Ocean Engineering & Oceanography 1

Series Editors

Manhar R. Dhanak
Florida Atlantic University
Boca Raton
FL, USA
E-mail: dhanak@oe.fau.edu

Nikolas Xiros
University of New Orleans
LA, USA
E-mail: nxiros@uno.edu

For further volumes:
http://www.springer.com/series/10524

Edmund Wittbrodt, Marek Szczotka,
Andrzej Maczyński, and Stanisław Wojciech

Rigid Finite Element Method in Analysis of Dynamics of Offshore Structures

Springer

Authors
Prof. Edmund Wittbrodt
Gdańsk University of Technology
Poland

Dr. Andrzej Maczyński
University of Bielsko-Biała
Poland

Dr. Marek Szczotka
University of Bielsko-Biała
Poland

Prof. Stanisław Wojciech
University of Bielsko-Biała
Poland

ISSN 2194-6396 e-ISSN 2194-640X
ISBN 978-3-642-29885-1 e-ISBN 978-3-642-29886-8
DOI 10.1007/978-3-642-29886-8
Springer Heidelberg New York Dordrecht London

Library of Congress Control Number: 2012937235

Printed on acid-free paper

Springer is part of Springer Science+Business Media (www.springer.com)

Contents

1 Introduction

Following the depletion of land natural resources (oil, gas, minerals etc.), methods of obtaining them from beds of seas and oceans are gaining more and more importance. However, exploiting undersea resources poses a number of challenging technical problems pertaining to their extraction, transport and processing in a specific, inhospitable environment. Important groups of machines used in offshore engineering are cranes applied to reloading and assembly works and specialized devices for laying pipes which transport oil and gas. A characteristic feature of their working conditions is sea waves causing significant movement of the base on which offshore structures are installed. This is a phenomenon which must be taken into consideration in design of such machines.

The structures which are used in offshore engineering, very often have flexible links and joints. This causes problems especially in the case then links deflections are large. That is the reason why the computer methods are used in design process of offshore machines and devices. On the market there is a significant number of general commercial packages, mostly based on finite element method, such as Nastran, Ansys, Abaqus. They have special modulus for modelling multibody systems with flexible links, and enable to take into consideration mentioned above phenomena, like base motion and large deflection of beam like links. Commercial computer packages are mainly used in large designing and production centers because they require from users special knowledge and experience. The small and middle firms very often prefer less general computer models and programs. Less general but better fitted to specific kind of designed structures, with limited data necessary for performing calculations and dedicated interface. That is the reason why some designers are still looking for models, methods and programs oriented for their specific machines and structures. The examples of methods that can be successfully applied in these cases are presented in the book.

The joint coordinates and homogenous transformations are applied to modelling multibody system presented herein. In order to take into account flexibility of beam like links, the rigid finite element method is proposed. These methods for many years have been developed at the Gdańsk University of Technology and the University of Bielsko-Biała. The base and detailed description of these methods is given in the book [Wittbrodt E., et al., 2006]. Hence in this monograph the base of homogenous transformations and rigid finite element method are presented succinctly, as necessary to understand the way in which the

E. Wittbrodt et al.: Rigid Finite Element Method, OEO 1, pp. 1–3.
springerlink.com

models are formulated. More attention is devoted to new models and methods developed for modelling offshore structures.

The present book contains new formulation of the rigid finite element method which allows to take into consideration nonlinear physical relations, as well as large deflections of links. The floating reference frames are introduced in this case. As in base formulation, the beam like links are divided into rigid elements, which reflect inertial features of links, and mass less spring-damping elements. Each rigid element has six degrees of freedom (three translations and three rotations). However, the flexible features of spring-damping elements are introduced as external forces and moments calculated according to physical dependencies. It allows to take into consideration different physical models of materials: linear, elasto-plastic, visco-plastic and others.

The models and methods presented in the book are applied to description of dynamics of some offshore structures, mainly cranes, ramps and pipes. Results of numerical calculations concern: crane used for transportation of BOP, column crane, A-frame. Also the results of numerical simulations for different methods of laying pipes on the see bottom are presented. The J-lay, S-lay and reel method are considered. The models, which allow their static and dynamic analysis to perform, are developed. The models take into account large base motion of ships or platform, on which considered structures are mounted, caused by sea waves. Important problems related to the control of offshore structures ensuring their safe operation are also discussed. Control enables compensation for the movement of the base caused by waves and reduction of dynamic loads of the elements of considered systems.

For selected models, their accuracy is verified by comparing their results with those yielded by other methods, including commercial finite element method packages.

In chapter 2 selected offshore engineering problems are briefly presented. Most prominent elements of the infrastructure necessary for the extraction of oil and gas as well as methods of offshore pipelines installation are introduced. Also given are the specifics of operation of offshore cranes and their basic typology.

Chapter 3 is devoted to the description of the impact of water on offshore structures. It includes basics of mechanics of the wave motion of water, the problem of determination of the values of forces acting on elements immersed in water and methods of simplified description of the motion of the base of offshore structures.

Preliminary information about defining the positions and orientations of coordinate systems, transformations of vectors and joint coordinates are contained in chapter 4. It also deals with application of homogeneous transformations and joint coordinates to describe the geometry of multibody systems.

In chapter 5, a method is presented to determine the elements of a Lagrange equation of the second order in the case of description of the dynamics of open kinematic chains consisting of rigid links. The equations of motion are obtained taking into account the kinetic and potential energy of gravity forces as well as external forces and moments acting on the multibody system. It is further assumed that the system is located on a movable base.

The next chapter treats the elements connecting links belonging to different branches of the system (in particular, one of the links may be the base). Two approaches to modelling constraints are also presented: the classical one consisting in the introduction of constraint equations and another one whereby the constraints are modelled as spring-damping elements with large stiffness and damping coefficients.

Basic information related to elasto-plastic and visco-elastic models of materials can be found in chapter 7.

The focus of chapter 8 is the rigid finite element method (the RFE method) which serves to discretise flexible links. Its concepts are presented along with the way of defining the generalized forces, formulating the equations of motion and determining the following energies: kinetic, potential of the gravity forces and elastic deformation of a link. The RFE method is described in two forms: classical and modified. A new approach to modelling large deflections based on this method is also discussed.

Chapter 9 contains examples of dynamic models of selected offshore structures. In addition to basic assumptions and mathematical descriptions, results of selected numerical simulations are also given. Presented are models of the following machines: a gantry for transporting BlowOut Preventor valves, a column crane with a shock absorber and the devices for laying pipes.

Problems pertaining to control are shown in chapter 10. The task of dynamic optimization in device control is discussed. A method is presented enabling vertical stabilization of the load of an offshore crane and its stabilization in three directions using a specialized auxiliary system. Analyses related to a system of active compensation for waves in the reel drive of the device for laying pipes are also included.

The book leverages some previous results of the authors' work, especially those present in Marek Szczotka's monograph entitled "The rigid finite element method in modelling of nonlinear offshore systems" [Szczotka M., 2011 b].

The authors wish to thank the co-authors of their publications from which some analyses presented in this volume are drawn, especially Dr. Paweł Fałat and Dr. Andrzej Urbaś.

2 Overview of Selected Problems in Offshore Technology

Extraction of undersea natural resources, particularly oil and gas, has expedited the progress in offshore technology for a few decades, including the construction of platforms as well as the development of new extraction techniques and methods of laying underwater pipelines. Various types of cranes are an important aid in the construction of extraction infrastructure as well as its operation and servicing. The current chapter describes some most important elements of the infrastructure necessary for extracting oil and gas and methods of installation of offshore pipelines. Specific conditions pertaining to offshore cranes' operation and their basic typology are presented.

2.1 Platforms as One of the Main Features of Offshore Infrastructure

The idea of extracting oil and gas from sea beds occurred over 100 years ago. The first wooden drilling platform was applieded at the end of 19th century off the coast of California. In 1911, the first installation to extract oil was engaged. It was located on the Caddo lake (on the border between the states of Louisiana and Texas) and yielded daily about 450 barrels [Wilson J. F., 2003]. The period following the World War II was marked by unceasing growth in number of offshore installations and as a consequence also in the amount of oil and gas being extracted. At that time, the areas where deposits were exploited and sought widened considerably. Presently, in many areas traditionally valued for such opportunities the deposits are running out or even already have. This applies particularly to the North Sea and the Norwegian Sea, where the existing deposits consist mainly of natural gas [Dokka A., Midttun O., 2006]. This situation forces companies to reach for resources located in less accessible areas featuring harder weather or greater depths. In Europe, the Barents Sea and other polar seas offer sample natural resources still remaining to be exploited. This opportunity doesn't come without difficulties though, namely: depth, heavy weather conditions and low temperatures. There has recently been a notable increase in amounts of resources yielded from waters surrounding South America (especially Brazil), West Africa, India, Australia and Oceania. Most of these endeavours are fairly new, thus

E. Wittbrodt et al.: Rigid Finite Element Method, OEO 1, pp. 5–25.
springerlink.com

employing post-2000 technology enabling extraction of oil and gas from large depths. Brazil in particular has shown considerable development of technology related to extracting resources from 2000 m and deeper, due primarily to their oil and gas search and extraction tycoon Petrobras. It is however worthy of notice that the rate of discovering new deposits is dropping, whereas just the opposite is the case with the demand for sources of energy and the pace at which known resources are exploited. Some forecasts thus state that between 2010 and 2015 we are at the peak amount of oil extracted from sea and ocean beds. Despite this, modern technologies created for the extraction of hard to reach resources and experience acquired may underlie new projects, including ones related to the production of energy. It is in seas and oceans where vast supplies of energy are to be found (waves, sea currents). Works aimed at exploiting them are gaining momentum.

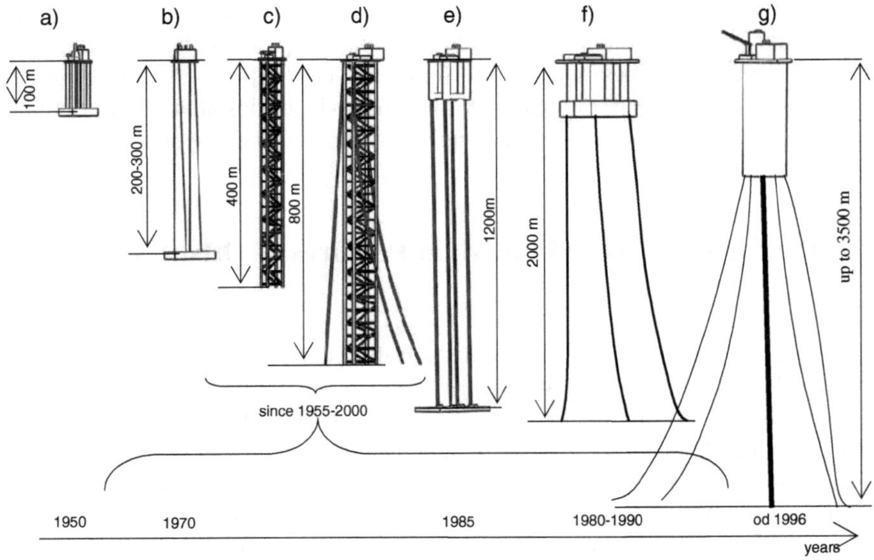

Fig. 2.1. Basic types of platforms

Drilling and extraction platforms are the most characteristic offshore structures related to the extraction of oil and gas (Fig. 2.1). A drilling platform is a floating structure equipped with a drilling rig, suited for making wells in the bottom of a sea. Platforms of this type are sporadically used simultaneously to exploit the deposits. An extraction (production) platform is suited for extraction of gas and oil and their preliminary purification. From there the stock is loaded onto tankers or transferred further by an underwater pipeline. These are usually structures supported off the bottom, although recently they have more and more often been constructed as floating or semi-floating.

Platforms were initially built on shallow seas, up to about 100 m in depth (Fig. 2.1a). They had a steel foundation lying on the bottom to which legs (three to six) were attached. Gravity platforms (Fig. 2.1b) started to be used shortly afterwards. Their foundations as well as legs were made of concrete. They enabled reaching depths of 200 m to 300 m. The type of structure based on a truss is most common nowadays (Fig. 2.1c). Of these the first to appear was a 30 m tall one in 1955. Their contemporary capability of operation in terms of depth reaches 400 m. A variation thereof are truss platforms equipped with additional mooring lines (Fig. 2.1d). In extraction industry they have been in use since the beginning of 1980s [Chakrabarti S. K., 2005]. They allow for extraction of deposits on seas with depths up to about 800 m, simultaneously admitting significant movement of the platform due to waves. Since the second half of 1980s, platforms lacking fixed connection with the bottom and instead kept afloat by buoyancy forces have started to be widely introduced. Their positioning with respect to the sea bottom requires additional elements to be installed in the bottom (Fig. 2.1e). Platforms of this type, called Tension Leg Platform (TLP), are installed mainly in areas 800 m to 1200 m deep. As their advantages count stability and immunity to vertical and rotary movements. Fig. 2.1f shows a scheme of a multihull semi-submersible platform used with depths in the range of 2000 m (in extreme cases even up to 3000 m). Such platforms are usually equipped with own propulsion assuring continual operation and stability even under intense waves and avoidance of personnel evacuation. Similarly constructed SPAR type platforms are shaped as vertical cylinders (Fig. 2.1g) with profile diameters up to 30 m or 40 m and heights of hundreds of meters.

In many cases when depth exceeds 1000 m, an additional infrastructure aiding the extraction of resources from a seabed is placed on the bottom which takes over some of the platform's tasks. The necessity to built an expensive production platform is thus avoided. Its role is fulfilled by specialized ships moored above the extraction field, or the stock is transferred by pipelines to land stations or nearby platforms.

2.2 Offshore Pipelines: Applications and Sample Installations

For over 100 years, transport of oil, gas, derivative compounds and water has been performed by means of pipelines. They are the most efficient way to transfer continuous medium to large distances. Their role has greatly increased with the development of extraction of resources from undersea deposits. Underwater pipelines presently constitute a considerable part of the entire infrastructure existing for oil and gas transportation. Because of the working environment, their structure, installation and extraction are much different to the case of overland and underground ones.

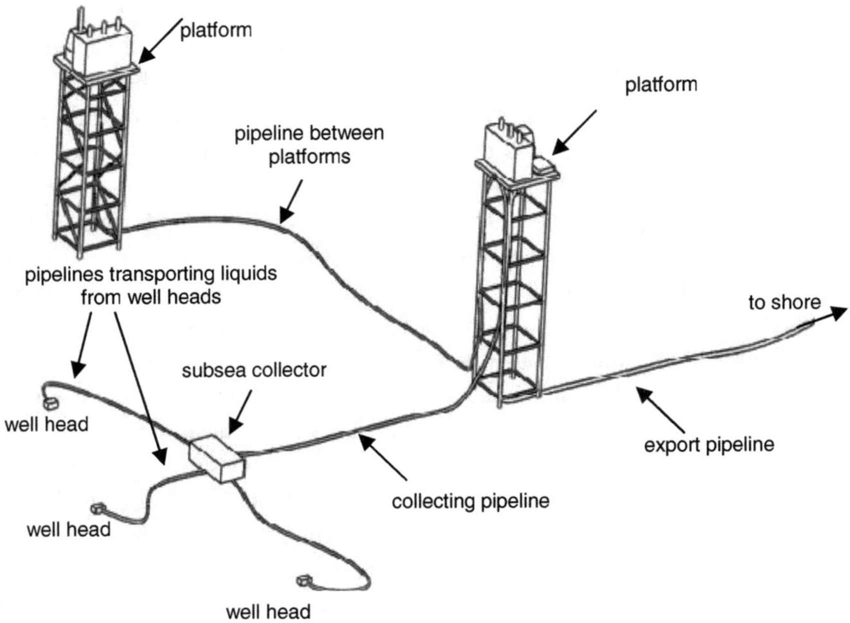

Fig. 2.2. Sample offshore infrastructure with miscellaneous pipelines

A sample gas field is shown in Fig. 2.2. The role of the production unit is fulfilled by an extraction platform. On the platform the stock is processed into the ready product, which is subsequently transferred by exporting pipelines to a receiving station.

Offshore pipelines may be basically divided as follows (Fig. 2.2):

- connecting heads of wells with collectors,
- collective – transferring resources from collectors to production platforms,
- transfer – connecting production platforms within a single oil or gas field or neighbouring ones,
- export – transporting products from production units to receiving stations (land bases, customer's receptive infrastructure),
- serving to transport water or other chemical compounds from production units (platforms or land facilities) to drilling heads,
- otherwise construed, often in the form of bundles of pipes or cables.

Large oil fields often have many platforms providing for different needs: accommodation, production, storage. The Ekofisk field in the North Sea built in 1970s may serve as an example.

Extraction and processing of oil and gas has lately more and more often been supported by Floating Production Storage & Offloading (FPSO) vessels (Fig. 2.3). Due to large dimensions (typically 300m to 500m in length), ships of this kind are very stable. An FSPO vessel is held in the desired position by means of mooring lines. It departs for a safe seclusion only under extreme storms. The extraction technique employing FSPO vessels is a fairly new development. Its main application is on smaller gas fields where periodic relocation of a vessel is desirable. Usually individual wells are connected by special pipelines to risers, which are vertical segments of pipes connecting collectors with a turret placed in the hull of an FSPO ship. Risers are highly flexible and feature special loops to compensate for FSPO movements due to waves and wind. In case of an FSPO vessel's emergency departure from a field the turret with risers and mooring lines is lowered to the bottom. Its reinstallation requires using a specialized winch with high capacity (800 T to 1000 T).

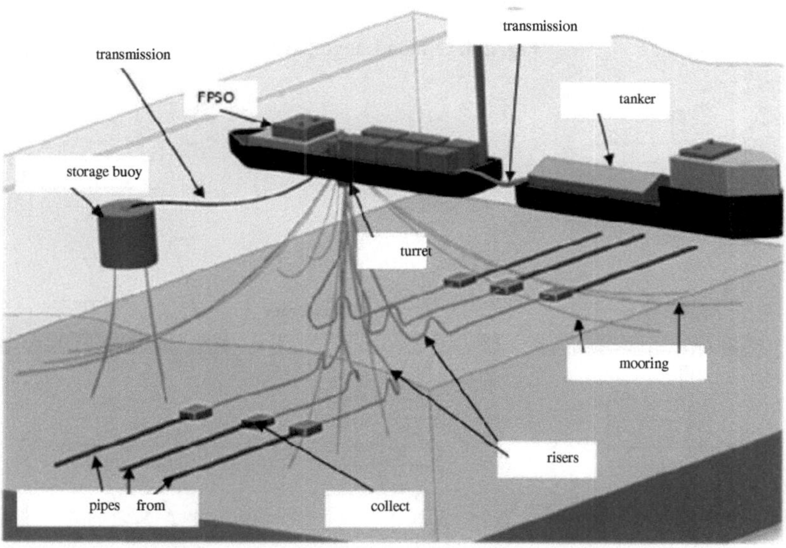

Fig. 2.3. Extraction and processing of oil or gas using an FPSO ship

In some cases only a processing station located on shore is used in the extraction of a deposit. The stock is then transferred by pipelines directly from the well to the land. The Ormen Lange gas field situated on the west coast of Norway (Fig. 2.4) is one such example.

As already mentioned, risers (Fig. 2.3) are used to transport oil or gas, as well as water and other substances, between the seabed and a drilling or production unit on the surface of the sea (platform, ship, FSPO). Risers may be drilling risers used for transport of fluids utilized in the process of making wells, or production risers which transport the stock from wells to vessels or platforms. It ought to be mentioned that the cost of risers may be comparable to that of a ship or a platform

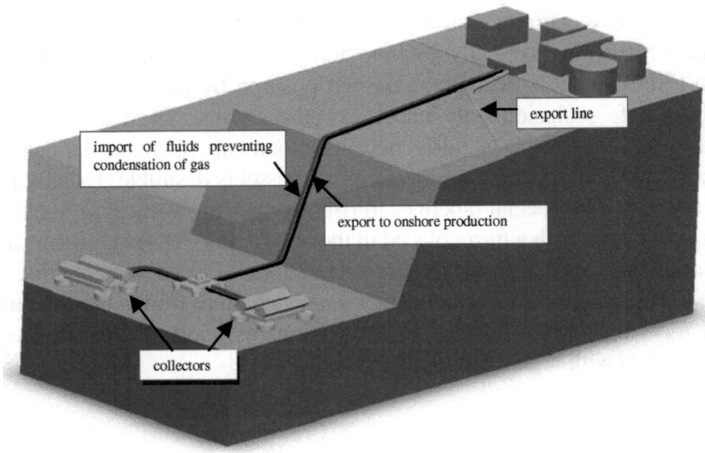

Fig. 2.4. Ormen Lange gas field

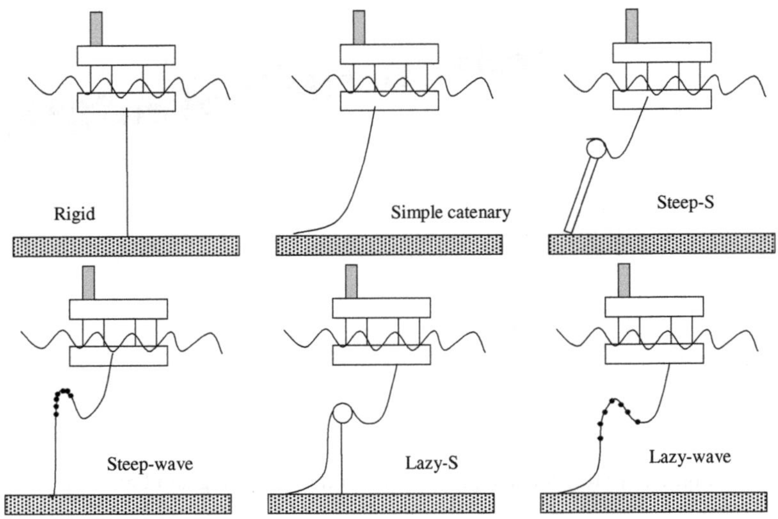

Fig. 2.5. Systems of risers used in oil and gas extraction

in case of deep sea extraction [Bai Y., Bai Q., 2005], [Chakrabarti S. K., 2005]. Basic configurations of risers are schematically depicted in Fig. 2.5. The choice of a given system depends on multiple factors: depth of the sea, structure and functionality of the vessel (particularly its manoeuvrability under wave action), number and setup of mooring lines, conditions present in the extraction area, including the intensity of sea currents.

A wide overview of the technology and elements of the infrastructure used in offshore industry can be found among others in [Gerwick B. D., 2000], [Bai Y., Bai Q., 2005] and [Chakrabarti S. K., 2005].

2.3 Offshore Pipelines Installation Techniques

Installation of underwater pipelines is a technically difficult operation, which differs substantially from the case of land ones. It requires using a separate set of often innovative methods. Currently the following basic methods of laying offshore pipelines may be distinguished:

- S-lay,
- J-lay,
- reel method,
- tow method.

Each of them has both advantages and disadvantages and limitations. The criteria for choosing the method for a particular case are following:

- sea depth,
- length of the segment to install,
- diameter of the pipeline,
- time allowed for the installation,
- total budget of the operation.

2.3.1 The S-Lay Method

The S-lay method is one of the oldest methods of laying underwater pipelines. It was used mainly for installations in shallow seas. It can be performed from either a specially equipped vessel or a platform. Low amplitudes of motion of multihull semi-submersible platforms have made them an often preferred utility for laying pipelines with the S-lay method. Such a solution is especially popular on seas with intense waves for the most part of the year (e.g. the North Sea and the Norwegian Sea). The schematic concept of the S-lay method is shown in Fig. 2.6. It is named after the shape taken by the pipeline being laid on the segment between the unit and the bottom. It resembles the letter *S*. A special structure called a stringer is used to support the pipeline suspended from the deck.

For installing long and high capacity pipelines in seas up to 600 m deep the S-lay method is most popular. It is applicable to pipelines of greatest diameters, even exceeding 1m.

The length and geometry of the ramp guiding the pipe depend primarily on the depth of the sea and the diameter of the pipeline. Control of the inclination angle of the ramp, thus also the shape of the pipeline being laid, is provided by two means:

- built-in buoyant elements whose filling appropriately with water makes it possible to regulate the immersion of individual segments of the ramp – such are the multi-modular buoyancy ramps,
- leveraging systems of ropes and a winch or other mechanisms controlling the inclination angle of the ramp.

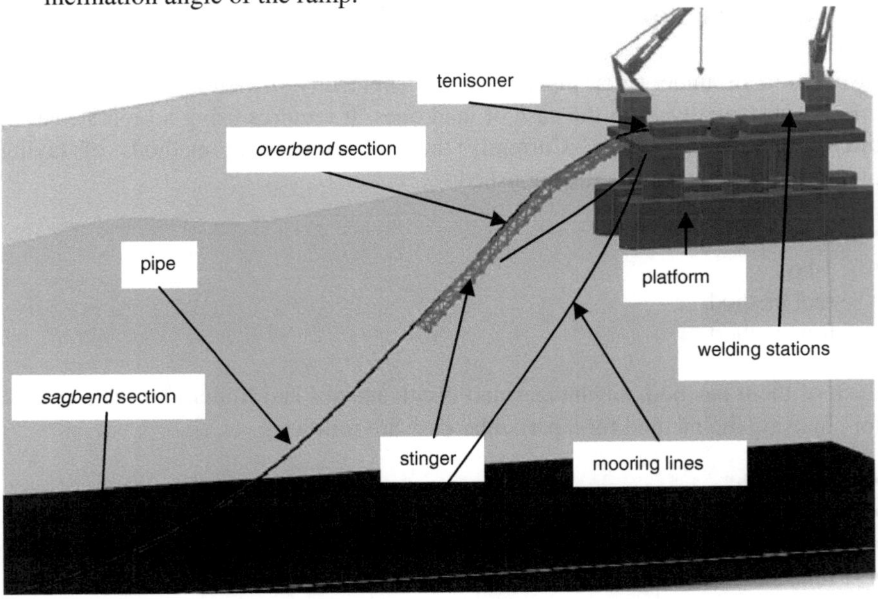

Fig. 2.6. Application of a semi-submersible platform to install o pipeline with the S-lay method

Tension systems are used to prevent buckling of the pipe being laid in the lower deflection and to keep the deformation of the material within the desired limits. These are specialized mechanisms placed in front of the entry point of the pipe onto the ramp guiding the pipeline and a set of anchor winches. Additional vessels control the anchors. Appropriate tension may also be created with thrusters along with sufficiently powerful engines. The necessity of creating an axial force in the pipeline being installed is a rather significant disadvantage of the method discussed. Its impact is more evident with greater depths, therefore the S-lay method is limited to shallower seas. Laying pipelines with the S-lay method may also be performed from ships or barges.

2.3.2 The J-Lay Method

Usage limitations of the S-lay method and simultaneously growing demand to lay pipelines at greater depths motivated the development of the J-lay method. Its name also reflects the shape of the pipe being laid. When suspended between the unit and the bottom, it resembles the letter *J*.

Similarly to the previous method, laying pipes with the J-lay method can be performed from decks of multihull semi-submersible platforms as well as monohull barges and ships. Its characteristic feature is a vertical guiding ramp. Setting the ramp in an almost vertical position eliminates the problems with exceeding admissible tensions in the material of the pipe in the area of the upper deflection. Additionally the J-lay method allows for considerable reduction of necessary forces to be exerted by tensioners and the unit itself. A disadvantage of the J-lay method is the ability to use a single welding station only, thus limiting the efficiency of laying pipelines. It typically reaches 1.5 km to 2 km per day, whereas with the S-lay method 5 km to 6 km per day are usual values.

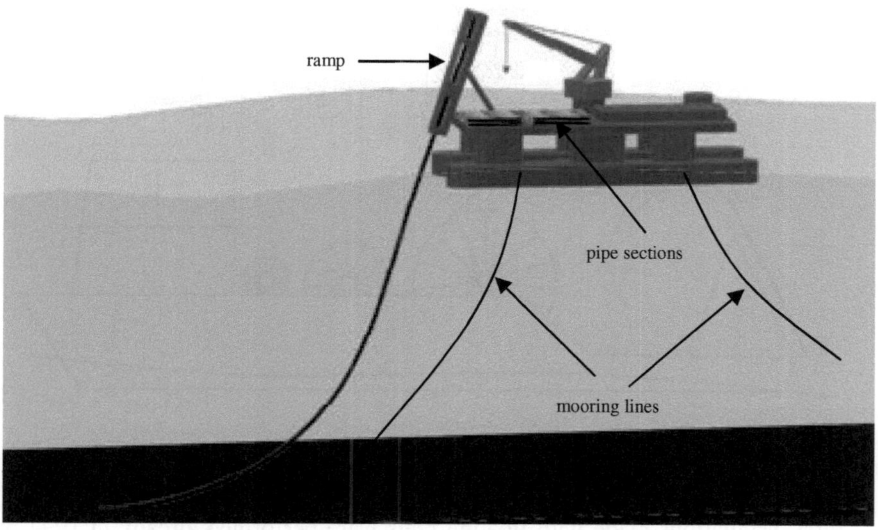

Fig. 2.7. Scheme of laying a pipeline with the J-lay method

In Fig. 2.7, a scheme of a platform installing a pipeline with the J-lay method is shown. The platform Saipem 7000 is one of the largest such units in the world. It was used, among other things, to install a high capacity pipeline running across the Black Sea with depths reaching 2200 m.

Whenever the inclination angle of the ramp can be controlled, the J-lay method may also be used in shallower seas. Usually, the angle can be set between about 65° and 90°, however in some ramps it can even be 30°.

2.3.3 The Reel Method

Among the major drawbacks of the S-lay and J-lay methods is the necessity to connect pipes on the platform or the vessel prior to laying them. Hence they require using transport ships supplying segments of pipes from the land, usually of lengths between 12 m and 50 m. Those must be subsequently reloaded with an offshore crane to the main unit. Therefore, in the case of pipelines with lower diameters, the reel method is popularly chosen. A vessel used in this method features a specialized reel. The pipe is wound onto the reel on the land, including connecting its segments by welding. Next, the vessel transports the entire pipe to its destination where it is installed. The reel method is used mainly on the North Sea. Its primary usage domain is laying relatively short segments of pipelines, e.g. those constituting elements of oil and gas fields infrastructure (Fig. 2.2). One of the foremost advantages of the method is its high efficiency reaching 2 km per hour.

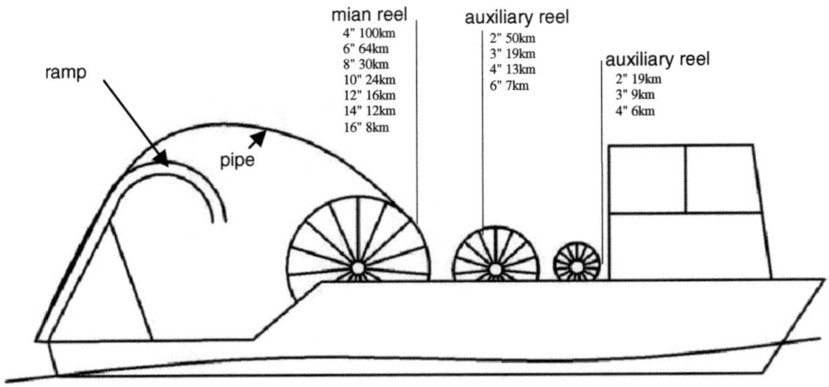

Fig. 2.8. A ship with a reel to wind pipelines

First installations made with floating reels were performed already in 1994 by the Allies during their invasion of France. In Fig. 2.8, a contemporary ship Apache[1] is shown, whose purpose is to work with the reel method. It features a main reel onto which 8 km to 100 km of pipes (depending on the diameter) can be wound and two smaller ones for short fragments of pipes and risers.

Winding of pipes onto the reel may cause permanent plastic deformations. It is the case with pipes with large outer diameters and small winding diameters of reels. This may substantially and adversely influence the properties of the installed pipeline. Therefore, only slight permanent plastic deformation of pipes being laid is admissible. The magnitudes of these deformations depend on: dimensions of the reel, outer and inner diameter of the pipes, tension during the winding.

At present, large reels are used, which can accommodate between 2000 T and 3500 T of a pipeline. Diameters of such reels exceed 30 m and the forces generated by their drives surpass 200 T. Smaller reels may form sets (of two or three pieces) which are supplied by transport vessels. An integral part of a ship suitable for laying

[1] The informations were taken from the operator's web page, the Technim company.

pipelines with the reel method is a guiding ramp, usually installed on the stern. Its purpose is to give shape to the pipe leaving the ship, whereby typically the reel method is combined with the J-lay method (Fig. 2.9). The ramp simultaneously serves as the supporting structure for elements guiding and straightening the pipe. In its upper part there is an aligner wheel, and following it a device straightening the pipe, eliminating its permanent deformations caused during winding. Additionally on the ramp, a set of devices is installed which control the speed at which the pipe slides from the ramp and hold the weight of the pipeline suspended in the water.

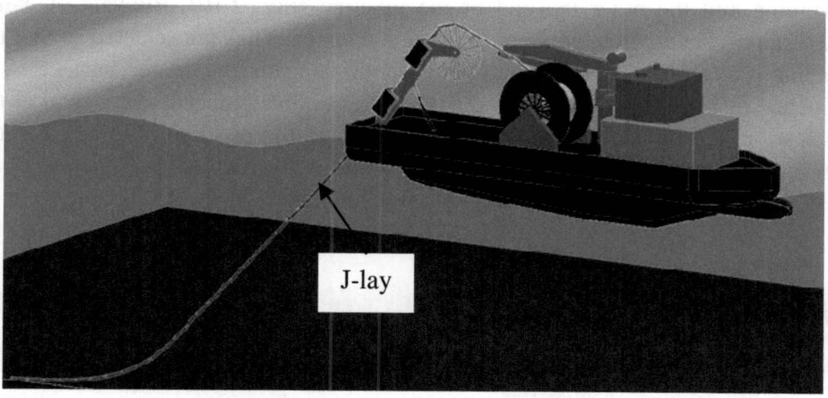

Fig. 2.9. A ship unwinding a pipe from a reel and laying it with the J-lay method

A similar solution is a structure with a reel whose rotation axis is vertical. Systems of this type are mainly used for installing bundles and cables. They are unwound from the reel (called a carousel) and passing through a system of tensioners are laid on the bottom. Another system in existence is one with the reel's axis parallel to the ship's longitudinal axis, in which case the laying is carried out from the vessel's side.

Limitation of the maximal outer diameter of the pipeline to be laid is the primary disadvantage of the reel method related to plastic deformations. Pipes installed using this technique have diameters up to 28 in. The reel method, moreover, introduces relatively large deformations in the material of the pipe (up to 5%) which may weaken the welds and deteriorate the pipeline's stability, including the occurrence of a spiral line. Furthermore, there exists a risk of ovalization of the pipeline's section leading to local instability. It may also occur that a pipeline needs to contain segments with different diameters or other components (e.g. valves, splitters and so on). These add to the difficulty of the method discussed. When the base in which the winding of pipes onto the reel takes place is distant from the destination, considerable growth of cost and time of the operation is to be expected. Additionally, huge mass of the reel and pipes or cables wound onto it exacerbates the dynamic forces caused by waves. Stability of the device and the whole ship is thus worsened. Despite all of these, the reel method is, as mentioned, eagerly used.

2.3.4 Methods of Towing Pipelines

In cases of short segments of pipelines, a few kilometres in length, it is possible to assemble them on the land and next to tow them wholly to the location where they are installed. The longest segment to have been installed in this way had 7km [Bai Y., Bai Q., 2005]. Towing is usually performed by two vessels, one towing the beginning and the other the end of the pipeline (Fig. 2.10). Buoyant elements placed along the pipeline are used to prevent damage. They are selected so that the pipeline stays at a certain controlled depth beneath the water surface. The influence of hydrodynamic forces due to waves is thus reduced. On arrival at the destination, the buoyant modules are removed and the pipeline is lowered to the bottom. Sometimes other towing techniques are used in which the pipeline floats on the surface, or is dragged on the bottom [Chakrabarti S. K., 2005].

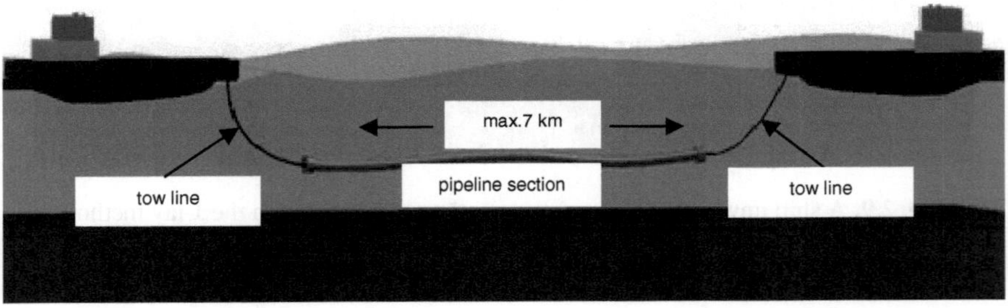

Fig. 2.10. Installing a pipeline using the towing method

Ease of preparation on the ground of the elements to be laid (which may be complex bundles of various cables and pipes installed together) is definitely an advantage here. The problems and costs related to the production of a pipeline or bundle in offshore conditions are avoided. An undeniable disadvantage is the difficulty of constructing longer pipelines with this method (control of forces and displacements during towing is a problem) and great amount of work and costs in cases of laying curved segments. For these reasons, the total number of kilometres of pipelines laid with the towing method is modest.

2.3.5 Other Operations in Service of Pipeline

Other than the very process of laying a pipeline, there exists a multitude of important operations forming comprehensive construction and extraction of exploration infrastructure. Among them are the following:

- securing a pipeline on the bottom,
- winding of pipes onto reels in specialized bases,
- operations of lifting a pipeline from the bottom, repairing, de-installing etc.

Securing pipelines is done with specialized ploughs. After the pipe has been laid on the bottom, the plough is lowered from the vessel and then towed by it. The plough forms a ditch in the seabed and simultaneously inserts the pipe into it. Another plough is used to bury the pipeline. When the diameter of the pipeline is small, devices cutting the ditch in the seabed may be used directly before laying it. The same devices usually install the pipe itself, provided that it is elastic enough. Yet another solution is to use specialized machines powered with high pressure. In some cases, the pipelines installed are buried with a layer of material supplied by ships (pebbles, gravel).

An example spool base in which offshore pipelines are produced is Orkanger base, Norway. Such bases must have the ability to store ready fragments of pipes of length up to a few kilometres. They are located in areas of intense extraction, where further works are planned for several years. Their advantages are low production costs and immunity to weather conditions.

Repairs and servicing of underwater pipelines is performed using specialized vessels capable of lifting pipes from the bottom. They are equipped with multiple reels onto which pipes or cables can be wound when they are damaged or being removed.

2.4 Reloading and Assembly Works Using Cranes: Tasks, Environmental Conditions, Types

Reloading and assembly works realised using various types of cranes are among widely performed and highly important operations in offshore engineering. One of the main features distinguishing offshore cranes from land ones are significant movements of the base caused by sea waves. In the case whereby a load is lifted from a supply ship also the load is in such motion. As a result, offshore cranes are far more exposed to dynamic overloads than their land counterparts. Those overloads have significant influence on the permissible operating range of the device. Constructors aim at designing a device in such a way that it can operate under wave action as intense as possible. Offshore cranes are therefore equipped with specialized anti-overload systems which minimize load oscillations and increase safety. It is also worth noting that weights of loads carried by offshore cranes often reach hundreds of tonnes. Winds, which are common in maritime areas, as well as extreme temperatures add further difficulty to their operation.

Taking the criterion of construction into account, the following types of offshore cranes may be distinguished:

- gantries,
- A-frames,
- boom cranes.

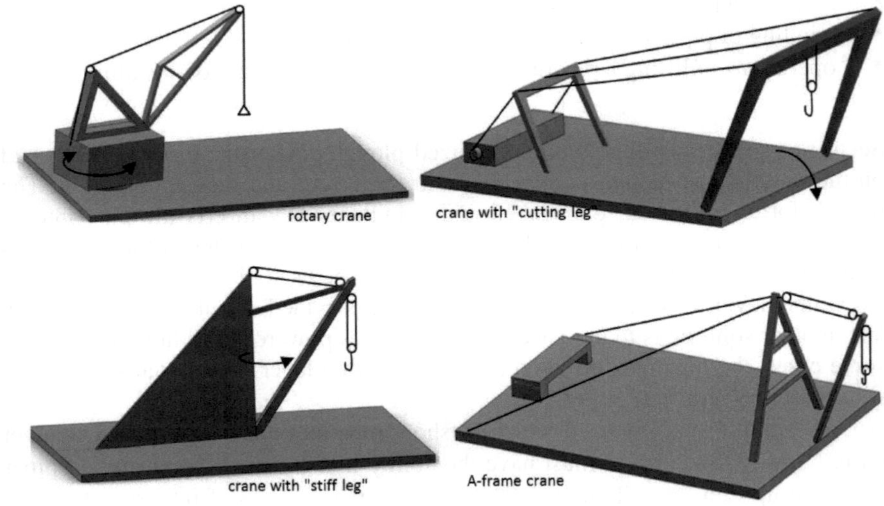

Fig. 2.11. Examples of cranes installed on vessels

Offshore cranes are installed both on platforms and sea vessels (ships, specialized barges). If they are equipped with a boom, it may be fixed or telescopic.

A common solution is to install offshore cranes on a special column. Depending on the criterion, column cranes may be classified as featuring:

- rope overhang control system
- hydraulic overhang control system
- truss boom
- box boom
- telescopic boom
- knuckle boom

Sample structures of offshore column cranes are depicted in Fig. 2.12.

Similarly to land technology, gantries are a popular choice. They often appear on large container ships. They are also installed on other sea vessels and platforms where they are used for assembly and service works. A gantry installed on platforms and used to relocate and lower the valves of a BlowOut Preventor (BOP) may serve as an example. Schemes of different installation possibilities of offshore gantries are in Fig. 2.13.

The multitude of construction solutions of offshore cranes is a result of the variety of their applications. Those range from reloading goods transported overseas, constructing and operating offshore infrastructure, to scientific research etc. In many cases such devices are constructed one-off for a particular order.

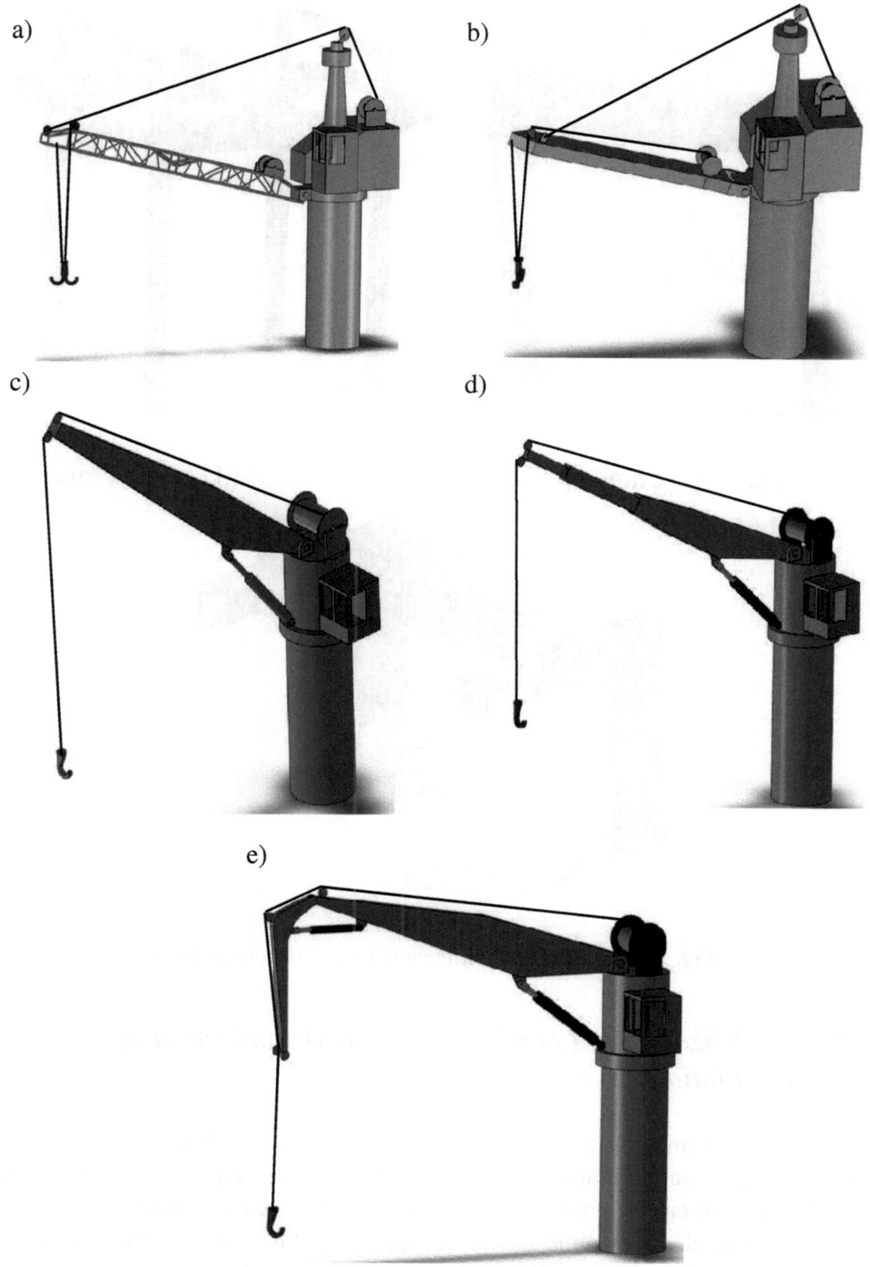

Fig. 2.12. Column cranes: a) with a rope overhang control system and a truss boom, b) with a rope overhang control system and a box boom, c) with a hydraulic overhang control system and a box boom, d) with a telescopic boom, e) with a knuckle boom

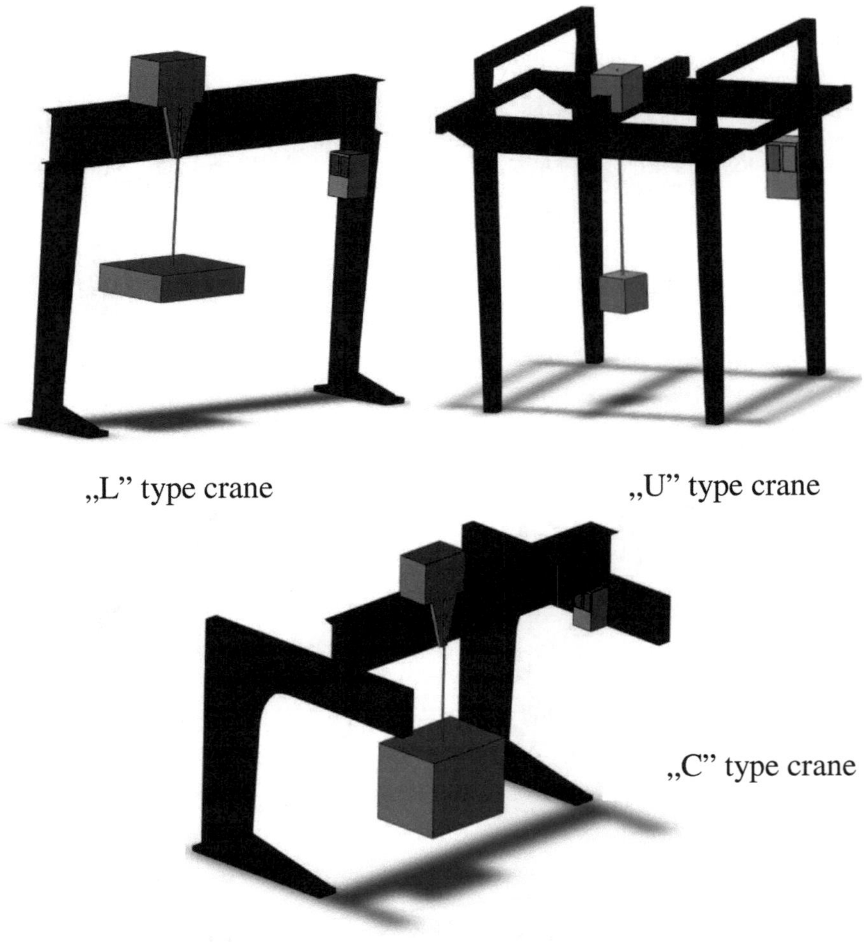

Fig. 2.13. Structures of different types of offshore gantries

2.4.1 Stabilization of Load Position and Minimization of Its Oscillations

When operating an offshore crane, the problem of load oscillations is of special importance. Those oscillations, caused mainly by sea waves, not only make reloading and assembly works more difficult, but also create an immediate danger for the personnel. In extreme cases, the load may hit a side of the sea vessel carrying the load or the supply vessel. Thus, market leaders endeavour to equip their products with specialized anti-oscillation systems. A system of this type, the SmartCrane™ Anti-Sway Crane Control for Rotating Boom Cranes, is offered by SmartCrane. Its working principle is to move the suspension point of the rope at the end of the boom. Li Y. and Balachandrana B. of the University of Maryland presented it at the symposiums MURI on Nonlinear Active Control of Dynamical Systems (Virginia Polytechnic Institute and State University, 1998-2001) as well as in their papers [Balachandran B., et al., 1999], [Li Y. Y., Balachandran B., 2001]. This solution may be enhanced by a closed-loop control system. In the mentioned papers, the mathematical models applied omitted the flexibility of the supporting structure of the crane. At the MURI symposiums it was also proposed that there exists another possibility of solving the problem of load oscillations. It consists in adequately controlling (also by means of a closed-loop system) the rotary motion and raising of the boom. The method was verified numerically and experimentally on a test stand. [Masoud Z. N., 2000], [Nayfeh A., Masoud Z., 2001], [Masoud Z., et al., 2004], [Nayfeh A., et al., 2005a]. The analyzed problem required a spatial model of the crane. However, the created model ignored the flexibility of the system. A feature worth mentioning is an additional provision for minimizing the motions of both the supplying and the receiving vessel by means of a stabilizing system. It consists in tying them together with ropes once positioned appropriately against the waves and moving at a specified speed. The concept is discussed in detail in [Nayfeh A. H., et al., 2005b]. Another method of stabilizing the load position in an offshore crane was the topic of the following works, among others: [Maczyński A., 2005], [Maczyński A., 2006], [Maczyński A., Wojciech S., 2007]. It will be presented with details in chapter 10. It assumes the use of an additional unit suspended at the end of the boom, guiding the tow rope at a certain segment. Changing its deviation from the vertical is a way to influence the load's tangential and radial oscillations. This solution provides for a great deal of influence on the motion of the load, and in combination with the winch it enables stabilization of the load in three directions. In [Maczyński A., Wojciech S., 2009] it was shown that stabilization of the load also minimizes the undesirable effect whereby the tow rope is stressed and eased. The analyses presented in the above mentioned papers were carried out for an offshore crane with a hydraulic overhang control system and a telescopic boom. In [Spathopoulos M. P., Fragopoulos D., 2004] a similar solution was considered in the planar case based on a simplified model of a crane ignoring the flexibility of the system. Control methods for both linear and nonlinear objects were used. Two different control algorithms minimizing load oscillations were also discussed in [Schaub H., 2008]. One algorithm was based on current measurements only, whereas the other additionally performed computations on a model of the system. Due to the necessity of real-time operation, the model of a crane should in this case be very simplified. The boom was thus modelled as a rigid link, the load as a material point and the distance between the end of the boom and the load was constant.

In industry practice, also other systems have been in use for years, e.g. PDC 200 scanning the profile of the load with a laser and subsequently compensating for the oscillations electronically. They are produced by the company Cegelec-AEG. Yet other solutions are ABB's System CPC and Caillard's ESCAD [Cosstick H., 1996].

2.4.2 *Safety Systems: Systems Limiting Dynamic Overloads in Offshore Cranes*

As can be seen from the previously presented reasoning, offshore cranes are heavily exploited devices often exposed to extreme conditions. Their malfunction during operation may cause significant material losses and even pose threats to human health and life. According to the EN13852-1 norm, each offshore crane must feature the following safety systems:

- emergency operation – in case of power failure, means shall be provided for a controlled slew, luff down and load lowering operations, to land the load and boom safely,
- emergency stop – the emergency stop shall retain its function regardless of any malfunction of the programmable control system, if installed,
- lateral boom protection system – an automatic protection system shall be provided to prevent lateral overload of the boom or overload on the slew mechanism if sidelead loads occur outside the design limits,
- manual overload protection system (MOPS) – system, activated by the crane operator, that protects the crane against possible overload by reducing the load carrying capacity and allowing the hook to be pulled away from the crane in any direction,
- automatic overload protection system (AOPS) – system that automatically safeguards and protects the crane against the effects of a gross overload during operations by allowing the hook to be pulled away from the crane in downwards direction within specified offlead and sidelead angles, without causing significant damage to the crane.

Appendix J to the norm EN13852-1 establishes a hierarchy of importance of these systems and signalling components. It is summarized in tables 2.1 and 2.2.

Table 2.1. Normal ranking of safety measures

Order of precedence	Safety measure	Safety measure
1st priority	Emergency stop	Manual overload protection system (MOPS)
2nd priority	Automatic overload protection system (AOPS)	
3rd priority	Other limiters	
4th priority	Indicators	

Table 2.2. Ranking of safety measures when mode for personnel lifting is selected

Order of precedence	Safety measure
1st priority	Emergency stop
2nd priority	Mode section switch and other limiters
3rd priority	Indicators

The so-called shock absorbers are another type of systems limiting dynamic overloads used in offshore cranes. One is installed on the boom (Fig. 2.14), the other at the manifold (Fig. 2.16). The task of a shock absorber is to consume the energy of a momentary overload. In the case of the first solution, the dynamic overloads are minimized by passing the rope through an additional movable sheave connected to a hydraulic system. The concept of its operation is explained with a scheme (Fig. 2.15). It is a system consisting of an accumulator filled with gas and a hydraulic actuator. When the force S applied to the piston rod increases to the cutoff level (static load summed with flow resistance in the actuator is usually assumed), it starts moving and the oil starts flowing from the cylinder to the accumulator. The working stroke Δ_2 of the piston is reached for the maximal value of the dynamic force. That stroke is lower than the maximal stroke Δ_{max}, as for safety reasons the stroke Δ_{safe} should be maintained. The force S is balanced by

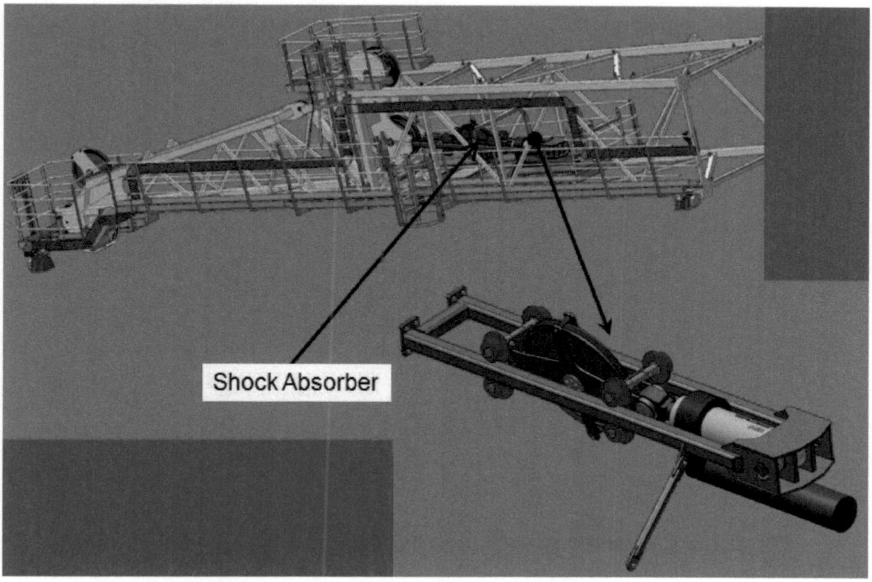

Fig. 2.14. Shock absorber installed on a boom[2]

[2] Picture published with the permission of National Oilwell Varco.

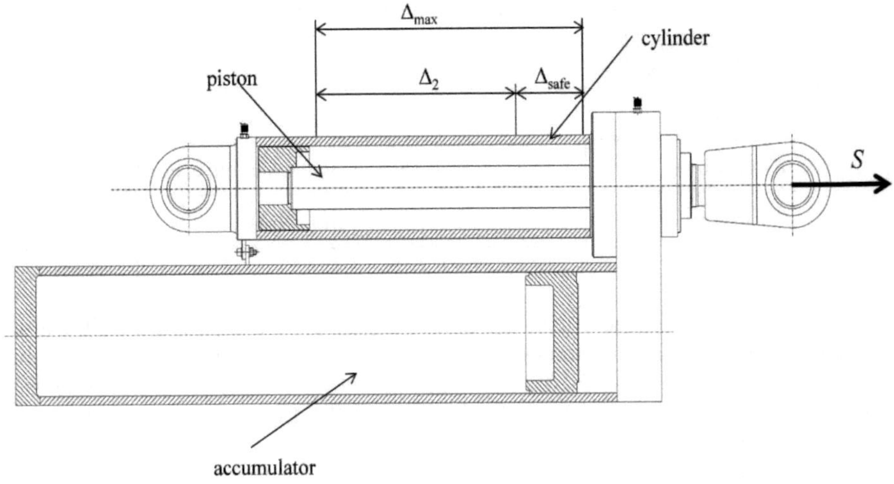

Fig. 2.15. Scheme of the system of a hydraulic shock absorber

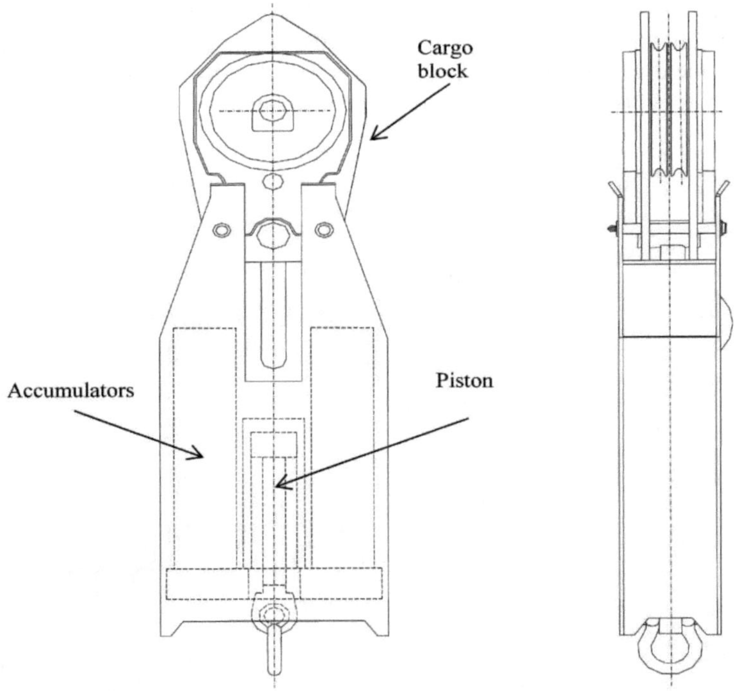

Fig. 2.16. Scheme of a shock absorber integrated with the manifold

the gas pressure in the accumulator. This type of shock absorber is especially recommended with single ratio in the lifting system because of the efficiency (the piston rod moves by the shortest distance).

The working principle of the second type of shock absorbers (Fig. 2.16) is analogous to the design described above, the difference being the placement of hydraulic accumulators consuming the energy in the manifold. This type of shock absorbers is particularly efficient with a multiple tackle in the lifting system of cranes. Its main disadvantage is the difficulty of supplying hydraulic installation to the manifold.

3 Impact of Water on Offshore Structures and Infrastructure

A characteristic phenomenon in offshore engineering is the impact of water on individual devices and elements of dynamic systems. This impact is a very complex phenomenon composed, among other things, of wave action, sea currents, hydrostatic and hydrodynamic forces. These processes are difficult to describe, and there exists a handful of approaches to modelling them, which differ in the level of idealization [Newman J. N., 1977], [Mei C. C., 1989], [Faltinsen O. M., 1990]. A possibly simple description of the motion of an offshore structure's base is often desirable, e.g. in the problems of control [Fossen T. I., 1994].

3.1 Basics of Water Wave Motion Mechanics

Hydrodynamic dependencies are at the foundation of any attempt at modelling objects situated in maritime environment. Formulas presented in this chapter allow us to determine the values which direct the motion of systems immersed in water, i.e. velocity and hydrodynamic pressure of the fluid.

Let us consider a point with coordinates given by the vector (Fig. 3.1):

$$\mathbf{x} = [x \quad y \quad z]^T, \tag{3.1}$$

in which the velocity of particles of the fluid in an inertial system equals:

$$\mathbf{v}(\mathbf{x},t) = [v_x(\mathbf{x},t) \quad v_y(\mathbf{x},t) \quad v_z(\mathbf{x},t)]^T. \tag{3.2}$$

In many cases it is prudent to assume that the density of the fluid is constant, hence the continuity equation for incompressible flow [Bukowski J., 1968], [Newman J. N., 1977]:

$$div(\mathbf{v}) = 0, \tag{3.3}$$

where $div(\mathbf{v}) = \nabla \mathbf{v} = \frac{\partial v_x}{\partial x} + \frac{\partial v_y}{\partial y} + \frac{\partial v_z}{\partial z}$.

E. Wittbrodt et al.: Rigid Finite Element Method, OEO 1, pp. 27–43.
springerlink.com

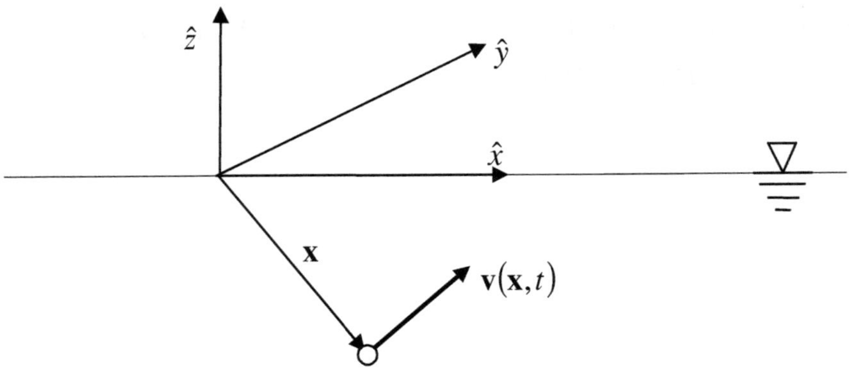

Fig. 3.1. Velocity of a particle of fluid in an inertial system

The momentum conservation principle for a Newtonian fluid is presented in the form of Navier-Stokes equations which, for an incompressible fluid, take the form [Bukowski J., 1968], [Newman J. N., 1977]:

$$\frac{\partial \mathbf{v}}{\partial t} + \mathbf{v}\nabla\mathbf{v} = \mathbf{F} - \frac{1}{\rho}\nabla p + \nu\nabla^2\mathbf{v}, \tag{3.4}$$

where $\mathbf{F} = [0 \quad 0 \quad -g]^T$ – mass forces,
$p = p(x, y, z, t)$ – pressure,
$\nabla = \frac{\partial}{\partial x}\vec{i} + \frac{\partial}{\partial y}\vec{j} + \frac{\partial}{\partial z}\vec{k}$,
$\mathbf{v}$ – velocity vector,
ρ, ν – density and viscosity of the fluid.

The Navier-Stokes equations (3.4) together with the continuity equation (3.3) form a system of nonlinear partial differential equations. No general solution is known (only numerical approximations are possible). To determine approximate pressure values lack of viscosity and irrotational flow are usually assumed. The irrotationality condition [White F. M., 2006]:

$$rot(\mathbf{v}) = \nabla \times \mathbf{v} = \mathbf{0} \tag{3.5}$$

ensures the existence of a velocity field potential. A function Φ called the potential is further sought, such that:

$$\mathbf{v} = \nabla\Phi. \tag{3.6}$$

Determining the potential Φ allows for calculation of velocity as the gradient of the potential. Applying the formula (3.6), the continuity equation (3.3) may be rewritten as a Laplace equation [Newman J. N., 1977], [El-Hawary F., ed., 2001], [White F. M., 2006]:

$$\frac{\partial^2\Phi}{\partial x^2}+\frac{\partial^2\Phi}{\partial y^2}+\frac{\partial^2\Phi}{\partial z^2}=0. \tag{3.7}$$

This equation should be completed with boundary conditions.

Omitting the last summand in the equation (3.4) standing for viscosity and taking the dependency $\mathbf{F} = -\nabla gz$ into account allows us to rewrite (3.4) in the form:

$$\frac{\partial \mathbf{v}}{\partial t}+\mathbf{v}\nabla\mathbf{v}=-\nabla\left(\frac{p}{\rho}+\Gamma\right) \tag{3.8}$$

where $\Gamma = gz$,

or, when the irrotationality condition is assumed, as a Bernoulli equation [Bukowski J., 1968], [Newman J. N., 1977], [El-Hawary F., ed., 2001]:

$$\frac{p}{\rho}+\frac{\partial\Phi}{\partial t}+\frac{1}{2}\left(\nabla^2\Phi\right)+\Gamma=const. \tag{3.9}$$

It makes it possible to determine the pressure if the potential Φ is known. Integrating the pressure over the surface of the body immersed in the fluid yields forces acting on it. In most practical cases, the hydrodynamics of ideal fluids is a sufficient theory for modelling the dynamics of systems occurring in offshore engineering. Viscosity, which was omitted from the equation (3.9), is sometimes re-added by stipulating additional empirical relations [Hoerner S. F., 1958], [Sarpkaya T., Isaacson M., 1981].

Equations of ideal fluid hydrodynamics enable determination of velocity and pressure fields for a regular wave (Fig. 3.2). In this case it is assumed that the profile of the wave (the free surface) is described by the formula:

$$z = \xi(x, y, t). \tag{3.10}$$

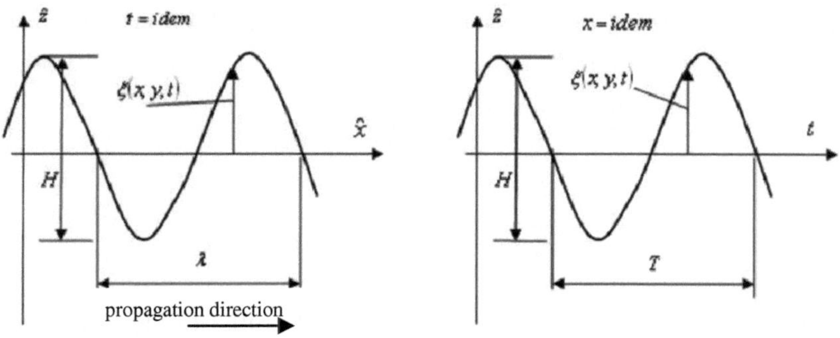

Fig. 3.2. Profile of a regular wave

Two boundary conditions for the free surface need to be also introduced. In the fist one, called the dynamic condition of free surface, the pressure of the fluid at the free surface is assumed to equal the atmospheric pressure. In the other, the so-called kinematic condition of free surface, the particles of the fluid on this surface are bound to it (they cannot break loose from it) [Newman J. N., 1977], [Faltinsen O. M., 1990].

3.1.1 Linear Model

Assuming the value of the quotient $\frac{p_0}{\rho}$ (p_0 – atmospheric pressure) in the Bernoulli equation (3.9) to be constant, taking the pressure on the free surface as p_0 and the profile of the wave to be described with the function [Newman J. N., 1977]:

$$\xi = \bar{\xi} \sin(\omega t - kx + \alpha), \tag{3.11}$$

a solution of the boundary problem is obtained in the form of a potential function (details are given, among others, in [Lighthill J., 1978], [Dean R. G., Dalrymple R. A., 1998]):

$$\Phi = \bar{\Phi} \cos(\omega t - kx), \tag{3.12}$$

where $\bar{\Phi} = \frac{\omega \bar{\xi}}{k} \frac{\cosh(k(d+z))}{\sinh(kd)}$,

$k = \frac{2\pi}{\lambda}$ – wave number,

λ – wave length.

Solving the Laplace equation with appropriate boundary conditions and using the definition of potential, the velocities of particles of the fluid may be calculated [DNV-RP-C205, 2007]:

$$v_x = \frac{\partial \Phi}{\partial x} = \omega \bar{\xi} \frac{\cosh(k(d+z))}{\sinh(kd)} \sin(\omega t - kx), \tag{3.13}$$

$$v_z = \frac{\partial \Phi}{\partial z} = \omega \bar{\xi} \frac{\sinh(k(d+z))}{\sinh(kd)} \cos(\omega t - kx). \tag{3.14}$$

Substituting the functions of potential (3.12) and profile (3.11) to the dynamic condition of free surface gives:

$$\omega^2 = gk \tanh(kd). \tag{3.15}$$

Pressure can be determined by using the potential obtained (3.12) and the Bernoulli equation (3.9):

$$p = -\rho g z + p_w \sin(\omega t - kx), \tag{3.16}$$

where $p_w = \rho g \bar{\xi} \dfrac{\cosh(k(d+z))}{\sinh(kd)}, \qquad z < 0.$

For large depths, i.e. for $kd \to \infty$, the above formulas take a simpler form:

velocity v_x, v_z:
$$v_x = v_z = \omega \bar{\xi} e^{kz}, \tag{3.17}$$

dispersion:
$$\omega = \sqrt{g k_0}\,,\ \lambda_0 = \frac{gT^2}{2\pi}, \tag{3.18}$$

pressure:
$$p = -\rho g z + \rho g \bar{\xi} e^{kz}. \tag{3.19}$$

Functions describing the trajectories of the fluid's particles are obtained by integration of velocity over time. The trajectories in local coordinate systems, when a wave profile conforming to (3.11) is assumed, change with sea depth (Fig. 3.3) forming ellipses. For $kd \to \infty$ the ellipses turn into circles whose radii may be approximated with the expression $\bar{\xi} e^{kz}$. For a shallow see the trajectories are flattened and tend to horizontal limes for $z \to -d$.

The linear model of waves described above (also called the Airy model) is adequate for waves with low amplitudes relatively to their length and sea depth. Its applicability is ruled by the following conditions:

$$\bar{\xi} k << 2\pi \text{ and } \bar{\xi} << d. \tag{3.20}$$

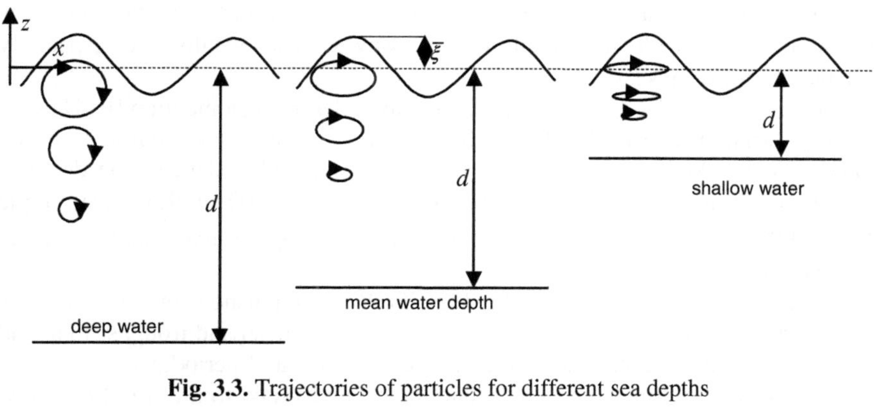

Fig. 3.3. Trajectories of particles for different sea depths

The limiting criteria allowing for the use of a given variant of the formulas produced, i.e. (3.13) – (3.16) or (3.17) – (3.19), depending on the depth d, are summarized in Table 3.1, following [Chakrabarti S. K., 2005].

Table 3.1. Criteria for depth and applicability of formulas in the linear model

Sea depth	Criterion	Wave length
large	$\frac{d}{\lambda} \geq \frac{1}{2}$	$\lambda = \lambda_0 = \frac{gT^2}{2\pi}$
medium	$\frac{1}{20} < \frac{d}{L} < \frac{1}{2}$	$\lambda = \lambda_0 \left[\tanh\left(\frac{2\pi d}{\lambda_0} \right) \right]^{\frac{1}{2}}$
small	$\frac{d}{\lambda} \leq \frac{1}{20}$	$\lambda = T\sqrt{gd}$

3.1.2 Stokes Model

The Stokes theory proposes models of waves which are nonlinear due to the nonlinearity of the dynamic condition of free surface in the variables $\mathbf{v}$ and p:

$$\frac{\partial p}{\partial t} + v_x \frac{\partial p}{\partial x} + v_z \frac{\partial p}{\partial z} = 0. \tag{3.21}$$

Depending on the number of components in the formula for velocity v_x, the Stokes theory is differentiated as first order, second order, third order, etc., see for example [Schwartz L., 1974], [Longuet-Higgins M. S., 1984]. Relations for the second order Stokes theory are shown in Table 3.2. In the description of waves this theory gives a fairly accurate approximation if the order of the method is increased along with the amplitude of waves. The results are especially satisfactory for deep seas.

There are numerous different models of waves deserving a mention [El-Hawary F., ed., 2001], [Webb D. J., 1978], [Komen G. J., et al., 1994], [Grue J., et al., 2003], [Tucker M. J., Pitt E. G., 2001]. Their applicability depends on depth d, wave period T and wave amplitude $\bar{\xi}$. Fig. 3.4, following [Dean R. G., Dalrymple R. A., 1998], shows a graph helpful in the choice of appropriate models (d and H are given in feet, T in seconds).

Trochoidal waves [Łomniewski K., 1969] distinguished in Fig. 3.4 are characterized by different shapes of crests (tall and narrow) and troughs (long and flat). They are an intermediate form between solitons and periodic waves (linear and nonlinear). Solitons lack troughs and their lengths tend to infinity. In shallow waters trochoidal waves transforms into solitons.

Table 3.2. Some dependencies in second order Stokes theory

potential Φ	$\Phi_l + \frac{3}{8}\frac{\pi H}{kT}\left(\frac{\pi H}{\lambda}\right)\frac{\cosh(2k(z+d))\sin(2\gamma)}{\sinh^4(kd)}$
velocity v_x	$v_{x,l} + \frac{3}{4}\frac{\pi H}{T}\left(\frac{\pi H}{\lambda}\right)\frac{\cosh(2k(z+d))}{\sinh^4(kd)}\cos(2\gamma)$
velocity v_z	$v_{z,l} + \frac{3}{4}\frac{\pi H}{T}\left(\frac{\pi H}{\lambda}\right)\frac{\sinh(2k(z+d))}{\sinh^4(kd)}\sin(2\gamma)$
pressure p	$p_l + \frac{3\rho g H}{4}\frac{\pi H}{\lambda\sinh(2kd)}\left[\frac{\cosh(2k(z+d))}{\sinh^2(kd)} - \frac{1}{3}\right]\cos(2\gamma) -$ $+ \frac{\rho g H}{4}\frac{\pi H}{\lambda\sinh(2kd)}\left[\cosh(2k(z+d)) - 1\right]$
where Φ_l, $v_{l,x}$, $v_{l,z}$, p_l are determined as in the linear wave theory (formulas (3.16) – (3.20)), λ is the wave length, $\gamma = \omega t - kx$	

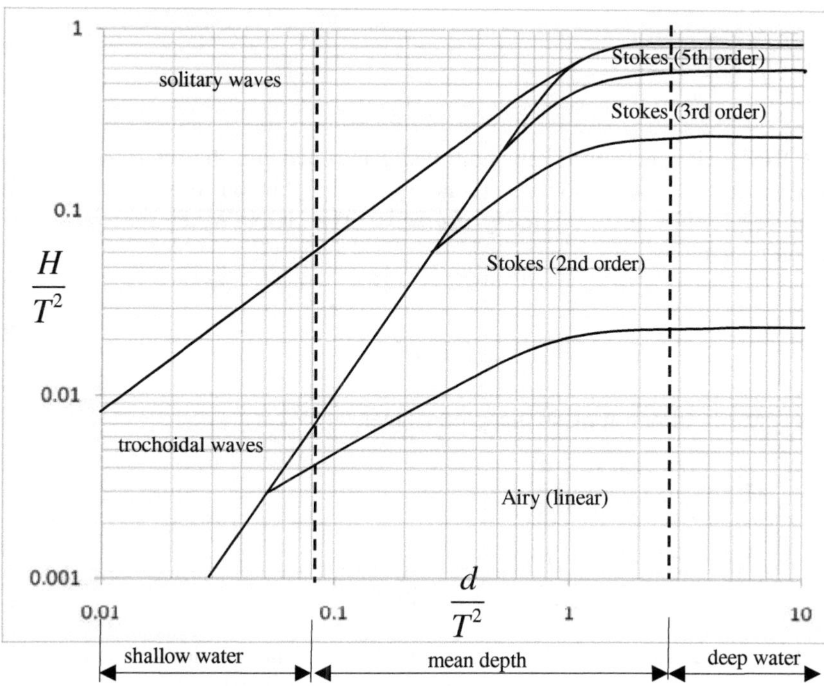

Fig. 3.4. Applicability ranges of different theories of sea waves

3.1.3 Statistical Description of Waves

In analyses and calculations, sometimes a spectral description of waves is used, which treats the phenomenon as a stationary problem within a finite time period (usually between 20 min and a few hours) [Massel S. R., 1996]. The spectrum $S(\omega)$ (spectral density of power) is defined as a function of some basic parameters of a wave, such as: significant wave height H_S, wave period T_p determined by the frequency of occurrence of function's $S(\omega)$ maxima and time T_z, calculated as mean wave period within the assumed time interval.

There are numerous methods to describe waves using a function of waves density distribution. Among the most often used are: the function devised by Pierson-Moskowitz [Pierson W. J., Moskowitz L., 1964] and its modified form, developed in the JONSWAP project [Hasselmann K., et al., 1973]. Examples of other, more complex functions are Ochi-Hubble [Ochi M. K., Hubble E. N., 1976] and Torsethaugen [Torsethaugen K., Haver S., 2004] distributions. Below, the forms proposed by Pierson-Moskowitz and JONSWAP are briefly presented, as they will be used in some mathematical models considered further herein.

The Pierson-Moskowitz spectrum is described thus:

$$S_{PM}(\omega) = \frac{5}{16} H_s^2 \omega_p^4 \omega^{-5} e^{\left(-\frac{5}{4}\left(\frac{\omega}{\omega_p}\right)^{-4}\right)} \tag{3.22}$$

where $\omega_p = \dfrac{2\pi}{T_p}$.

The JONSWAP spectrum takes additionally into account the intermediate states occurring at the onset of waves (contrary to the Pierson-Moskowitz spectrum which treats only waves already formed). The JONSWAP spectral density function is given by:

$$S_J(\omega) = A_\eta S_{PM}(\omega) \cdot \eta^{e^{\left(-0.5\left(\frac{\omega-\omega_p}{\sigma\omega_p}\right)^2\right)}} \tag{3.23}$$

where η – dimensionless shape parameter,

$$\sigma = \begin{cases} \sigma_a & \text{for } \omega \le \omega_p \\ \sigma_b & \text{for } \omega > \omega_p, \end{cases}$$

$A_\eta = 1 - 0.287 \ln(\eta)$,

σ_a, σ_b – parameters of the distribution.

For $\eta = 1$, the relation (3.23) reduces to (3.22). The method of selecting the parameters η, σ_a, σ_b is specified in norms, e.g. [DNV-RP-C205, 2007], and in literature [Clauss G. F., et al., 1992], [Holthuijsen L. H., 2007].

Spectral density $S(\omega)$ may be used to generate an irregular wave. Such waves are closer to actual waves than those considered by many authors in their papers. The simplest example of an irregular wave is a sum of a given number N of harmonic components:

$$\xi_r(t)=\sum_{k=1}^{N} A_k \cos(\omega_k t+\varepsilon_k), \tag{3.24}$$

where A_k – random amplitude, calculated as $A_k=\sqrt{2\cdot\Delta\omega\cdot S(\omega_k)}$,

$\varepsilon_k \in \langle 0,\ldots,2\pi\rangle$ – random initial phase with uniform distribution,

$\Delta\omega_k=\omega_k-\omega_{k-1}$.

The dependency (3.24) allows us to determine the waves parameters according to the formulas (3.12) – (3.16) or those given in Table 3.2. Summation is performed for every calculated value. For example, the velocity from the formula (3.13) is determined thus:

$$v_x=\sum_{i=1}^{N}\omega_i\bar{\xi}_i\frac{\cosh(k_i(d+z))}{\sinh(k_i d)}\sin(\omega_i t-k_i x+\varepsilon_i). \tag{3.25}$$

The number of components N may reach a few hundred and more (guidelines are specified in appropriate norms). A pre-filled array of values needed, combined with interpolation over time, is therefore desirable when integrating the equations of a system's dynamics. Such approach ensures much shorter times taken by computations.

3.2 Determination of Forces Acting on Objects Immersed in Water

Let us consider a body immersed in a liquid. Action of the liquid on the body results from the motion of the body and of the liquid itself. Diffraction and radiation phenomena must be taken into consideration in the general case, as the body may influence the motion of the liquid. However, in performing analyses of structures with small characteristic dimensions relatively to the wave length, a simplifying assumption may be introduced that the Morison equation [Morison J. R., et al., 1950] governs the forces acting on a body immersed in water [Faltinsen O. M., 1990], [Sarpkaya T., Isaacson M., 1981]. Their determination requires the knowledge of coefficients whose values are yielded from appropriate laboratory experiments. In examples contained in the current book, interaction between water and pipelines is analyzed. The pipelines are modelled with beam elements. It can be easily proven that the Morison equation's applicability criterion, defined in [DNV-RP-C205, 2007] as:

$$\lambda > 5D\ , \tag{3.26}$$

is satisfied even for pipelines with large diameters (D greater than 1 m). This justifies the assumption of the Morison equation's applicability in dynamic analyses of pipelines throughout the rest of the volume. The force of a liquid's action on a body in general is defined as a function of multiple dimensionless parameters [Faltinsen O. M., 1990]:

$$f = f\left(t, T, KC, \mathrm{Re}, d_{fr}, \hat{e}, \hat{u}\right), \tag{3.27}$$

where $KC = \dfrac{u_0 T}{D}$ – Keugelan-Carpenter number,

$\mathrm{Re} = \dfrac{u_0 D}{\nu}$ – Reynolds number,

$d_{fr} = \dfrac{\pi D}{\lambda}$ – diffraction parameter,

$\hat{e} = \dfrac{K}{D}$ – reduced body surface roughness,

$\hat{u} = \dfrac{u_0}{f_s D}$ – reduced velocity of the liquid,

f_s – angular frequency of the body's oscillations,

u_0 – amplitude of the liquid's velocity,

ν – viscosity of the liquid,

K – surface roughness,

D – diameter (characteristic dimension) of the body.

The force of the liquid's action on a segment of a given body may be decomposed into tangent and normal components (Fig. 3.5b). For an asymmetric segment, turbulence and other effects, a lifting force F_L may additionally occur, which acts in the direction perpendicular to the liquid's flow (Fig. 3.5c).

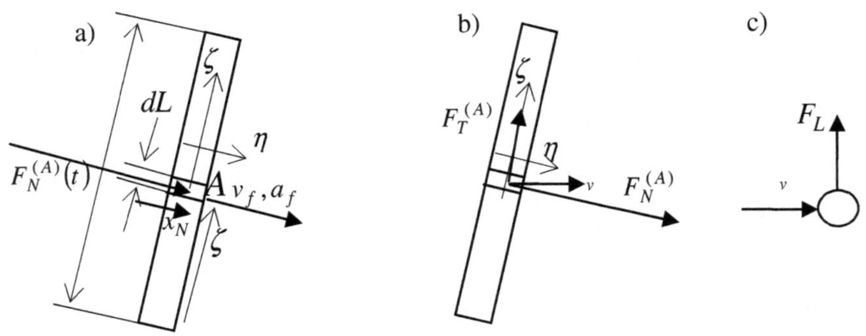

Fig. 3.5. Forces acting on a segment of a body: a) determination of the normal component, b) normal and tangential components, c) lifting force

The normal force acting on the segment determined by a coordinate ς (Fig. 3.5a) can be described with the Morison equation [DNV-RP-C205, 2007]:

$$F_N^{(A)}(t,\varsigma) = \rho A a_f + \rho C_A A a_r + \frac{1}{2}\rho C_D D v_r |v_r|, \tag{3.28}$$

where $F_N^{(A)}(t,\varsigma)$ – normal force acting on the segment of the body at the coordinate's value ς,

ρ – density of the liquid,

$A = A(\varsigma)$ – area of the body's segment determined by coordinat ς,

a_f – acceleration of the liquid,

$a_r = a_f - \ddot{x}_N$ – relative acceleration,

x_N – displacement of the body in the direction normal to the longitudinal axis,

$v_r = v_f - \dot{x}_N$ – velocity of the liquid relative to the body,

C_A – dimensionless coefficient of added mass,

C_D – dimensionless drag coefficient.

The Morison equation (3.28) may be used if the following relations hold [Chakrabarti S. K., 2005]:

$$KC > 6 \text{ and } d_{fr} << 0.5. \tag{3.29}$$

E.g., by substituting data of the largest diameters of pipelines, i.e. $D = 1$ m, with wave period $T = 10$ s and its amplitude $A = 1$ m, and taking into account the linear theory of waves (formula (3.13)), we obtain $KC = 6.28$, $d_{fr} = 0.056$. The condition (3.29) ought to be checked supplementing the general condition (3.26) due to the fact that the diffraction parameter determines the magnitude of wave dispersion as the result of meeting the object. When the number KC is small, the Morison equation should be replaced with calculations using e.g. potential theory or Froude-Krylov forces [Chakrabarti S. K., 2005]. The force given by (3.28) is particularly suitable for modelling structures such as ropes, pipes, beams. It may also yield satisfactory results for small 3D objects. On the other hand, it may not be successfully applied to large objects whose influence on the motion of water particles is significant enough to cause reflections of waves.

The net normal force and its application point x_h in the local coordinate system $0\varsigma\eta$ (Fig. 3.5) with origin in the middle of the item's length can be calculated from:

$$F_N = \int_{-0.5L}^{0.5L} F_N^{(A)}(t,\varsigma)\, d\varsigma, \tag{3.30}$$

$$x_h = \frac{\int_{-0.5L}^{0.5L} F_N^{(A)}(t,\varsigma)\cdot\varsigma\cdot d\varsigma}{F_N}, \tag{3.31}$$

where $F_N^{(A)}(t,\varsigma)$ – force described by (3.28),
L – length of the object.

Coefficients C_A and C_D which appear in the Morison equation are functions of dimensionless parameters from (3.27), whereby:

$$C_A = C_A(\text{Re}, KC, \hat{e}), \tag{3.32}$$

$$C_D = C_D(\text{Re}, KC, \hat{e}), \tag{3.33}$$

where $\text{Re}, KC, \hat{e}$ are defined in (3.27).

Further details and practical advice about how these coefficients depend on different variables are to be found, among other things, in [Sarpkaya T., Isaacson M., 1981], [API-RP-2A-LRFD, 1993], [Bai Y., Bai Q., 2005], [Chakrabarti S. K., 2005], [DNV-RP-F105, 2006], [DNV-RP-C205, 2007]. These works also include cases of calculations for items placed very close to the free surface, the bottom or another large object which changes the coefficient of the added mass.

The tangent force caused by hydrodynamic resistance should be considered mainly in analyses of long objects with rough surfaces. It may be formulated thus [DNV-RP-C205, 2007]:

$$F_T = \frac{1}{2}\rho C_{Dt} v^2, \tag{3.34}$$

where C_{Dt} – coefficient of hydrodynamic resistance in the tangent direction,
v – amplitude of the liquid's net velocity.

The resistance coefficient C_{Dt} may be described with the formula [Eames M.C., 1968]:

$$C_{Dt} = C_D(m + n\sin(\alpha))\cos(\alpha), \tag{3.35}$$

where α – angle between net velocity and the item's longitudinal axis,
m, n – coefficients from Table 3.3.

Table 3.3. Values of the coefficients *m*, *n* according to [Eames M. C., 1968]

Item type	*m*	*n*
Smooth cylindrical surfaces	0.02-0.03	0.04-0.05
Porous cables, pipes	0.25-0.5	0.25-0.50
Six-strand ropes	0.03	0.06

Also the seabed exerts forces on elements of a pipeline as it is being laid. The problem of finding a description of the seabed and its interaction with installed objects was the subject of many works. Behaviour of the seabed is highly dependent on the sea considered. It is thus difficult to formulate one universally applicable model. Empirical models are most commonly used, which reproduce approximately the character of a given type of seabed. In the present volume, a model developed by Verley and Lund [Verley R., Lund K. M., 1995] will be used. It gives the following formula for the value of seabed penetration in the normal direction:

$$\frac{\Delta u_n}{D} = 0.0071\left(Y Z^{0.3}\right)^{3.2} + 0.062\left(Y Z^{0.3}\right)^{0.7}, \tag{3.36}$$

where $Y = \dfrac{N_C}{D t_u}$,

$Z = \dfrac{t_u}{D\rho'}$,

Δu_n – seabed penetration in [m],

N_C – contact force per a pipeline's length link [kN/m],

t_u – seabed material's shear strength [kPa],

ρ' – density of seabed material (wet) [kN/m^3].

The dependency (3.36) holds for $Y Z^{0.3} < 2.5$. Otherwise, the following formula gives better accuracy:

$$\frac{\Delta u_n}{D} = 0.09\left(Y Z^{0.3}\right). \tag{3.37}$$

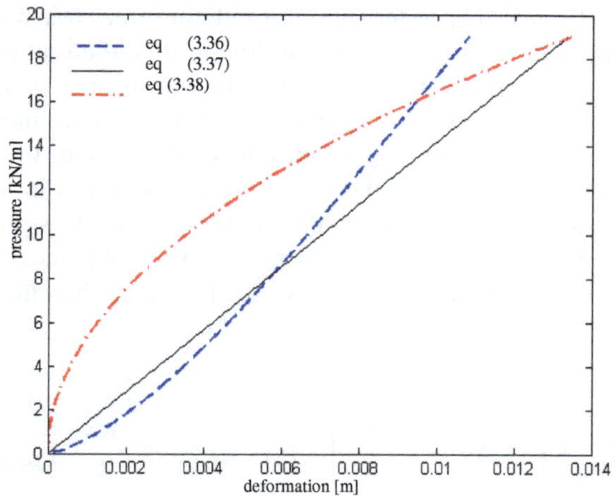

Fig. 3.6. Sample characteristics of sea beds according to Verley-Lund and bearing capacity models

Fig. 3.6 presents sample characteristics of sea beds described by (3.36) and (3.37) differentiated by bearing capacity model. In the last case, analogously to the model determining the bearing capacity, the following formulas are used [Bai Y., Bai Q., 2005]:

$$\frac{\Delta u_n}{D} = 0.5\left(1-\left(1-4\left(\frac{(b+0.5)f}{2\left(b+0.5f^3\right)}\right)^2\right)^{0.5}\right), \tag{3.38}$$

where $f = \dfrac{\alpha}{0.5(b+0.5)}$,

$\alpha = \dfrac{N_C}{2\rho' D^2}$,

$b = \dfrac{f_B t_u}{\rho' D}$,

f_b – coefficient (for large support lengths taking the value 5.14).

The graphs in Fig. 3.6 are plotted for the following parameters: $D = 0.3\,\text{m}$, $t_u = 35\,\text{kPa}$, $\rho' = 1.215\,\frac{\text{g}}{\text{cm}^3}$.

Forces caused by waves or sea currents which act on elements of pipes or other structures lying on a seabed result in their displacement in the forces' direction. The action of the seabed in the transverse (horizontal) direction needs therefore to be taken into consideration. More even so, given that the bodies in question often lie in sand or other material (e.g. clay). For objects penetrating the seabed's material only slightly, the Coulomb model of friction may be used [Bai Y., Bai Q., 2005], however, with soft materials or deeper penetration the description becomes more complex. Typical characteristics formulating the transversal forces as functions of displacement are nonlinear. Adding to this, the seabed in the contact area may become deformed and its material's consistence may vary due to displacement. Often a characteristic composed of two different linear parts is used. One of them describes the working conditions of the system up to the moment when a ground layer is shorn off. The other treats the forces in action when the object is moved along with seabed's material. This approach is presented, among other things, in the norm [DNV-RP-F105, 2006]. Guidelines concerning the selection of parameters and its applicability are also stated therein. The transversal force is calculated thus:

$$F_{lat} = \begin{cases} k_{l1}\Delta u_t & \text{for} \quad F_{lat} < \mu_t F_V \\ \mu_t F_V + k_{l2}\left(\Delta u_t - \dfrac{\mu_t F_V}{k_{l1}}\right) & \text{for} \quad \mu_t F_V \le F_{lat} < F_{lat,\max} \end{cases}, \tag{3.39}$$

where k_{l1} – equivalent transversal stiffness coefficient up to shearing off of the seabed's material,

k_{l2} – stiffness coefficient for motion developed in the transverse direction,

$\Delta\mu_t$ – transverse displacement,

F_V – vertical contact force per length of link,

μ_t – friction coefficient between the seabed and the body (transversal),

$F_{lat,\max} = F_{lat,\max}(\mu_t, F_V, D, t_u, \nu, \rho', G)$ – transverse force per length of link maximally transmissible by the seabed's material,

ν, G – Poisson number and shear modulus of the seabed's material.

There also exist other models of sea beds, including more complex ones. In [DNV-RP-F105, 2006] guidelines can be found on the use of different models and parameters of the seabed. The fact deserving a mention is that the models described above and used below omit in particular any changes of the seabed's properties due to the oscillatory impact of pipes and cables.

3.3 Methods of Simplified Description of Movement of Offshore Structures

The monograph [Adamiec-Wójcik I., et al., 2008] contains a short survey of papers related to the description of a vessel's motion under wave action. Those are most often fairly complex models used e.g. to determine strains to which the hull is exposed. However, they are hardly useful in quick dynamic analyses of offshore structures. Therefore, many works in the field of dynamics of such objects, particularly cranes, take the assumption that movements of the base can be described with relatively simple functions. Often it is further assumed that the vessel moves only in the vertical plane passing through the longitudinal symmetry axis of the deck. Such propositions seem reasonable for most offshore cranes that operate predominantly on vessels which are moored and properly positioned against the waves. Instead of making assumptions about the base's motion, some papers prefer to deal with forces acting on it.

It is common practice to assume that the motion of a vessel or any given point of a crane is harmonic or pseudo-harmonic. Sinusoidal waves with angular frequencies of 0.56 and 0.74 rad/s and height of 1 m directed along a ship's longitudinal axis are considered in [Das S. N., Das S. K., 2005]. In the papers [Balachandran B., et al., 1999] and [Li Y. Y., Balachandran B., 2001] two kinds of functions are used to describe the motion of the jib's head in a crane installed on a ship. These are:

- harmonic

$$x_e = (F \sin \omega t)\cos\psi \quad ; \quad y_e = (F \sin \omega t)\sin\psi\,, \tag{3.40}$$

- periodic

$$
\begin{aligned}
x_e &= F\left(\sin \omega t + \frac{1}{4}\sin 2\omega t + \frac{1}{9}\sin 3\omega t\right)\cos\psi ,\\
y_e &= F\left(\sin \omega t + \frac{1}{4}\sin 2\omega t + \frac{1}{9}\sin 3\omega t\right)\sin\psi ,
\end{aligned}
\tag{3.41}
$$

where F – excitation amplitude,
ω – angular frequency of excitation,
ψ – boom inclination angle.

The papers [Osiński M., Wojciech S., 1994], [Osiński M., Wojciech S., 1998] and [Osiński M., et al., 2004] focus on the planar problem of lifting a load from a ship's deck, whereby the motion of the deck is described with a harmonic function:

$$
y_d = F\sin(\omega t + \beta) , \tag{3.42}
$$

where β is the phase angle.

General motion of the base (3 displacements and 3 rotations) defined with pseudo-harmonic functions (reducing to harmonic when a single component is taken) can be considered for the model of a crane presented in [Maczyński A., 2005], [Maczyński A., Wojciech S., 2007]. Similarly for A-frames (harmonic functions) [Fałat P., 2004], [Adamiec-Wojcik I., et al., 2009] and BOP gantries for transportation [Urbaś A., et al., 2010], [Urbaś A., 2011].

In [Ellermann K., et al., 2002] and [Ellermann K., et al., 2003] two components are distinguished among the forces exerted by the waves: a periodically changing one and a constant one (related to drifting). These forces are determined from the following formulas:

$$
\mathbf{F}_w(t) = \begin{bmatrix} a\,er_x\cos(\omega t) - a\,ei_x\sin(\omega t) + a^2 p_{drag} \\ a\,er_\theta\cos(\omega t) - a\,ei_\theta\sin(\omega t) \\ a\,er_z\cos(\omega t) - a\,ei_z\sin(\omega t) \end{bmatrix} , \tag{3.43}
$$

where a – wave amplitude,
er_j, ei_j – coefficients empirically determined for a particular type of ships, whereby $j \in \{x, \theta, z\}$,
ω – wave angular frequency,
x, θ, z – generalized coordinates of the hull,
p_{drag} – drift force determined empirically.

In the article [Cha J. H., et al., 2010a], hydrodynamic forces are proposed to be present among those acting on the crane's base: one due to radiation and another excited by the wave. The forces stemming from sea waves acting on the vessel where the crane is installed are determined based on the spectrum of the wave in [Witz J. A., 1995].

Sometimes, the motion of the base is described using measurements already performed in real conditions. In [Masoud Z. N., 2000] it is assumed that a ship with a crane on board oscillates transversally and longitudinally and is subjected to the motions of heaving, swaying and rolling. Calculations are based on data obtained empirically [Fossen T. I., 1994] which describe transverse and longitudinal oscillations as well as heaving, swaying and rolling of a selected point of the ship (the reference point). Also in [Driscoll F. R., et al., 2000], measured displacements of an A-frame are used to study a model of a cage suspended in large depths (1730 m). In the paper [Pedrazzi C., Barbieri G., 1998] the ADAMS package is used to analyse the dynamics of a vessel with a crane. The sea is modelled as a massless object which moves vertically relatively to the bottom. Its motion is defined in two ways: as a spline in time based on real measurements of sea waves and as an analytic function constructed using a pseudostochastic model of a wave.

In many of the models and computer programmes discussed herein, a provision is made for defining the general motion of an offshore structure's base as a pseudo-harmonic function with arbitrary number of components.

4 Homogeneous Transformations and Joint Coordinates in the Description of Geometry of Multibody Systems

Basic models of bodies used in dynamic analysis of mechanical systems, including multibody systems, are a material point and a rigid body. They have, respectively, three and six degrees of freedom. To describe their positions either three or six independent coordinates must therefore be given. Usually, the position is given in a rectangular clockwise Cartesian system. It is then convenient to express the position of a point as a vector, also called a radius vector. To describe a body's position, an additional coordinate system is attached to it in a fixed way. The position of this coordinate system, thus also of the body, is defined by giving the position vector of a selected point of the body (usually coinciding with the origin of the coordinate system attached to the body) and additionally a 3×3 matrix called a rotation matrix. In classical mechanics, displacement of a body from one position to another is treated as a superposition of two motions: translation and rotation. As a consequence, if a position vector of a point in the movable coordinate system attached to the body is given, and a position vector of this point in the reference system is to be determined, two mathematical operations are necessary: multiplication of the rotation matrices and addition of two vectors. By introducing the method of homogeneous transformations, the notation can be simplified. Such transformations are described by 4×4 matrices and take into account both a translation of a coordinate system and its rotation. The convenience of such interpretation makes it highly popular in robotics [Craig J. J., 1988], [Morecki A., et al., 2002], [Spong M. W., et al., 2006], [Jezierski E., 2006], which is a domain where multibody systems commonly occur.

In the classical approach, the positions of links are expressed in a global static coordinate system. To describe a system with n links, $6{\times}n$ parameters have to be specified. These are called absolute coordinates. The use of joint coordinates which define motions of links relative to their predecessors in a kinematic chain enable a description of the positions of the system's consecutive links with far less parameters.

The current chapter offers a basic introduction to describing positions and orientations of coordinate systems, transformations of vectors and joint coordinates. Application of homogeneous transformations and joint coordinates to describe the geometry of multibody systems is also discussed.

E. Wittbrodt et al.: Rigid Finite Element Method, OEO 1, pp. 45–58.
springerlink.com

4.1 Position Vector: Rotation Matrix

To describe a point's position in a three-dimensional space, the following procedure is used. First, a coordinate system is defined in which the point's position will be determined. It is called the reference or base system. For the remaining part of this book that system is assumed to be Cartesian and clockwise. Next, the position vector is determined. The starting point of this vector coincides with the system's origin and the end is at the point in question (Fig. 4.1). If the origin is denoted with {0} and the point with P, the position vector may be given as:

$$ {}^{0}\mathbf{r}_P = \begin{bmatrix} {}^{0}x_P \\ {}^{0}y_P \\ {}^{0}z_P \end{bmatrix} = \begin{bmatrix} {}^{0}x_P & {}^{0}y_P & {}^{0}z_P \end{bmatrix}^T, \tag{4.1} $$

where the index T stands for the transposition operator of a vector (or matrix).

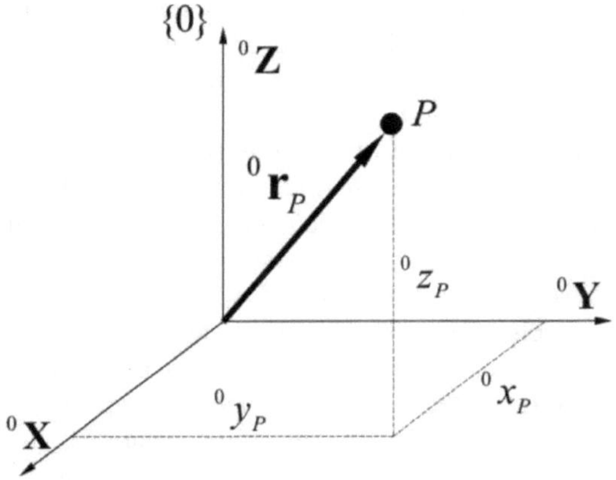

Fig. 4.1. Position vector of a point

Points will be denoted with capital letters and coordinate systems with either capital letters or bracketed digits (thus { }). The upper preceding index accompanying a vector symbol **r** (or a component thereof), will indicate the coordinate system in which the vector **r** is specified, whereas in the lower following index the point to which the given vector pertains will be given.

Determination of a rigid body's position in a three-dimensional space requires defining two coordinate systems: a reference one, relative to which the position will be specified, and another one attached in a fixed way to the given body. Hence, the description of the body's position can be construed as positioning two coordinate systems against each other (Fig. 4.2).

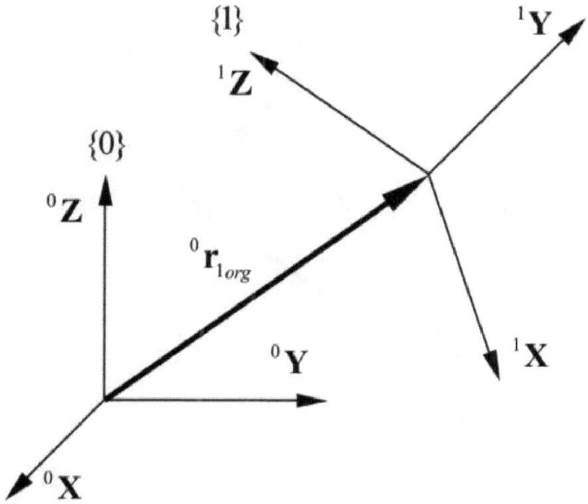

Fig. 4.2. Relative position of two coordinate systems

Unambiguous statement of a position may be done by giving the position vector ${}^0\mathbf{r}_{1org}$ of system {1}'s origin in system {0} and a rotation matrix ${}^0_1\mathbf{R}$ defining the orientation of system {1} relative to system {0}. The vector ${}^0\mathbf{r}_{1org}$ takes the form:

$$ {}^0\mathbf{r}_{1org} = \begin{bmatrix} {}^0x_{1org} & {}^0y_{1org} & {}^0z_{1org} \end{bmatrix}^T , \tag{4.2}$$

and the rotation matrix can be written as:

$$ {}^0_1\mathbf{R} = \begin{bmatrix} {}^1\mathbf{X}\,{}^0\mathbf{X} & {}^1\mathbf{Y}\,{}^0\mathbf{X} & {}^1\mathbf{Z}\,{}^0\mathbf{X} \\ {}^1\mathbf{X}\,{}^0\mathbf{Y} & {}^1\mathbf{Y}\,{}^0\mathbf{Y} & {}^1\mathbf{Z}\,{}^0\mathbf{Y} \\ {}^1\mathbf{X}\,{}^0\mathbf{Z} & {}^1\mathbf{Y}\,{}^0\mathbf{Z} & {}^1\mathbf{Z}\,{}^0\mathbf{Z} \end{bmatrix}. \tag{4.3}$$

Elements of the rotation matrix are the appropriate scalar products of versors of the axes of systems {1} and {0}. Since the scalar product of versors equals the cosine of the angle between them, this matrix is sometimes called the direction cosine matrix.

The vector ${}^0\mathbf{r}_{1org}$ and the rotation matrix ${}^0_1\mathbf{R}$ enable determining the position vector ${}^0\mathbf{r}_P$ of the point P in the system {0} (Fig. 4.3) given the position vector of the point P in the system {1}. It can verily be stated:

$$ {}^0\mathbf{r}_P = {}^0\mathbf{r}_{1org} + {}^0_1\mathbf{R}\,{}^1\mathbf{r}_P\,. \tag{4.4}$$

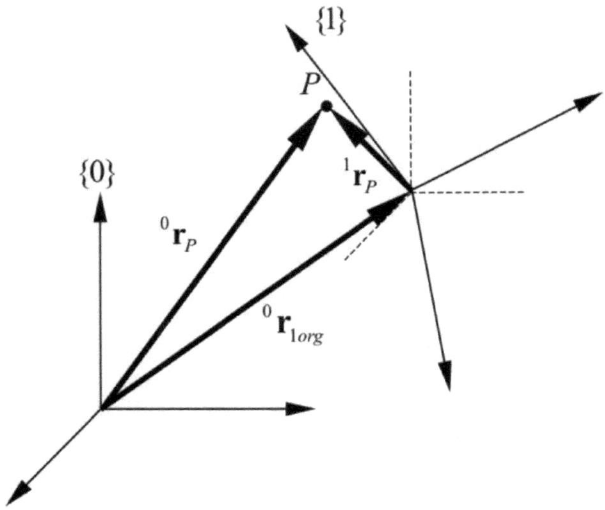

Fig. 4.3. Point P specified in coordinate systems {1} and {0}

It is a noteworthy observation that columns and rows of a rotation matrix are orthonormal vectors[1]. Whence follows an important dependency:

$$ {}^1_0\mathbf{R} = {}^0_1\mathbf{R}^{-1} = {}^0_1\mathbf{R}^T , \tag{4.5} $$

which states that the inverse of a rotation matrix equals its transpose. This property greatly simplifies transformations of formulas and improves numerical efficiency.

The rotation matrix (4.3) is specified with nine elements. However, as the versors of axes of a coordinate system must satisfy 6 conditions (three for their lengths and three for them to be perpendicular to each other), there are only three independent parameters defining the rotation matrix.

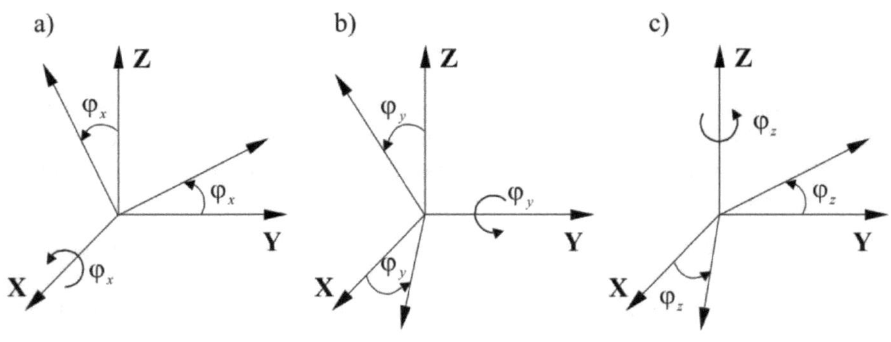

Fig. 4.4. Elementarny rotations around the axes **X**, **Y** and **Z**

[1] Column vectors $\mathbf{X}_1,...,\mathbf{X}_k$ are orthogonal if $\mathbf{X}_i^T\mathbf{X}_j = 0$ for each $i \neq j$ and they are orthonormal if they are orthogonal and furthermore $\mathbf{X}_i^T\mathbf{X}_i = \pm 1$ for $i = 1,...,k$

Let us define rotation matrices for the so-called elementarny rotations. These are rotations around a single axis **X**, **Y** or **Z** of the coordinate system. On schemes a), b) and c) in Fig. 4.4 rotations around the **X** axis by the angle φ_x, **Y** by φ_y and **Z** by φ_z are shown. Rotation matrices for elementarny rotations have the following forms:

$$\mathbf{R}_{\mathrm{X}}(\varphi_x)=\begin{bmatrix} 1 & 0 & 0 \\ 0 & \cos\varphi_x & -\sin\varphi_x \\ 0 & \sin\varphi_x & \cos\varphi_x \end{bmatrix}, \tag{4.6.1}$$

$$\mathbf{R}_{\mathrm{Y}}(\varphi_y)=\begin{bmatrix} \cos\varphi_y & 0 & \sin\varphi_y \\ 0 & 1 & 0 \\ -\sin\varphi_y & 0 & \cos\varphi_y \end{bmatrix}, \tag{4.6.2}$$

$$\mathbf{R}_{\mathrm{Z}}(\varphi_z)=\begin{bmatrix} \cos\varphi_z & -\sin\varphi_z & 0 \\ \sin\varphi_z & \cos\varphi_z & 0 \\ 0 & 0 & 1 \end{bmatrix}. \tag{4.6.3}$$

As mentioned before, of the nine elements of a rotation matrix (4.3) only three are independent. An arbitrary rotation of a given coordinate system relative to another one can thus be presented as a composition of three elementarny rotations. These rotations may be performed around axes of a previously fixed or the current coordinate system. In each case, there exist twelve different variations built from these rotations. The angles of rotation around the axes of the current system are called the Euler angles. This subject is further described, among other things, in [Craig J. J., 2004].

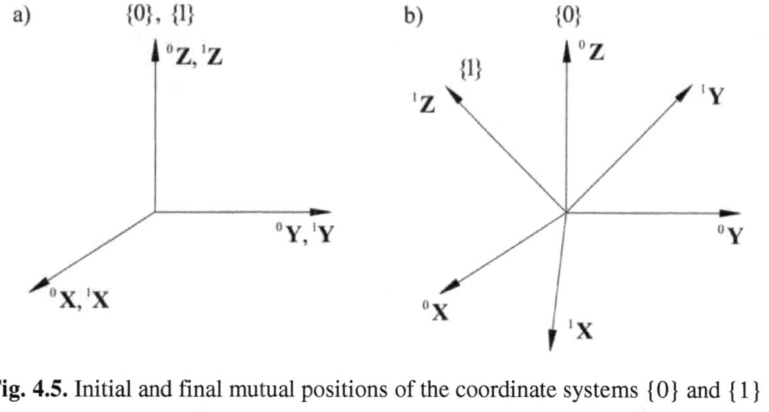

Fig. 4.5. Initial and final mutual positions of the coordinate systems {0} and {1}

In Fig. 4.5, a convention is shown for the so-called ZYX Euler angles. It is presented here because of its application further in this volume. Let us assume that initially the systems {0} and {1} coincide (Fig. 4.5a). To move the coordinate system {1} to the position shown in Fig. 4.5b, the following procedure may be applied:

- turn the system {1} about ${}^{1}\mathbf{Z}$ by the angle φ_z,[2]
- turn the system {1} about ${}^{1}\mathbf{Y}$ by the angle φ_y,
- turn the system {1} about ${}^{1}\mathbf{X}$ by the angle φ_x.

The above allows us to write:

$$ {}_{1}^{0}\mathbf{R} = \mathbf{R}_{{}^{1}\mathrm{Z}}(\varphi_z)\mathbf{R}_{{}^{1}\mathrm{Y}}(\varphi_y)\mathbf{R}_{{}^{1}\mathrm{X}}(\varphi_x). \tag{4.7} $$

Substituting the matrices of elementarny rotations (4.6) gives:

$$ {}_{1}^{0}\mathbf{R} = \begin{bmatrix} c\varphi_z c\varphi_y & c\varphi_z s\varphi_y s\varphi_x - s\varphi_z c\varphi_x & c\varphi_z s\varphi_y c\varphi_x + s\varphi_z s\varphi_x \\ s\varphi_z c\varphi_y & s\varphi_z s\varphi_y s\varphi_x + c\varphi_z c\varphi_x & s\varphi_z s\varphi_y c\varphi_x - c\varphi_z s\varphi_x \\ -s\varphi_y & c\varphi_y s\varphi_x & c\varphi_y c\varphi_x \end{bmatrix}, \tag{4.8} $$

where $c\varphi = \cos\varphi,\ s\varphi = \sin\varphi$.

Discussing the rotation matrix, we should add that sometimes the rotation angles, φ_x and φ_y in particular, are small. This allows the form of the matrix (4.8) to be simplified. However, such step usually leads to the loss of orthonormality by the matrix, i.e. the equation (4.5) no longer holds.

4.2 Homogeneous Transformation

Transformation of coordinates with the formula (4.4) requires carrying out two operations: multiplication of the matrix ${}_{1}^{0}\mathbf{R}$ by the vector ${}^{1}\mathbf{r}_P$ and addition of the result to the vector ${}^{0}\mathbf{r}_{1org}$. Notation following (4.4) is inconvenient, especially in cases of more complex transformations. Hence in [Craig J. J., 2004] rewriting of the relation (4.4) is proposed:

$$ {}^{0}\mathbf{r}_{P,4} = {}_{1}^{0}\mathbf{T}\,{}^{1}\mathbf{r}_{P,4}, \tag{4.9} $$

where ${}_{1}^{0}\mathbf{T}$ – homogeneous transformation matrix with dimensions 4×4,

${}^{0}\mathbf{r}_{P,4}, {}^{1}\mathbf{r}_{P,4}$ – position vectors of the point P with dimensions 4×1 in systems {0} and {1}, respectively.

---

[2] In order to simplify notation, in symbols φ_x, φ_y, φ_z, we omit the upper index that indicates the coordinate system.

Taking (4.4) the equation (4.9) may be put in this way:

$$\underbrace{\begin{bmatrix} {}^{0}\mathbf{r}_P \\ \hline 1 \end{bmatrix}}_{{}^{0}\mathbf{r}_{P,4}} = \underbrace{\left[\begin{array}{c|c} {}^{0}_{1}\mathbf{R} & {}^{0}\mathbf{r}_{1org} \\ \hline 0\ \ 0\ \ 0 & 1 \end{array}\right]}_{{}^{0}_{1}\mathbf{T}} \underbrace{\begin{bmatrix} {}^{1}\mathbf{r}_P \\ \hline 1 \end{bmatrix}}_{{}^{1}\mathbf{r}_{P,4}}. \tag{4.10}$$

This can be easily seen to be the equation (4.4) supplemented with the identity 1=1. A homogeneous transformation matrix ${}^{0}_{1}\mathbf{T}$ will, in short, be called a homogeneous transformation. It combines the operations of a rotation and a translation. It also necessitates the representation of position as a 4×1 vector.

The notion of joint coordinates[3] is introduced in [Craig J. J., 2004]. To use these coordinates to describe a point's position in a three-dimensional space, four values need to be given: (x_1, x_2, x_3, x_4), where x_4 must not equal zero. Given joint coordinates, Cartesian coordinates may be determined according to:

$$x = \frac{x_1}{x_4}, \qquad y = \frac{x_2}{x_4}, \qquad z = \frac{x_3}{x_4}. \tag{4.11}$$

In the present book (likewise in most papers on robotics, e.g. [Morecki A., et al., 2002], [Spong M. W., et al., 2006], [Jezierski E., 2006]) $x_4 = 1$ is assumed, hence in (4.10) the following change is introduced:

$$\mathbf{r}_{P,4} = \begin{bmatrix} x_P \\ y_P \\ z_P \\ 1 \end{bmatrix} = \begin{bmatrix} \mathbf{r}_P \\ 1 \end{bmatrix}. \tag{4.12}$$

In further considerations, we will omit the additional index 4, which indicates a position vector with dimensions 4×1. The dimension of a vector is determined by whether it is pertinent to a rotation matrix, or to a homogeneous transformation. It is usually clear enough. Therefore, the relation (4.9) will be written in the short form:

$${}^{0}\mathbf{r}_P = {}^{0}_{1}\mathbf{T}\,{}^{1}\mathbf{r}_P. \tag{4.13}$$

As is the case with rotation matrices, elementarny homogeneous transformations may also be distinguished. They are defined for rotations about the coordinate system's axes and a translation:

---

[3] The method of homogeneous transformations and coordinates is widely used in computer graphics. The most general form of a homogeneous transformation is $\left[\begin{array}{c|c} \text{rotation} & \text{tramslation} \\ \hline \text{perspective} & \text{scaling} \end{array}\right]$.

$$\mathbf{T}_X(\phi_x)=\begin{bmatrix}1 & 0 & 0 & 0\\ 0 & \cos\phi_x & -\sin\phi_x & 0\\ 0 & \sin\phi_x & \cos\phi_x & 0\\ 0 & 0 & 0 & 1\end{bmatrix} \text{ rotation by the angle } \varphi_x \text{ around the axis } \mathbf{X}, \qquad (4.14.1)$$

$$\mathbf{T}_Y(\phi_y)=\begin{bmatrix}\cos\phi_y & 0 & \sin\phi_y & 0\\ 0 & 1 & 0 & 0\\ -\sin\phi_y & 0 & \cos\phi_y & 0\\ 0 & 0 & 0 & 1\end{bmatrix} \text{ rotation by the angle } \varphi_y \text{ around the axis } \mathbf{Y}, \qquad (4.14.2)$$

$$\mathbf{T}_Z(\phi_z)=\begin{bmatrix}\cos\phi_z & -\sin\phi_z & 0 & 0\\ \sin\phi_z & \cos\phi_z & 0 & 0\\ 0 & 0 & 1 & 0\\ 0 & 0 & 0 & 1\end{bmatrix} \text{ rotation by the angle } \varphi_z \text{ around the axis } \mathbf{Z}, \qquad (4.14.3)$$

$$\mathbf{T}(\mathbf{a})=\begin{bmatrix}1 & 0 & 0 & a_x\\ 0 & 1 & 0 & a_y\\ 0 & 0 & 1 & a_z\\ 0 & 0 & 0 & 1\end{bmatrix} \text{ translation by a vector } \mathbf{a}=\begin{bmatrix}a_x\\ a_y\\ a_z\end{bmatrix}. \qquad (4.14.4)$$

Let us observe that the homogeneous transformation (4.14.4) can be expressed as a composition of three simpler translations. It is due to the fact that the translation by the vector **a** can be replaced by three translations along individual axes of the coordinate system.

If ZYX Euler angles (Fig. 4.6) are used to define a homogeneous transformation (4.9), it takes the form:

$${}^0_1\mathbf{T}=\begin{bmatrix}c\phi_z c\phi_y & c\phi_z s\phi_y s\phi_x - s\phi_z c\phi_x & c\phi_z s\phi_y c\phi_x + s\phi_z s\phi_x & {}^0x_{1org}\\ s\phi_z c\phi_y & s\phi_z s\phi_y s\phi_x + c\phi_z c\phi_x & s\phi_z s\phi_y c\phi_x - c\phi_z s\phi_x & {}^0y_{1org}\\ -s\phi_y & c\phi_y s\phi_x & c\phi_y c\phi_x & {}^0z_{1org}\\ 0 & 0 & 0 & 1\end{bmatrix}. \qquad (4.15)$$

In addition to the coordinate systems {0} and {1} let us consider another system {2}, whose position and orientation relative to the system {1} is given by a matrix ${}^1_2\mathbf{T}$ (Fig. 4.7). If the position vector ${}^2\mathbf{r}_P$ is known and furthermore ${}^0_1\mathbf{T}$, the following is obtained:

$${}^0_2\mathbf{T}={}^0_1\mathbf{T}\,{}^1_2\mathbf{T} \qquad (4.16)$$

allowing us to determine:

$${}^0\mathbf{r}_p={}^0_2\mathbf{T}\,{}^2\mathbf{r}_P={}^0_1\mathbf{T}\,{}^1_2\mathbf{T}\,{}^2\mathbf{r}_P\,. \qquad (4.17)$$

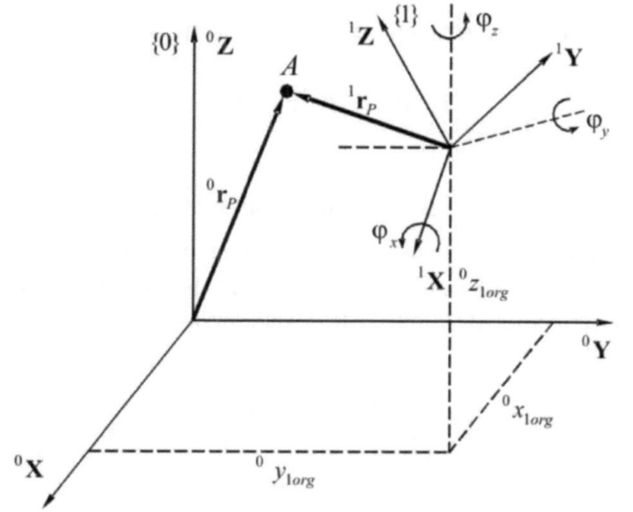

Fig. 4.6. ZYX Euler angles

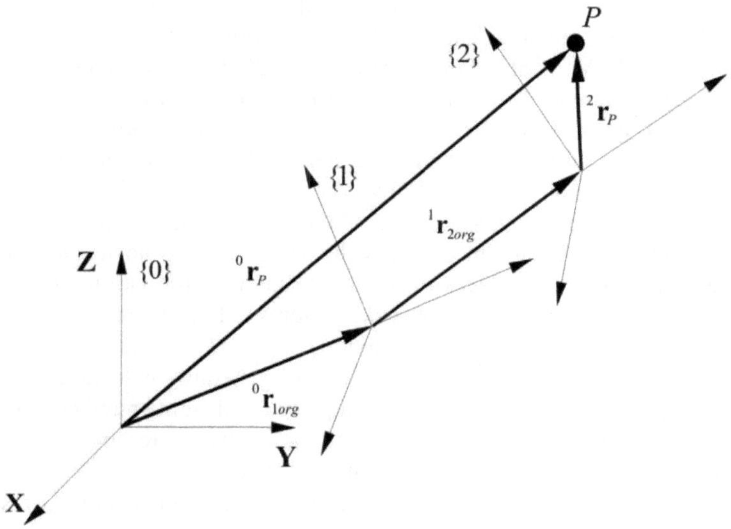

Fig. 4.7. Three coordinate systems: {0}, {1} and {2}

Attention should be drawn to the fact that equation (4.17) is much more convenient than its form that would follow from (4.4):

$$^0\mathbf{r}_P = {}^0_1\mathbf{R}\left({}^1\mathbf{r}_{2org} + {}^1_2\mathbf{R}\,{}^2\mathbf{r}_P\right) + {}^0\mathbf{r}_{1org}\,. \tag{4.18}$$

The homogeneous transformation ${}^{0}_{2}\mathbf{T}$ may be written as:

$$ {}^{0}_{2}\mathbf{T}=\left[\begin{array}{ccc|c} & {}^{0}_{1}\mathbf{R}\,{}^{1}_{2}\mathbf{R} & & {}^{0}_{1}\mathbf{R}\,{}^{1}\mathbf{r}_{2org}+{}^{0}\mathbf{r}_{1org} \\ \hline 0 & 0 & 0 & 1 \end{array}\right]. \tag{4.19} $$

Inverting a homogeneous transformation is an often needed step. It means that when a transformation ${}^{0}_{1}\mathbf{T}$ is given, the transformation ${}^{0}_{1}\mathbf{T}^{-1}={}^{1}_{0}\mathbf{T}$ is to be determined. To invert the matrix, any method may be used, e.g. that of elementary operations described in [Dziubiński I., Świątkowski T., 1982]. They can also be calculated in this way:

$$ {}^{0}_{1}\mathbf{T}^{-1}={}^{1}_{0}\mathbf{T}=\left[\begin{array}{ccc|c} & {}^{0}_{1}\mathbf{R}^{T} & & -{}^{0}_{1}\mathbf{R}^{T}\,{}^{0}\mathbf{r}_{1org} \\ \hline 0 & 0 & 0 & 1 \end{array}\right]. \tag{4.20} $$

4.3 Denavit-Hartenberg Notation: Joint Coordinates

In the analysis of multibody system modelled as open kinematic chains, the Denavit-Hartenberg notatin [Denavit J., Hartenberg R. S., 1955], [Craig J. J., 2004], [Spong M. W., et al., 2006], [Jezierski E., 2006] is prevalent. This notation minimizes the number of values necessary to describe the positions of a mechanism's links. Positions of subsequent links of the kinematic chain are expressed in the current coordinate systems. Their positions in the global coordinate system are obtained by means of appropriate homogeneous transformations presented in section 4.2. Methods of description of the geometry of links and their interconnections will be discussed next. Defining local coordinate systems and matrices of homogeneous transformations that allow for switching between consecutive coordinate systems will also be treated.

Further discussion assumes that individual links of the system are connected into kinematic joints of the 5th class[4]. No loss of generality is incurred by this, for a kinematic joint of a lower class may be replaced with some number of kinematic joints of the 5th class with zero lengths[5]. In the analyses presented, two types of kinematic joints will appear: revolute joints and sliders.

From the kinematics point of view, the task of a rigid link p (Fig. 4.8) is to keep the axis of $p+1$ in fixed position relative to the axis of p. For kinematic considerations, it is therefore sufficient to describe the geometry of the link p with

---

[4] Most papers, e.g. [Morecki A., at al., 2002], define the class number as the number of degrees of freedom reduced by a given joint. However, some monographs ([Parczewski Z., 1978], [Gronowicz A., 2003]) the rank of a class is defined to be the number of independent motions allowed by a given joint. A joint of the 5th class is then described as one of the 1st class.

[5] A joint of the 4th class can be replaced by two joints of the 5th class, one of the 3rd class with three of them and so on.

two parameters: its length a_p, i.e. the distance between the axes of p and p+1, and the angle α_p between those axes. In Fig. 4.8 these values are shown unambiguously in space. The angle α_p is taken around the line a_p according to the right-hand rule (it is assumed that the line a_p is oriented from the axis of p towards the axis of p+1).

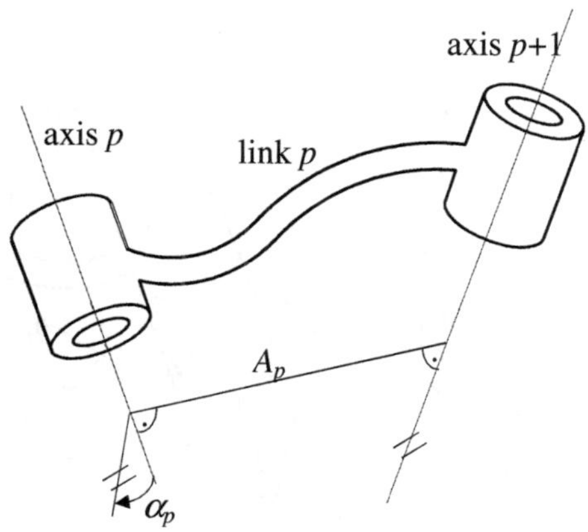

Fig. 4.8. A link p

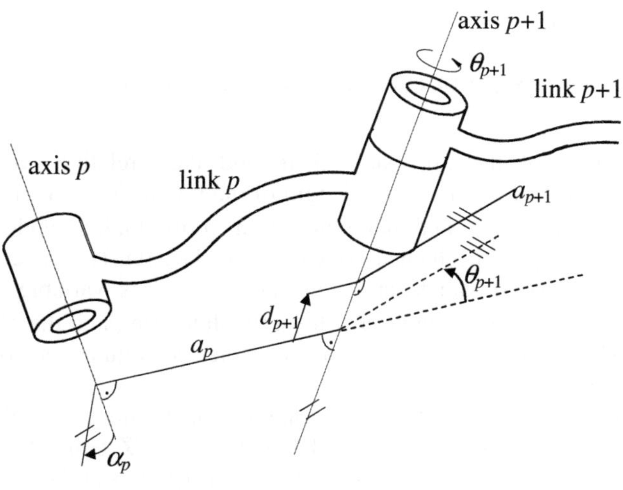

Fig. 4.9. Connection of links p and p+1

As stated previously, two consecutive links are assumed to form a revolute joint or a slider of the 5th class. Links p and p+1 are connected by an axis denoted by p+1 (Fig. 4.9). The interconnection p+1 requires adding two values to the description: the offset d_{p+1} of the link and the configuration angle θ_{p+1} (Fig. 4.9). A more detailed discussion of the problem of describing the link and its connection with another link from the motions' points of view can, among other things, be found in [Craig J. J., 2004]. Let us, however, remark that in a revolute joint the configuration angle θ_{p+1} is a joint variable (the offset d_{p+1} of the link is constant), whereas in a slider the offset d_{p+1} is variable (with the angle θ_{p+1} being constant).

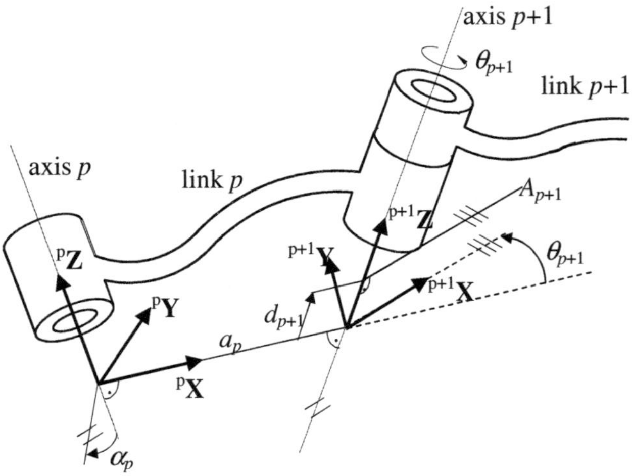

Fig. 4.10. Coordinate system of the link p

Individual links of the kinematic chain and their relative positions may therefore be described by four values: length of the link a_p, twist angle of the link α_p, offset of the link d_{p+1} and interconnection configuration angle θ_{p+1}. This method of describing a mechanism is called the Denavit-Hartenberg notation. Three of the mentioned values are constant and one is variable – it is the interconnection configuration angle θ_{p+1} for a revolute joint, and the offset d_{p+1} of the link for a slider. The set of all the joint variables constitutes the configuration space of the chain.

To the Denavit-Hartenberg notation, a method of defining a local coordinate system for a given link is related (Fig. 4.10). The axis $^p\mathbf{Z}$ coincides with the p connecting axis. The origin of the coordinate system $\{p\}$ lies at the intersection of the axis of p and the line a_p. In the case of $a_p = 0$ the origin of $\{p\}$ lies at the intersection of the axes of p and p+1, and the axis $^p\mathbf{X}$ lies on the common normal to the connecting axes of p and p+1. When the axes of p and p+1 do not intersect, the axis $^p\mathbf{X}$ is directed towards the connecting axis of p+1. The axis $^p\mathbf{Y}$ is taken in the way making the coordinate system clockwise. The coordinate system $\{0\}$ attached to the foundation and the system $\{n\}$ attached to the last link of the chain may be chosen arbitrarily. They are usually chosen in a way maximizing the number of Denavit-Hartenberg parameters and variables which equal zero.

Knowing the Denavit-Hartenberg parameters and variables for a link p allows to determine the form of the homogeneous transformation mapping the system $\{p+1\}$ onto $\{p\}$:

$$
{}^{p}_{p+1}\mathbf{T} = \begin{bmatrix} c\theta_{p+1} & -s\theta_{p+1} & 0 & a_p \\ s\theta_{p+1}c\alpha_p & c\theta_{p+1}c\alpha_p & -s\alpha_p & s\alpha_p d_{p+1} \\ s\theta_{p+1}s\alpha_p & c\theta_{p+1}s\alpha_p & c\alpha_p & c\alpha_p d_{p+1} \\ 0 & 0 & 0 & 1 \end{bmatrix}. \qquad (4.21)
$$

The transformation (4.21) can be easily obtained observing that it may be treated as a product of four elementarny transformations: two of them due to rotations by the angles α_p and θ_p around the axes ${}^{p}\mathbf{X}$ and ${}^{p+1}\mathbf{Z}$, respectively, and two due to translations by the distances a_p and d_{p+1} along the axes ${}^{p}\mathbf{X}$ and ${}^{p+1}\mathbf{Z}$, respectively. [Craig J. J., 2004] offers a more detailed explanation.

Coordinates from the system $\{p+1\}$ are mapped to the system $\{p\}$ according to the formula:

$$
{}^{p}\mathbf{r} = {}^{p}_{p+1}\mathbf{T}\ {}^{p+1}\mathbf{r}\,, \qquad (4.22)
$$

and to the inertial system with:

$$
\mathbf{r} = {}^{0}\mathbf{r} = \left[\prod_{i=0}^{p} {}^{i}_{i+1}\mathbf{T}(q_{i+1})\right] {}^{p+1}\mathbf{r}\,, \qquad (4.23)
$$

where q_{i+1} – joint (configuration) variable of the joint $i+1$.

Let us again underline that using the Denavit-Hartenberg notation causes each of the matrices ${}^{i}_{i+1}\mathbf{T}$ depend only on a single variable ($q_{i+1} = \theta_{i+1}$ if the kinematic joint is a revolute or $q_{i+1} = d_{i+1}$ if it is a slider).

In case of the connection between the links p and $p+1$ being of a lower class than 5th, it can be, as already mentioned, replaced by an appropriate series of revolute joints or sliders. Thus, where k is the class number of the connection between the links p and $p+1$, the following may be assumed:

$$
{}^{p}_{p+1}\mathbf{T}_{(k)} = {}^{p}_{p+1}\mathbf{T}_{(k)}\left(q_{p+1,1},\ldots,q_{p+1,6-k}\right) = \prod_{j=1}^{6-k} {}^{p}_{p+1}\mathbf{T}_{(k,j)}\left(q_{p+1,j}\right). \qquad (4.24)
$$

To bring more clarity into the matter, let us consider the case of the links p and $p+1$ being connected with a spherical joint (Fig. 4.11). Then $k = 3$ and:

$$
q_{p+1,1} = \psi_{p+1}, \quad q_{p+1,2} = \theta_{p+1}, \quad q_{p+1,3} = \varphi_{p+1}\,, \qquad (4.25)
$$

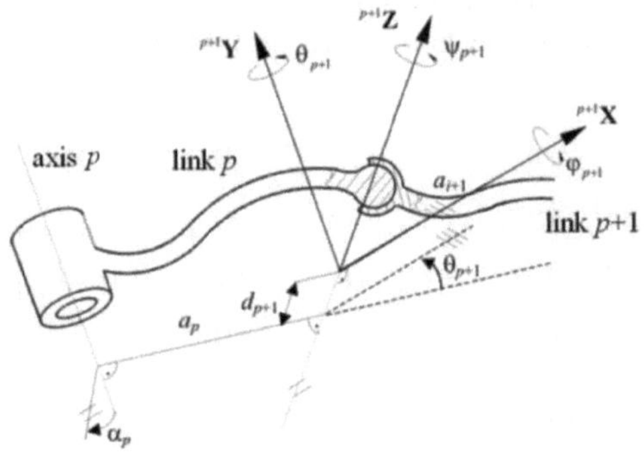

Fig. 4.11. A spherical joint of links p and $p+1$

so the motion of the link $p+1$ relative to p is described by three independent variables $\psi_{p+1}, \theta_{p+1}, \varphi_{p+1}$ being ZYX Euler angles. The matrices ${}^{p}_{p+1}\mathbf{T}_{(k,j)}$ in (4.24) take these forms:

$$
{}^{p}_{p+1}\mathbf{T}_{(k,1)} = \begin{bmatrix} c\psi_{p+1} & -s\psi_{p+1} & 0 & a_p \\ s\psi_{p+1}c\alpha_p & c\psi_{p+1}c\alpha_p & -s\alpha_p & s\alpha_p d_{p+1} \\ s\psi_{p+1}s\alpha_p & c\psi_{p+1}s\alpha_p & c\alpha_p & c\alpha_p d_{p+1} \\ 0 & 0 & 0 & 1 \end{bmatrix}, \tag{4.26.1}
$$

$$
{}^{p}_{p+1}\mathbf{T}_{(k,2)} = \begin{bmatrix} c\theta_{p+1} & 0 & s\theta_{p+1} & 0 \\ 0 & 1 & 0 & 0 \\ -s\theta_{p+1} & 0 & c\theta_{p+1} & 0 \\ 0 & 0 & 0 & 1 \end{bmatrix}, \tag{4.26.2}
$$

$$
{}^{p}_{p+1}\mathbf{T}_{(k,3)} = \begin{bmatrix} 1 & 0 & 0 & 0 \\ 0 & c\varphi_{p+1} & -s\varphi_{p+1} & 0 \\ 0 & s\varphi_{p+1} & c\varphi_{p+1} & 0 \\ 0 & 0 & 0 & 1 \end{bmatrix}. \tag{4.26.3}
$$

5 Equations of Motion of Systems with Rigid Links

In the current chapter the main steps of determining the components of the equation of motion for open kinematic chains consisting of rigid links are presented [Wittbrodt E., et al., 2006]. The method is based on the Lagrange equations of the second order, homogeneous transformations and joint coordinates.

The Lagrange equations of the second order may be written as:

$$\boldsymbol{\varepsilon}_{\mathbf{q}}(E)+\frac{\partial V}{\partial \mathbf{q}}+\frac{\partial D}{\partial \dot{\mathbf{q}}}=\mathbf{Q}, \tag{5.1}$$

where $\boldsymbol{\varepsilon}_{\mathbf{q}}(E)=\left(\frac{d}{dt}\frac{\partial E}{\partial \dot{q}_k}-\frac{\partial E}{\partial q_k}\right)_{k=1,\dots,n}$

$$\frac{\partial V}{\partial \mathbf{q}}=\left(\frac{\partial V}{\partial q_k}\right)_{k=1,\dots,n}, \qquad \frac{\partial D}{\partial \dot{\mathbf{q}}}=\left(\frac{\partial D}{\partial \dot{q}_k}\right)_{k=1,\dots,n}, \qquad \mathbf{Q}=(Q_k)_{k=1,\dots,n}.$$

$\mathbf{q}=[q_1 \ \dots \ q_k \ \dots \ q_n]^T$ – vector of generalized coordinates,

$\dot{\mathbf{q}}=[\dot{q}_1 \ \dots \ \dot{q}_k \ \dots \ \dot{q}_n]^T$ – vector of generalized velocities,

E – kinetic energy,

V – potential energy,

D – function of dissipation energy,

Q_k – non-potential generalized force corresponding to the k-th generalized coordinate,

n – number of generalized.

In the following reasoning, the dissipation of energy is omitted (D=0) and the multibody system is assumed to be situated on a movable base $\{A\}$ (Fig. 5.1) whose motion relative to the inertial (global) system $\{0\} = \{\}$ is known.

E. Wittbrodt et al.: Rigid Finite Element Method, OEO 1, pp. 59–73.
springerlink.com

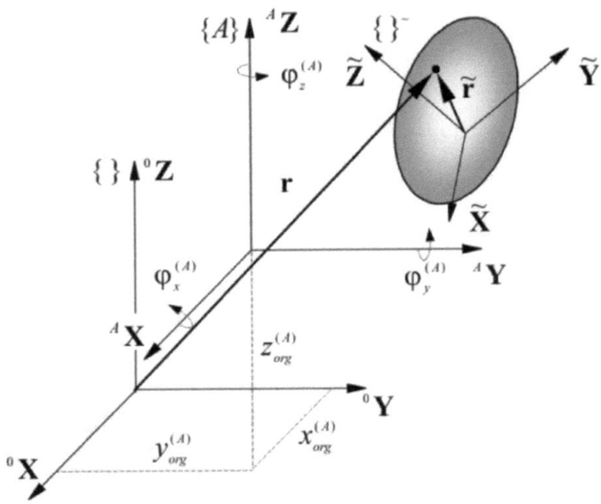

Fig. 5.1. Coordinate systems: {} – stationary (inertial) global one, $\{A\}$ – that of the movable base, $\{\}^\sim$ – local one attached to the considered link

For the sake of notation's clarity, the coordinate system {0} will for the remaining part be identified with the inertial system {}. Additionally, the following notation will be assumed:

$$ {}_{p}^{0}\mathbf{T} = \mathbf{T}^{(p)}, \tag{5.2}$$

where p – number of the link in the kinematic chain.

Let us introduce the following denotations:

$$ x_{org}^{(p)},\ y_{org}^{(p)},\ z_{org}^{(p)}, \tag{5.3}$$

for the origin of the system $\{p\}$ in the coordinate system of the preceding link and:

$$ \varphi_x^{(p)}, \varphi_y^{(p)}, \varphi_z^{(p)}, \tag{5.4}$$

for ZYX Euler angles determining the orientation of the axes of the system $\{p\}$ relative to the axes of the preceding system.

Matrix of the homogeneous transformation ${}_{A}^{0}\mathbf{T}$ taking into account the motion of the system $\{A\}$ relative to the system { } may be represented as a product of six matrices, each of which being a function of a single time-dependent variable only:

$$ {}_{A}^{0}\mathbf{T}(t) = {}_{A}^{0}\mathbf{T}_1\, {}_{A}^{0}\mathbf{T}_2\, {}_{A}^{0}\mathbf{T}_3\, {}_{A}^{0}\mathbf{T}_6\, {}_{A}^{0}\mathbf{T}_5\, {}_{A}^{0}\mathbf{T}_4\,, \tag{5.5}$$

where

$$ {}^{0}_{A}\mathbf{T}_1 = {}^{0}_{A}\mathbf{T}_1\left(x_{org}^{(A)}\right) = \begin{bmatrix} 1 & 0 & 0 & x_{org}^{(A)} \\ 0 & 1 & 0 & 0 \\ 0 & 0 & 1 & 0 \\ 0 & 0 & 0 & 1 \end{bmatrix}, \quad {}^{0}_{A}\mathbf{T}_4 = {}^{0}_{A}\mathbf{T}_4\left(\varphi_x^{(A)}\right) = \begin{bmatrix} 1 & 0 & 0 & 0 \\ 0 & c\varphi_x^{(A)} & -s\varphi_x^{(A)} & 0 \\ 0 & s\varphi_x^{(A)} & c\varphi_x^{(A)} & 0 \\ 0 & 0 & 0 & 1 \end{bmatrix}, $$

$$ {}^{0}_{A}\mathbf{T}_2 = {}^{0}_{A}\mathbf{T}_2\left(y_{org}^{(A)}\right) = \begin{bmatrix} 1 & 0 & 0 & 0 \\ 0 & 1 & 0 & y_{org}^{(A)} \\ 0 & 0 & 1 & 0 \\ 0 & 0 & 0 & 1 \end{bmatrix}, \quad {}^{0}_{A}\mathbf{T}_5 = {}^{0}_{A}\mathbf{T}_5\left(\varphi_y^{(A)}\right) = \begin{bmatrix} c\varphi_y^{(A)} & 0 & s\varphi_y^{(A)} & 0 \\ 0 & 1 & 0 & 0 \\ -s\varphi_y^{(A)} & 0 & c\varphi_y^{(A)} & 0 \\ 0 & 0 & 0 & 1 \end{bmatrix}, $$

$$ {}^{0}_{A}\mathbf{T}_3 = {}^{0}_{A}\mathbf{T}_3\left(z_{org}^{(A)}\right) = \begin{bmatrix} 1 & 0 & 0 & 0 \\ 0 & 1 & 0 & 0 \\ 0 & 0 & 1 & z_{org}^{(A)} \\ 0 & 0 & 0 & 1 \end{bmatrix}, \quad {}^{0}_{A}\mathbf{T}_6 = {}^{0}_{A}\mathbf{T}_6\left(\varphi_z^{(A)}\right) = \begin{bmatrix} c\varphi_z^{(A)} & -s\varphi_z^{(A)} & 0 & 0 \\ s\varphi_z^{(A)} & c\varphi_z^{(A)} & 0 & 0 \\ 0 & 0 & 1 & 0 \\ 0 & 0 & 0 & 1 \end{bmatrix}, $$

$$ x_{org}^{(A)} = x_{org}^{(A)}(t), \qquad y_{org}^{(A)} = y_{org}^{(A)}(t), \qquad z_{org}^{(A)} = z_{org}^{(A)}(t), $$

$$ \varphi_x^{(A)} = \varphi_x^{(A)}(t), \qquad \varphi_y^{(A)} = \varphi_y^{(A)}(t), \qquad \varphi_z^{(A)} = \varphi_z^{(A)}(t). $$

The order of rotations included in the matrix ${}^{0}_{A}\mathbf{T}$ conforms to the convention for ZYX Euler angles presented in section 4.1.

If $\tilde{\mathbf{r}} = \begin{bmatrix}\tilde{x} & \tilde{y} & \tilde{z} & 1\end{bmatrix}^T$ is a vector determining the coordinates of a mass dm in the local system $\{\}^{\sim}$ attached to given link of the system, then the coordinates of this mass in the system $\{\}$ can be given with this formula:

$$ \mathbf{r} = {}^{0}_{A}\mathbf{T}(t)\,\overline{\mathbf{T}}(\mathbf{q})\,\tilde{\mathbf{r}} = \mathbf{T}\,\tilde{\mathbf{r}}\,, \tag{5.6} $$

where $\overline{\mathbf{T}}(\mathbf{q}) = {}^{A}_{\{\}^{\sim}}\mathbf{T}(q_1,\ldots,q_n)$ – matrix of coordinate transformation from the local system $\{\}^{\sim}$ to the system $\{A\}$, dependent on the generalized coordinates of the link,

$\mathbf{T} = {}^{0}_{A}\mathbf{T}(t)\overline{\mathbf{T}}(\mathbf{q})$.

In a particular case whereby the base $\{A\}$ of a multibody system is motionless, the following may be assumed:

$$ {}^{0}_{A}\mathbf{T}(t) = \mathbf{I}\,, \tag{5.7.1} $$

where $\mathbf{I}$ is the identity matrix.

Then:

$$ \mathbf{T} = \overline{\mathbf{T}}(\mathbf{q})\,. \tag{5.7.2} $$

5.1 Kinetic Energy of a Link

Kinetic energy E of a link with mass m can be calculated using the trace of a matrix [Paul R. P., 1981], [Jurewič E. I., 1984]. The kinetic energy of an elementary mass dm with coordinates (x, y, z) can then be represented as:

$$dE = \frac{1}{2}\mathrm{tr}\left\{\dot{\mathbf{r}}\,\dot{\mathbf{r}}^T\right\}dm = \frac{1}{2}\mathrm{tr}\left\{\begin{bmatrix} \dot{x}\dot{x} & \dot{x}\dot{y} & \dot{x}\dot{z} & 0 \\ \dot{y}\dot{x} & \dot{y}\dot{y} & \dot{y}\dot{z} & 0 \\ \dot{z}\dot{x} & \dot{z}\dot{y} & \dot{z}\dot{z} & 0 \\ 0 & 0 & 0 & 0 \end{bmatrix}\right\}dm = \tag{5.8}$$
$$= \frac{1}{2}(\dot{x}^2 + \dot{y}^2 + \dot{z}^2)dm = \frac{1}{2}v^2 dm,$$

where $\mathrm{tr}(\mathbf{A}) = \sum_{i=1}^{u} a_{ii}$ – trace of the matrix $\mathbf{A}_{u\times u} = (a_{ij})_{i,\,j=1,\ldots,u}$,

$v^2 = \dot{x}^2 + \dot{y}^2 + \dot{z}^2$.

Since the vector $\tilde{\mathbf{r}}$ which determines the position of the elementary mass dm in the local coordinate system has constant coordinates (in time), then:

$$\dot{\mathbf{r}} = \dot{\mathbf{T}}\,\tilde{\mathbf{r}}, \tag{5.9}$$

and the expression giving the kinetic energy of the considered link takes the form:

$$E = \frac{1}{2}\int_m \mathrm{tr}\left\{\dot{\mathbf{r}}\,\dot{\mathbf{r}}^T\right\}dm = \frac{1}{2}\int_m \mathrm{tr}\left\{\dot{\mathbf{T}}\,\dot{\tilde{\mathbf{r}}}\,\dot{\tilde{\mathbf{r}}}^T\,\dot{\mathbf{T}}^T\right\}dm = \tag{5.10}$$
$$= \frac{1}{2}\mathrm{tr}\left\{\dot{\mathbf{T}}\left[\int_m \dot{\tilde{\mathbf{r}}}\,\dot{\tilde{\mathbf{r}}}^T dm\right]\dot{\mathbf{T}}^T\right\} = \frac{1}{2}\mathrm{tr}\{\dot{\mathbf{T}}\,\mathbf{H}\,\dot{\mathbf{T}}^T\}.$$

The matrix $\mathbf{H}$ occurring in the above formula is the matrix of inertia of the link whose elements may be calculated thus:

$$\mathbf{H} = \int_m \dot{\tilde{\mathbf{r}}}\,\dot{\tilde{\mathbf{r}}}^T dm = \begin{bmatrix} J_{(\tilde{X}\tilde{X})} & J_{\tilde{X}\tilde{Y}} & J_{\tilde{X}\tilde{Z}} & J_{\tilde{X}} \\ J_{\tilde{X}\tilde{Y}} & J_{(\tilde{Y}\tilde{Y})} & J_{\tilde{Y}\tilde{Z}} & J_{\tilde{Y}} \\ J_{\tilde{X}\tilde{Z}} & J_{\tilde{Y}\tilde{Z}} & J_{(\tilde{Z}\tilde{Z})} & J_{\tilde{Z}} \\ J_{\tilde{X}} & J_{\tilde{Y}} & J_{\tilde{Z}} & m \end{bmatrix}, \tag{5.11}$$

where $J_{\tilde{X}\tilde{X}} = \int\limits_m \tilde{x}^2 dm,\quad J_{\tilde{Y}\tilde{Y}} = \int\limits_m \tilde{y}^2 dm,\quad J_{\tilde{Z}\tilde{Z}} = \int\limits_m \tilde{z}^2 dm$ – planar moments of inertia in the coordinate system $\{\}^{\sim}$,

$J_{\tilde{X}\tilde{Y}} = \int\limits_m \tilde{x}\tilde{y}\, dm,\quad J_{\tilde{X}\tilde{Z}} = \int\limits_m \tilde{x}\tilde{z}\, dm,\quad J_{\tilde{Y}\tilde{Z}} = \int\limits_m \tilde{y}\tilde{z}\, dm$ – centrifugal (deviatoric) moments of inertia in the coordinate system $\{\}^{\sim}$,

$J_{\tilde{X}} = \int\limits_m \tilde{x}\, dm,\quad J_{\tilde{Y}} = \int\limits_m \tilde{y}\, dm,\quad J_{\tilde{Z}} = \int\limits_m \tilde{z}\, dm$ – static moments of inertia of the link in the coordinate system $\{\}^{\sim}$,

m – mass of the link.

The following relations hold:

$$J_{(\tilde{X}\tilde{X})} = \frac{1}{2}\left(\bar{J}_{\tilde{Y}} + \bar{J}_{\tilde{Z}} - \bar{J}_{\tilde{X}}\right), \tag{5.12.1}$$

$$J_{(\tilde{Y}\tilde{Y})} = \frac{1}{2}\left(\bar{J}_{\tilde{X}} + \bar{J}_{\tilde{Z}} - \bar{J}_{\tilde{Y}}\right), \tag{5.12.2}$$

$$J_{(\tilde{Z}\tilde{Z})} = \frac{1}{2}\left(\bar{J}_{\tilde{X}} + \bar{J}_{\tilde{Y}} - \bar{J}_{\tilde{Z}}\right), \tag{5.12.3}$$

where $\bar{J}_{\tilde{X}} = \int\limits_m \left(\tilde{y}^2 + \tilde{z}^2\right) dm,\ \bar{J}_{\tilde{Y}} = \int\limits_m \left(\tilde{x}^2 + \tilde{z}^2\right) dm,\ \bar{J}_{\tilde{Z}} = \int\limits_m \left(\tilde{x}^2 + \tilde{y}^2\right) dm$ are mass moments of inertia of the link relative to the axes $\tilde{\mathbf{X}}, \tilde{\mathbf{Y}}, \tilde{\mathbf{Z}}$, respectively.

Taking (5.6) into account, we may write the matrix $\dot{\mathbf{T}}$ as:

$$\dot{\mathbf{T}} = \frac{d\mathbf{T}}{dt} = \frac{d}{dt}\left[{}^0_A\mathbf{T}(t)\,\overline{\mathbf{T}}(\mathbf{q})\right] = {}^0_A\dot{\mathbf{T}}\,\overline{\mathbf{T}} + {}^0_A\mathbf{T}\,\dot{\overline{\mathbf{T}}}. \tag{5.13}$$

Since:

$$\dot{\overline{\mathbf{T}}} = \frac{d\overline{\mathbf{T}}}{dt} = \sum_{i=1}^{n} \frac{\partial \overline{\mathbf{T}}}{\partial q_i}\dot{q}_i = \sum_{i=1}^{n} \overline{\mathbf{T}}_i\, \dot{q}_i, \tag{5.14}$$

where $\overline{\mathbf{T}}_i = \dfrac{\partial \overline{\mathbf{T}}}{\partial q_i}$, the following is obtained:

$$\dot{\mathbf{T}} = {}^0_A\dot{\mathbf{T}}\,\overline{\mathbf{T}} + \sum_{i=1}^{n} \mathbf{T}_i\,\dot{q}_i\,, \tag{5.15}$$

where $\mathbf{T}_i = {}^0_A\mathbf{T}\,\overline{\mathbf{T}}_i$.

A reasoning analogous to that in [Wittbrodt E., et al., 2006] leads to:

$$\varepsilon_i(E) = \frac{d}{dt}\frac{\partial E}{\partial \dot{q}_i} - \frac{\partial E}{\partial q_i} = \mathrm{tr}\left\{\mathbf{T}_i\mathbf{H}\ddot{\mathbf{T}}^T\right\} \quad i{=}1,\ldots,n. \tag{5.16}$$

The matrix $\ddot{\mathbf{T}}$ may be calculated by differentiating (5.13):

$$\ddot{\mathbf{T}} = {}^0_A\ddot{\mathbf{T}}\,\overline{\mathbf{T}} + 2\,{}^0_A\dot{\mathbf{T}}\,\dot{\overline{\mathbf{T}}} + {}^0_A\mathbf{T}\,\ddot{\overline{\mathbf{T}}}\,, \tag{5.17.1}$$

where $\dot{\overline{\mathbf{T}}}$ is defined in (5.14).

The matrix $\ddot{\overline{\mathbf{T}}}$ is obtained by differentiating by time the formula (5.14), giving:

$$\ddot{\overline{\mathbf{T}}} = \sum_{i=1}^{n}\left(\frac{d\overline{\mathbf{T}}_i}{dt}\dot{q}_i + \overline{\mathbf{T}}_i\,\ddot{q}_i\right) = \sum_{i=1}^{n}\left[\sum_{j=1}^{n}\left(\frac{\partial\overline{\mathbf{T}}_i}{\partial q_j}\dot{q}_j\right)\dot{q}_i + \overline{\mathbf{T}}_i\,\ddot{q}_i\right] = \sum_{i=1}^{n}\sum_{j=1}^{n}\overline{\mathbf{T}}_{i,j}\,\dot{q}_i\,\dot{q}_j + \sum_{i=1}^{n}\overline{\mathbf{T}}_i\,\ddot{q}_i\,, \tag{5.17.2}$$

where $\overline{\mathbf{T}}_{i,j} = \dfrac{\partial\overline{\mathbf{T}}_i}{\partial q_j} = \dfrac{\partial^2\overline{\mathbf{T}}}{\partial q_i\partial q_j}$.

Taking (5.17) into account, we may rewrite the relation (5.16) as:

$$\begin{aligned}\varepsilon_i(E) &= \mathrm{tr}\left\{\mathbf{T}_i\,\mathbf{H}\left[{}^0_A\ddot{\mathbf{T}}\,\overline{\mathbf{T}} + 2\,{}^0_A\dot{\mathbf{T}}\,\dot{\overline{\mathbf{T}}} + \sum_{l=1}^{n}\sum_{j=1}^{n}\mathbf{T}_{l,j}\,\dot{q}_l\,\dot{q}_j + \sum_{l=1}^{n}\mathbf{T}_l\,\ddot{q}_l\right]^T\right\} = \\ &= \sum_{l=1}^{n} a_{i,l}(\mathbf{q})\,\ddot{q}_l + e_i(\mathbf{q}) \qquad \text{for} \quad i = 1,2,\ldots,n,\end{aligned} \tag{5.18}$$

where $a_{i,l}(\mathbf{q}) = \mathrm{tr}\left\{\mathbf{T}_i\;\mathbf{H}\;\mathbf{T}_l^{\,T}\right\}$,

$$\begin{aligned}e_i(\mathbf{q}) &= \mathrm{tr}\left\{\mathbf{T}_i\,\mathbf{H}\left[{}^0_A\ddot{\mathbf{T}}\,\overline{\mathbf{T}} + 2\,{}^0_A\dot{\mathbf{T}}\,\dot{\overline{\mathbf{T}}} + \sum_{l=1}^{n}\sum_{j=1}^{n}\mathbf{T}_{l,j}\,\dot{q}_l\,\dot{q}_j\right]^T\right\} = \\ &= \mathrm{tr}\left\{\mathbf{T}_i\,\mathbf{H}\left[{}^0_A\ddot{\mathbf{T}}\,\overline{\mathbf{T}} + 2\,{}^0_A\dot{\mathbf{T}}\sum_{j=1}^{n}\overline{\mathbf{T}}_j\,\dot{q}_j + \sum_{l=1}^{n}\sum_{j=l}^{n}\delta_{l,j}\,\mathbf{T}_{l,j}\,\dot{q}_l\,\dot{q}_j\right]^T\right\},\end{aligned}$$

$$\delta_{l,j} = \begin{cases}1 & \text{when} \quad l = j \\ 2 & \text{when} \quad l \neq j\end{cases}.$$

The above equation may be presented in a matrix form:

$$\varepsilon(E) = \mathbf{A}\ddot{\mathbf{q}} + \mathbf{e}\,, \tag{5.19}$$

where $\mathbf{A} = \mathbf{A}(t,\mathbf{q}) = \left(a_{i,l}(t,\mathbf{q})\right)_{i,l=1,\ldots,n}$,

$\mathbf{e} = \mathbf{e}(t,\mathbf{q},\dot{\mathbf{q}}) = \left(e_i(t,\mathbf{q},\ddot{\mathbf{q}})\right)_{i=1,\ldots,n}$.

An important property of the matrix **A** is its symmetry. The dependencies (5.18) and (5.19) will be used to formulate the equations of motion of analysed multibody systems.

5.2 Potential Energy of Gravity Forces of a Link

Let the coordinates of the centre of mass of a given link in its local coordinate system $\{\}^{\sim}$ be specified by the vector:

$$\tilde{\mathbf{r}}_C = \begin{bmatrix}\tilde{x}_C & \tilde{y}_C & \tilde{z}_C & 1\end{bmatrix}^T . \tag{5.20}$$

Assuming the axis ${}^0\mathbf{Z}$ of the global (inertial) coordinate system $\{\}$ to be perpendicular to the Earth's surface, we obtain the following formula that gives the potential energy of the gravity forces of the link:

$$V^g = m\,g\,z_C\,, \tag{5.21}$$

where g – acceleration due to gravity,

z_C – component in the direction of the axis ${}^0\mathbf{Z}$ of the vector $\mathbf{r}_C = \begin{bmatrix}x_C & y_C & z_C & 1\end{bmatrix}^T$ specifying the position of the centre of mass of the link in the inertial system.

By knowing the transformation matrix **T** from the local coordinate system $\{\}^{\sim}$ to the global one $\{\}$ the following may be obtained from equation (5.21):

$$\frac{\partial V^g}{\partial \mathbf{q}} = \mathbf{G}\,, \tag{5.22}$$

where $\mathbf{G} = \mathbf{G}(\mathbf{q}) = \left(g_i\right)_{i=1,\ldots,n}$,

$$g_i = \frac{\partial V}{\partial q_i} = m\,g\,\boldsymbol{\theta}_3\,\mathbf{T}_i\,\mathbf{r}'_C\,,$$

$$\boldsymbol{\theta}_3 = \begin{bmatrix}0 & 0 & 1 & 0\end{bmatrix}.$$

Elements of the vector **G** depend therefore on the matrix $\mathbf{T}_i$ and hence on time t and the vector of generalized coordinates **q**.

5.3 Generalized Forces: Equations of Motion of a Link

If non-potential forces or moments thereof act on a given link, they must be taken into account in the equations of motion as generalized forces. When the convention of homogeneous transformations and coordinates is applied, vectors of forces and their moments, unlike those of positions, have zero as their fourth coordinate:

$$\tilde{\mathbf{F}} = \begin{bmatrix} \tilde{F}_x & \tilde{F}_y & \tilde{F}_z & 0 \end{bmatrix}^T, \tag{5.23}$$

$$\tilde{\mathbf{M}} = \begin{bmatrix} \tilde{M}_x & \tilde{M}_y & \tilde{M}_z & 0 \end{bmatrix}^T. \tag{5.24}$$

Let us assume that a force $\tilde{\mathbf{F}}$ is applied to the link at the point N (Fig. 5.2).

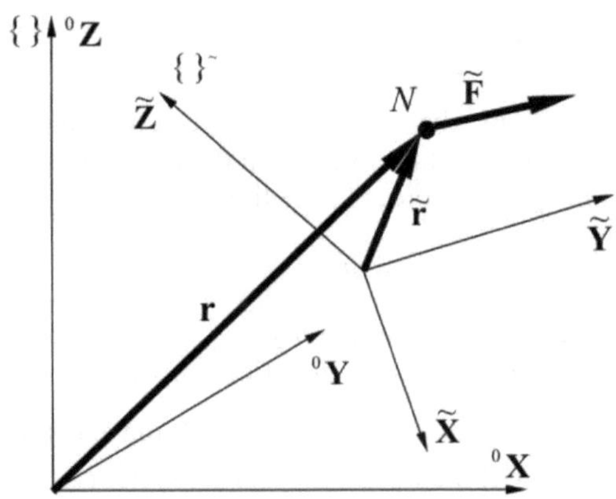

Fig. 5.2. Force acting in the local coordinate system

The force $\tilde{\mathbf{F}}$ is described in the inertial system by:

$$\mathbf{F} = \mathbf{T}\tilde{\mathbf{F}}. \tag{5.25}$$

The generalized force corresponding to the i-th generalized coordinate [Leyko J., 1996] may be written thus:

$$Q_i(\mathbf{F}) = \mathbf{F}^T \frac{\partial \mathbf{r}}{\partial q_i} = \mathbf{F}^T \mathbf{T}_i \tilde{\mathbf{r}} \quad \text{for } i = 1,\ldots,n. \tag{5.26}$$

Using (5.25), we may transform the formula (5.26) to obtain:

$$Q_i(\tilde{\mathbf{F}}) = \tilde{\mathbf{F}}^T\, \mathbf{T}^T\, \mathbf{T}_i\, \tilde{\mathbf{r}}. \tag{5.27}$$

If an external moment of force $\tilde{\mathbf{M}}$ specified by (5.24) is applied to a given link, it is possible, by representing its components as pairs of forces [Grzegożek W., et al., 2003] and performing appropriate transformations, obtain the formula for the generalized force corresponding to the i-th generalized coordinate which is due to the moment $\tilde{\mathbf{M}}$:

$$Q_i(\tilde{\mathbf{M}}) = \tilde{M}_x \sum_{l=1}^{3} t_{l,3} t_{i,l,2} + \tilde{M}_y \sum_{l=1}^{3} t_{l,1} t_{i,l,3} + \tilde{M}_z \sum_{l=1}^{3} t_{l,2} t_{i,l,1}, \tag{5.28}$$

where $\left(t_{m,l}\right)_{m,l=1...3}$, $\left(t_{i,m,l}\right)_{\substack{i=1,...,n \\ m,l=1,...,3}}$ are the corresponding elements of matrices $\mathbf{T}$ and $\mathbf{T}_i$, respectively.

Finally, using the Lagrange equations of the second kind, the equations of motion of the link concerned are put in this form:

$$\mathbf{A}\ddot{\mathbf{q}} = \mathbf{Q} - \mathbf{G} - \mathbf{e}, \tag{5.29}$$

where $\mathbf{A}$ – matrix of inertia defined in (5.19),
$\mathbf{G}$ – vector of gravity forces defined in (5.22),
$\mathbf{e}$ – vector of nonlinear forces defined in (5.19),
$\mathbf{Q}$ – vector of non-potential forces,
$\mathbf{Q} = \left(Q_i\right)_{i=1,...,n}$,
$Q_i = Q_i\left(\tilde{\mathbf{F}}\right) + Q_i\left(\tilde{\mathbf{M}}\right)$.

5.4 Generalization of the Procedure

The equations of motion for a single link having been determined, the equations of motion of an arbitrary open kinematic chain (Fig. 5.3) can be formulated.

Since joint coordinates are used to describe motion, the motion of a link p depends on its generalized coordinates, of which there are $\tilde{n}_p$, and on the generalized coordinates of its predecessor s in the chain. The total number of generalized coordinates for a link p (including all the generalized coordinates of preceding links) will be denoted by n_p. The vector of generalized coordinates of a link p may therefore be written:

$$\mathbf{q}^{(p)} = \begin{bmatrix} \mathbf{q}^{(s)} \\ \tilde{\mathbf{q}}^{(p)} \end{bmatrix}, \tag{5.30}$$

where $\mathbf{q}^{(s)}$ – vector of generalized coordinates describing the motion of the link s preceding the link p,

$\tilde{\mathbf{q}}^{(p)} = [\tilde{q}_1^{(p)} \quad \ldots \quad \tilde{q}_{\tilde{n}_p}^{(p)}]^T$ – vector of generalized coordinates of the link p describing its motion relative to the link s,

$\tilde{q}_i^{(p)} \in \left\{x^{(p)} \quad y^{(p)} \quad z^{(p)} \quad \varphi_x^{(p)} \quad \varphi_y^{(p)} \quad \varphi_z^{(p)}\right\}$,

$\mathbf{q}^{(p)} = [q_1^{(p)} \quad \ldots \quad q_{n_p}^{(p)}]^T$,

$n_p = n_s + \tilde{n}_p$.

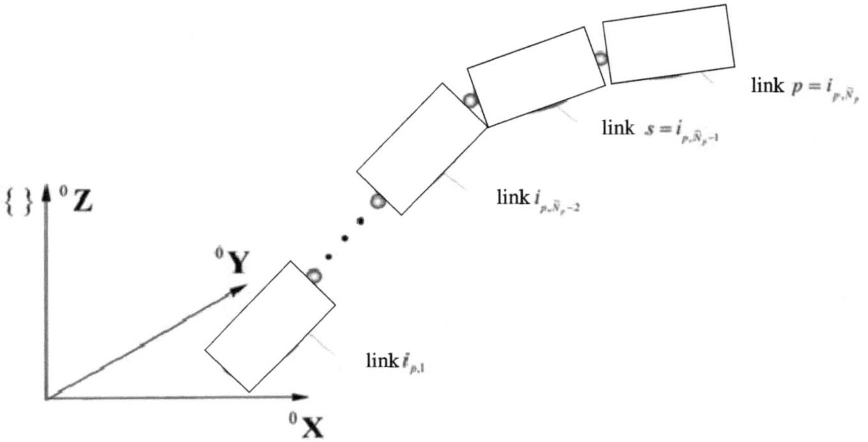

Fig. 5.3. A link p and links preceding it in a kinematic chain

The presented procedure takes the tree structure of kinematic chains into consideration. Therefore, consecutive links in a chain need not be assigned consecutive ordinal numbers. In such cases, one needs to define an ordered set of indices of the preceding links in the kinematic chain along with the index of the concerned link p:

$$N_p = \{i_{p,1}, \ldots, i_{p,l}, \ldots, i_{p,\tilde{N}_p}\}, \tag{5.31}$$

whereas:

$$\begin{aligned} &\tilde{N}_p && \text{– number of elements of the set } N_p, \\ &i_{p,\tilde{N}_p} && = p, \\ &i_{p,\tilde{N}_p-1} && = s, \\ &N_p && = N_s \cup \{p\}, \\ &\tilde{N}_p && = \tilde{N}_s + 1. \end{aligned} \tag{5.32}$$

The vector $\mathbf{q}^{(p)}$ from the formula (5.30) may now be written as:

$$\mathbf{q}^{(p)} = \begin{bmatrix} \tilde{\mathbf{q}}^{(i_{p,1})^T} & \ldots & \tilde{\mathbf{q}}^{(i_{p,l})^T} & \ldots & \tilde{\mathbf{q}}^{(i_{p,\tilde{N}_p})^T} \end{bmatrix}^T, \tag{5.33}$$

and the transformation from the coordinate system of the point p of the link to the inertial system { } may be expressed as follows:

$$\mathbf{r}^{(p)} = \mathbf{T}^{(p)}\tilde{\mathbf{r}}^{(p)} = \mathbf{T}^{(s)}\tilde{\mathbf{T}}^{(p)}\tilde{\mathbf{r}}^{(p)} = {}^{0}_{A}\mathbf{T}\,\overline{\mathbf{T}}^{(p)}\tilde{\mathbf{r}}^{(p)}, \tag{5.34}$$

where $\mathbf{T}^{(s)} = {}^{0}_{A}\mathbf{T}(t)\,\overline{\mathbf{T}}^{(s)}\left(\mathbf{q}^{(s)}\right)$,

$\mathbf{T}^{(p)} = {}^{0}_{A}\mathbf{T}(t)\,\overline{\mathbf{T}}^{(s)}\left(\mathbf{q}^{(s)}\right)\tilde{\mathbf{T}}^{(p)}\left(\tilde{\mathbf{q}}^{(p)}\right) = {}^{0}_{A}\mathbf{T}(t)\,\overline{\mathbf{T}}^{(p)}\left(\mathbf{q}^{(p)}\right)$,

$\overline{\mathbf{T}}^{(p)} = \overline{\mathbf{T}}^{(s)}\tilde{\mathbf{T}}^{(p)}$,

$\tilde{\mathbf{r}}^{(p)}$ – vector of coordinates of the point in the local coordinate system $\{p\}$.

Following the reasoning in section 5.1, thus is the kinetic energy of the link p:

$$E_p = \frac{1}{2}\operatorname{tr}\{\dot{\mathbf{T}}^{(p)}\mathbf{H}^{(p)}\dot{\mathbf{T}}^{(p)^T}\}, \tag{5.35}$$

where $\mathbf{H}^{(p)}$ – defined as in (5.11),

and the Lagrange operator for the link p takes the form:

$$\boldsymbol{\varepsilon}_{\mathbf{q}^{(p)}}(E_p) = \tilde{\mathbf{A}}^{(p)}\ddot{\mathbf{q}}^{(p)} + \tilde{\mathbf{e}}^{(p)}, \tag{5.36}$$

where $\tilde{\mathbf{A}}^{(p)} = (\tilde{a}_{k,j}^{(p)})_{k,j=1,\ldots,n_p}$,

$\tilde{\mathbf{e}}^{(p)} = (\tilde{e}_k^{(p)})_{k=1,\ldots,n_p}$,

$\tilde{a}_{k,j}^{(p)} = \operatorname{tr}\left\{\mathbf{T}_k^{(p)}\mathbf{H}^{(p)}\mathbf{T}_j^{(p)^T}\right\}$,

$$\tilde{e}_k^{(p)} = \operatorname{tr}\left\{\mathbf{T}_k^{(p)}\mathbf{H}^{(p)}\left[{}^{0}_{A}\ddot{\mathbf{T}}\,\overline{\mathbf{T}}^{(p)} + 2\,{}^{0}_{A}\dot{\mathbf{T}}\dot{\overline{\mathbf{T}}}^{(p)} + \sum_{l=1}^{n_p}\sum_{j=1}^{n_p}\mathbf{T}_{l,j}^{(p)}\dot{q}_l^{(p)}\dot{q}_j^{(p)}\right]^T\right\},$$

$$\mathbf{T}_k^{(p)} = \frac{\partial\mathbf{T}^{(p)}}{\partial q_k^{(p)}} = {}^{0}_{A}\mathbf{T}(t)\frac{\partial\overline{\mathbf{T}}^{(p)}}{\partial q_k^{(p)}} = \begin{cases} {}^{0}_{A}\mathbf{T}(t)\overline{\mathbf{T}}_k^{(s)}\tilde{\mathbf{T}}^{(p)} & \text{for } k = 1,\ldots,n_s, \\ {}^{0}_{A}\mathbf{T}(t)\overline{\mathbf{T}}^{(s)}\tilde{\mathbf{T}}_{k-n_s}^{(p)} & \text{for } k = n_s+1,\ldots,n_p, \end{cases}$$

$$\mathbf{T}_{l,j}^{(p)} = \frac{\partial\mathbf{T}_l^{(p)}}{\partial q_j^{(p)}} = {}^{0}_{A}\mathbf{T}(t)\frac{\partial^2\overline{\mathbf{T}}^{(p)}}{\partial q_l^{(p)}\partial q_j^{(p)}} = \begin{cases} {}^{0}_{A}\mathbf{T}\overline{\mathbf{T}}_{l,j}^{(s)}\tilde{\mathbf{T}}^{(p)} & \text{for } l,j = 1,\ldots,n_s, \\ {}^{0}_{A}\mathbf{T}\overline{\mathbf{T}}_l^{(s)}\tilde{\mathbf{T}}_{j-n_s}^{(p)} & \text{for } \begin{cases} l = 1,\ldots,n_s, \\ j = n_s+1,\ldots,n_p, \end{cases} \\ {}^{0}_{A}\mathbf{T}\overline{\mathbf{T}}^{(s)}\tilde{\mathbf{T}}_{l-n_s,j-n_s}^{(p)} & \text{for } l,j = n_s+1,\ldots,n_p. \end{cases}$$

This relation may also be written with a matrix and vector blocks:

$$\boldsymbol{\varepsilon}_{\mathbf{q}^{(p)}}(E^{(p)})=\begin{bmatrix}\tilde{\mathbf{A}}^{(p)}_{i_{p,1},i_{p,1}} & \cdots & \tilde{\mathbf{A}}^{(p)}_{i_{p,1},i_{p,k}} & \cdots & \tilde{\mathbf{A}}^{(p)}_{i_{p,1},i_{p,\tilde{N}_p}}\\ \vdots & & \vdots & & \vdots\\ \tilde{\mathbf{A}}^{(p)}_{i_{p,j},i_{p,1}} & \cdots & \tilde{\mathbf{A}}^{(p)}_{i_{p,j},i_{p,k}} & \cdots & \tilde{\mathbf{A}}^{(p)}_{i_{p,j},i_{p,\tilde{N}_p}}\\ \vdots & & \vdots & & \vdots\\ \tilde{\mathbf{A}}^{(p)}_{i_{p,\tilde{N}_p},i_{p,1}} & \cdots & \tilde{\mathbf{A}}^{(p)}_{i_{p,\tilde{N}_p},i_{p,k}} & \cdots & \tilde{\mathbf{A}}^{(p)}_{i_{p,\tilde{N}_p},i_{p,\tilde{N}_p}}\end{bmatrix}\begin{bmatrix}\ddot{\tilde{\mathbf{q}}}^{(i_{p,1})}\\ \vdots\\ \ddot{\tilde{\mathbf{q}}}^{(i_{p,j})}\\ \vdots\\ \ddot{\tilde{\mathbf{q}}}^{(i_{p,\tilde{N}_p})}\end{bmatrix}+\begin{bmatrix}\tilde{\mathbf{e}}^{(p)}_{i_{p,1}}\\ \vdots\\ \tilde{\mathbf{e}}^{(p)}_{i_{p,j}}\\ \vdots\\ \tilde{\mathbf{e}}^{(p)}_{i_{p,\tilde{N}_p}}\end{bmatrix}, \tag{5.37}$$

where $\tilde{\mathbf{A}}^{(p)}_{i_{p,j},i_{p,k}}=\left(\tilde{a}^{(p)}_{n_{\alpha_j}+l,n_{\alpha_k}+m}\right)_{\substack{l=1,\dots,\tilde{n}_{i_{p,j}}\\ m=1,\dots,\tilde{n}_{i_{p,k}}}}$,

$$\tilde{\mathbf{e}}^{(p)}_{i_{p,j}}=\left(\tilde{e}^{(p)}_{n_{\alpha_j}+l}\right)_{l=1,\dots,\tilde{n}_{i_{p,j}}},$$

$$n_{\alpha_j}=\sum_{\nu=1}^{j-1}\tilde{n}_{i_{p,\nu}}$$

or:

$$\boldsymbol{\varepsilon}_{\mathbf{q}^{(p)}}(E^{(p)})=\begin{bmatrix}\hat{\mathbf{A}}^{(p)}_{s,s} & \hat{\mathbf{A}}^{(p)}_{s,p}\\ \hat{\mathbf{A}}^{(p)}_{p,s} & \hat{\mathbf{A}}^{(p)}_{p,p}\end{bmatrix}\begin{bmatrix}\ddot{\mathbf{q}}^{(s)}\\ \ddot{\tilde{\mathbf{q}}}^{(p)}\end{bmatrix}+\begin{bmatrix}\hat{\mathbf{e}}^{(p)}_{s}\\ \hat{\mathbf{e}}^{(p)}_{p}\end{bmatrix}, \tag{5.38}$$

where $\hat{\mathbf{A}}^{(p)}_{s,s}=\begin{bmatrix}\tilde{\mathbf{A}}^{(p)}_{i_{p,1},i_{p,1}} & \cdots & \tilde{\mathbf{A}}^{(p)}_{i_{p,1},s}\\ \vdots & \vdots & \vdots\\ \tilde{\mathbf{A}}^{(p)}_{i_{p,1},s} & \cdots & \tilde{\mathbf{A}}^{(p)}_{s,s}\end{bmatrix}$, $\hat{\mathbf{A}}^{(p)}_{s,p}=\begin{bmatrix}\tilde{\mathbf{A}}^{(p)}_{i_{p,1},i_{p,\tilde{N}(p)}}\\ \vdots\\ \tilde{\mathbf{A}}^{(p)}_{s,i_{p,\tilde{N}(p)}}\end{bmatrix}$,

$$\hat{\mathbf{A}}^{(p)}_{p,s}=\begin{bmatrix}\tilde{\mathbf{A}}^{(p)}_{p,i_{p,1}} & \cdots & \tilde{\mathbf{A}}^{(p)}_{p,s}\end{bmatrix}, \qquad \hat{\mathbf{A}}^{(p)}_{p,p}=\tilde{\mathbf{A}}^{(p)}_{p,p},$$

$$\hat{\mathbf{e}}^{(p)}_{s}=\begin{bmatrix}\tilde{\mathbf{e}}^{(p)}_{i_{p,1}}\\ \vdots\\ \tilde{\mathbf{e}}^{(p)}_{s}\end{bmatrix}, \qquad \hat{\mathbf{e}}^{(p)}_{p}=\left[\tilde{\mathbf{e}}^{(p)}_{p}\right],$$

s, p – defined in (5.32).

Making use of the dependency (5.22) obtained in section 5.2, we may express the derivative of potential energy of gravity forces of the link p with respect to generalized coordinates as:

$$\frac{\partial V^{g}_{p}}{\partial \mathbf{q}^{(p)}}=\left(\tilde{g}^{(p)}_{l}\right)_{l=1,\dots,n_p}, \tag{5.39}$$

where $\tilde{g}^{(p)}_{l}=m^{(p)}\,g\,\boldsymbol{\theta}_3\,\mathbf{T}^{(p)}_{l}\,\tilde{\mathbf{r}}^{(p)}_{c}$.

The dependency (5.39) may thus be rewritten:

$$\frac{\partial V_p^g}{\partial \mathbf{q}^{(p)}} = \tilde{\mathbf{G}}^{(p)} = \begin{bmatrix} \tilde{\mathbf{G}}_1^{(p)} \\ \vdots \\ \tilde{\mathbf{G}}_{i_{p,j}}^{(p)} \\ \vdots \\ \tilde{\mathbf{G}}_{i_{p,\tilde{N}_p}}^{(p)} \end{bmatrix}, \tag{5.40}$$

where $\tilde{\mathbf{G}}_j^{(p)} = \left(\tilde{g}_{n_{\alpha_j}+l}^{(p)}\right)_{l=1,\ldots,\tilde{n}_{i_{p,j}}}$,

equivalently:

$$\frac{\partial V_p^g}{\partial \mathbf{q}^{(p)}} = \tilde{\mathbf{G}}^{(p)} = \begin{bmatrix} \hat{\mathbf{G}}_s^{(p)} \\ \hat{\mathbf{G}}_p^{(p)} \end{bmatrix}, \tag{5.41}$$

where $\hat{\mathbf{G}}_s^{(p)} = \begin{bmatrix} \tilde{\mathbf{G}}_{i_{p,1}}^{(p)} \\ \vdots \\ \tilde{\mathbf{G}}_s^{(p)} \end{bmatrix}$, $\hat{\mathbf{G}}_p^{(p)} = \tilde{\mathbf{G}}_p^{(p)}$.

The generalized forces due to external forces and moments thereof are calculated like in section 5.3 giving:

$$\tilde{\mathbf{Q}}^{(p)} = \begin{bmatrix} \tilde{\mathbf{Q}}_{i_{p,1}}^{(p)} \\ \vdots \\ \tilde{\mathbf{Q}}_{i_{p,j}}^{(p)} \\ \vdots \\ \tilde{\mathbf{Q}}_{i_{p,\tilde{N}_p}}^{(p)} \end{bmatrix}, \tag{5.42}$$

where $\tilde{\mathbf{Q}}_{i_{p,j}}^{(p)} = \left(\tilde{Q}_{n_{\alpha_j}+l}^{(p)}(\tilde{\mathbf{F}}^{(p)}) + \tilde{Q}_{n_{\alpha_j}+l}^{(p)}(\tilde{\mathbf{M}}^{(p)})\right)_{l=1,\ldots,\tilde{n}_{i_{p,j}}}$,

n_{α_j} – defined in (5.37),

or:

$$\tilde{\mathbf{Q}}^{(p)} = \begin{bmatrix} \hat{\mathbf{Q}}_s^{(p)} \\ \hat{\mathbf{Q}}_p^{(p)} \end{bmatrix}, \tag{5.43}$$

where $\hat{\mathbf{Q}}_s^{(p)} = \begin{bmatrix} \tilde{\mathbf{Q}}_{i_{p,1}}^{(p)} \\ \vdots \\ \tilde{\mathbf{Q}}_s^{(p)} \end{bmatrix}$, $\hat{\mathbf{Q}}_p^{(p)} = \tilde{\mathbf{Q}}_p^{(p)}$.

Finally, the equations of motion of the link p may be written thus:

$$\tilde{\mathbf{A}}^{(p)}\ddot{\mathbf{q}}^{(p)}=\tilde{\mathbf{f}}^{(p)}, \tag{5.44}$$

where $\tilde{\mathbf{f}}^{(p)}=\tilde{\mathbf{Q}}^{(p)}-\tilde{\mathbf{e}}^{(p)}-\tilde{\mathbf{G}}^{(p)}$,

$\tilde{\mathbf{A}}^{(p)},\tilde{\mathbf{e}}^{(p)},\tilde{\mathbf{G}}^{(p)},\tilde{\mathbf{Q}}^{(p)}$ – defined by formulas (5.36), (5.40) and (5.42), respectively.

The equations (5.44) describing the motion of the link p also indicate how its motion depends on the generalized coordinates of the preceding links (i.e. the coordinates $\tilde{\mathbf{q}}^{(i_{p,1})},\ldots,\tilde{\mathbf{q}}^{(s)}$ of the vector $\mathbf{q}^{(s)}$) and its own coordinates, i.e. $\tilde{\mathbf{q}}^{(p)}$. Therefore, those equations may be written in the form:

$$\begin{bmatrix}\hat{\mathbf{A}}_{s,s}^{(p)} & \hat{\mathbf{A}}_{s,p}^{(p)}\\ \hat{\mathbf{A}}_{p,s}^{(p)} & \hat{\mathbf{A}}_{p,p}^{(p)}\end{bmatrix}\begin{bmatrix}\ddot{\mathbf{q}}^{(s)}\\ \ddot{\tilde{\mathbf{q}}}^{(p)}\end{bmatrix}=\begin{bmatrix}\hat{\mathbf{f}}_{s}^{(p)}\\ \hat{\mathbf{f}}_{p}^{(p)}\end{bmatrix}, \tag{5.45}$$

where $\hat{\mathbf{f}}_s^{(p)}=\hat{\mathbf{Q}}_s^{(p)}-\hat{\mathbf{e}}_s^{(p)}-\hat{\mathbf{G}}_s^{(p)}$, $\hat{\mathbf{f}}_p^{(p)}=\hat{\mathbf{Q}}_p^{(p)}-\hat{\mathbf{e}}_p^{(p)}-\hat{\mathbf{G}}_p^{(p)}$.

The above equations of motion are obtained taking into account the kinetic energy and the potential energy of gravity forces of a single link p as well as the force $\tilde{\mathbf{F}}^{(p)}$ and the moment of force $\tilde{\mathbf{M}}^{(p)}$ acting upon this link.

If a kinematic chain has links numbered 1 to p, the energies: kinetic and potential of gravity forces of the system are given by the expressions:

$$E=\sum_{i=1}^{p}E_i, \tag{5.46}$$

$$V^g=\sum_{i=1}^{p}V_i^g. \tag{5.47}$$

Let us assume that the equations of motion of the links 1 to p-1, which take into account the kinetic energy E_i, the potential energy V_i^g, the forces $\tilde{\mathbf{F}}^{(i)}$ and the moments of forces $\tilde{\mathbf{M}}^{(i)}$ $(i=1,\ldots,p-1)$, have the form:

$$\mathbf{A}^{(p-1)}\ddot{\mathbf{q}}^{(p-1)}=\mathbf{f}^{(p-1)}, \tag{5.48}$$

where $\mathbf{A}^{(p-1)}$ is a matrix of dimension $n_{p-1}\times n_{p-1}$, and $\mathbf{q}$ and $\mathbf{f}$ are $n_{p-1}=\sum_{i=1}^{p-1}\tilde{n}_i$ -element vectors.

Adding a link p so that it connects with the link $s \le p-1$ belonging to the considered kinematic chain makes the equations of motion of the entire system of links 1 to p expressible as:

$$\mathbf{A}^{(p)}\ddot{\mathbf{q}}^{(p)} = \mathbf{f}^{(p)}, \tag{5.49.1}$$

or:

$$\left[\begin{array}{c:c} \mathbf{A}^{(p-1)} + \breve{\mathbf{A}}_{s,s}^{(p)} & \breve{\mathbf{A}}_{s,p}^{(p)} \\ \hdashline \breve{\mathbf{A}}_{p,s}^{(p)} & \hat{\mathbf{A}}_{p,p}^{(p)} \end{array}\right] \begin{bmatrix} \ddot{\mathbf{q}}^{(p-1)} \\ \ddot{\tilde{\mathbf{q}}}^{(p)} \end{bmatrix} = \begin{bmatrix} \mathbf{f}^{(p-1)} + \breve{\mathbf{f}}_s^{(p)} \\ \hat{\mathbf{f}}_p^{(p)} \end{bmatrix}, \tag{5.49.2}$$

where $\breve{\mathbf{A}}_{s,s}^{(p)}, \breve{\mathbf{A}}_{s,p}^{(p)}, \breve{\mathbf{A}}_{p,s}^{(p)}, \breve{\mathbf{f}}_s^{(p)}$ – matrices of dimensions $n_{p-1} \times n_{p-1}$, $n_{p-1} \times \tilde{n}_p$, $\tilde{n}_p \times n_{p-1}$ in which the appropriate submatrices with indices $i, j \in N_p$ are calculated according to:

$$\left(\breve{\mathbf{A}}_{s,s}^{(p)}\right)_{k,l} = \begin{cases} \tilde{\mathbf{A}}_{k,l}^{(p)} & \text{when } k,l \in N_p, \\ \mathbf{0} & \text{otherwise,} \end{cases} \qquad \text{for } k,l = 1,\dots,n_{p-1},$$

$$\left(\breve{\mathbf{A}}_{s,p}^{(p)}\right)_{k} = \begin{cases} \tilde{\mathbf{A}}_{k,p}^{(p)} & \text{when } k \in N_p, \\ \mathbf{0} & \text{otherwise,} \end{cases} \qquad \text{for } k = 1,\dots,n_{p-1},$$

$$\left(\breve{\mathbf{A}}_{p,s}^{(p)}\right)_{k} = \begin{cases} \tilde{\mathbf{A}}_{p,k}^{(p)} & \text{when } k \in N_p, \\ \mathbf{0} & \text{otherwise,} \end{cases} \qquad \text{for } k = 1,\dots,n_{p-1},$$

$$\left(\breve{\mathbf{f}}_{s}^{(p)}\right)_{k} = \begin{cases} \tilde{\mathbf{f}}_{k}^{(p)} & \text{when } k \in N_p, \\ \mathbf{0} & \text{otherwise,} \end{cases} \qquad \text{for } k = 1,\dots,n_{p-1}.$$

6 Modelling of Joining Elements: Constraint Equations

Individual links of a kinematic chain are often interconnected by elastic or damping (or both) elements. Among these are mainly: springs, dampers, absorbers, actuators. Components expressing the potential energy accumulated in such elements and its dissipation need to be introduced to the system's equations of motion. The present chapter discusses a method of modelling spring-damping elements treated as massless objects. Constraint equations occurring when kinematic subchains are joined in certain systems are also presented.

6.1 Spring-Damping Connecting Elements

The considerations presented below are under the assumption that the modelled element connects movable links i and j (Fig. 6.1). In the general case, the links i and j may belong to different branches of the chain.

Point A is the attachment location of the connecting element to the link i, whereas B the point where the element is attached to the link j (Fig. 6.1). The deformation of the connecting element is given by:

$$\Delta d_{i,j} = |\mathbf{r}_{AB}| - d_{i,j}^0, \tag{6.1}$$

where $d_{i,j}^0$ – free length of the element connecting the links i and j,

$\mathbf{r}_{AB}$ – vector starting at A and ending at B expressed in the inertial system { }.

The energy of elastic deformation of the connecting element may be determined from:

$$V_{i,j}^s = \frac{1}{2} c_{i,j} \left(\Delta d_{i,j}\right)^2, \tag{6.2}$$

where $c_{i,j}$ – stiffness coefficient.

The vector $\mathbf{r}_{AB}$ may be calculated in this way:

$$\mathbf{r}_{AB} = \mathbf{r}_B - \mathbf{r}_A, \tag{6.3}$$

E. Wittbrodt et al.: Rigid Finite Element Method, OEO 1, pp. 75–89.
springerlink.com

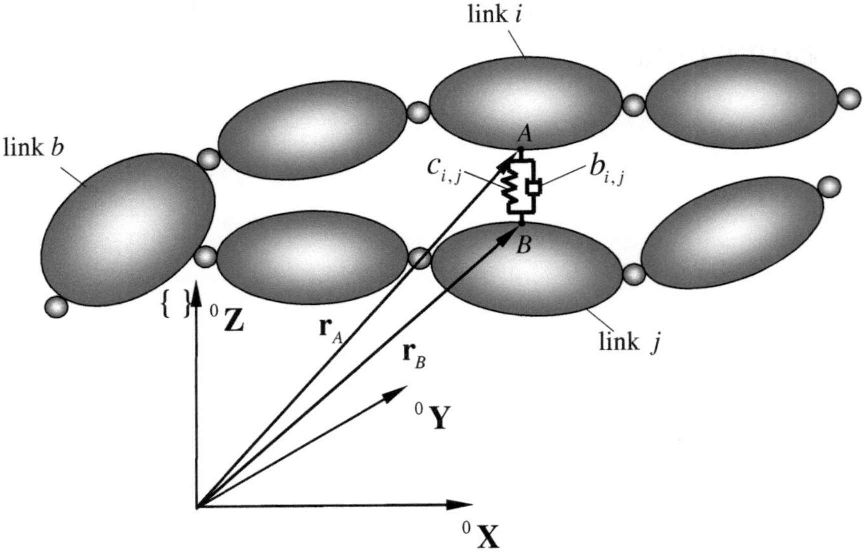

Fig. 6.1. The element connecting the links i and j

whereas:

$$\mathbf{r}_A = \mathbf{T}^{(i)}\,\tilde{\mathbf{r}}_A^{(i)}, \tag{6.4.1}$$

$$\mathbf{r}_B = \mathbf{T}^{(j)}\,\tilde{\mathbf{r}}_B^{(j)}, \tag{6.4.2}$$

where $\tilde{\mathbf{r}}_A^{(i)}, \tilde{\mathbf{r}}_B^{(j)}$ – the vectors of coordinates of the points A and B in the systems $\{i\}$ and $\{j\}$, respectively,

$\mathbf{r}_A, \mathbf{r}_B$ – the vectors of coordinates of the points A and B in the inertial system $\{\,\}$.

Considering the case in which the branches containing respectively the links i and j have as a common base a link b and a coordinate system $\{b\}$ attached to it, the homogeneous transformations in (6.4) may be specified as:

$$\mathbf{T}^{(i)} = \mathbf{T}^{(i)}\left(t, q_i^{(i)}, \ldots, q_{n_i}^{(i)}\right) = \mathbf{T}^{(i)}\left(t, \mathbf{q}^{(b)}, \tilde{\mathbf{q}}^{(b,i)}\right) = \mathbf{T}^{(i)}\left(t, q_1^{(b)}, \ldots, q_{n_b}^{(b)}, \tilde{q}_1^{(b,i)}, \ldots, \tilde{q}_{\tilde{n}_{b,i}}^{(b,i)}\right), \tag{6.5.1}$$

$$\mathbf{T}^{(j)} = \mathbf{T}^{(j)}\left(t, q_i^{(j)}, \ldots, q_{n_j}^{(j)}\right) = \mathbf{T}^{(i)}\left(t, \mathbf{q}^{(b)}, \tilde{\mathbf{q}}^{(b,j)}\right) = \mathbf{T}^{(j)}\left(t, q_1^{(b)}, \ldots, q_{n_b}^{(b)}, \tilde{q}_1^{(b,j)}, \ldots, \tilde{q}_{\tilde{n}_{b,j}}^{(b,j)}\right), \tag{6.5.2}$$

where $\mathbf{q}^{(b)} = \left[q_1^{(b)} \ \ldots \ q_{n_b}^{(b)}\right]^T$ – vector of generalized coordinates describing the motion of the common base $\{b\}$ of the links i and j,

$\tilde{\mathbf{q}}^{(b,i)} = \left[\tilde{q}_1^{(b,i)} \quad \dots \quad \tilde{q}_{\tilde{n}_{b,i}}^{(b,i)}\right]^T$ – vector of generalized coordinates describing the motion of the link i relative to the common base $\{b\}$,

$\tilde{\mathbf{q}}^{(b,j)} = \left[\tilde{q}_1^{(b,j)} \quad \dots \quad \tilde{q}_{\tilde{n}_{b,j}}^{(b,j)}\right]^T$ – vector of generalized coordinates describing the motion of the link j relative to the common base $\{b\}$.

A fact worthy of remark is that, in special cases, the indices $n_b, \tilde{n}_{b,i}, \tilde{n}_{b,j}$ may equal zero. And so, when:

$n_b = 0$ – vectors of generalized coordinates of the links i and j have no common elements: $\mathbf{q}^{(i)} \cap \mathbf{q}^{(j)} = \varnothing$, ($\varnothing$ denotes the empty set),

$\tilde{n}_{b,i} = 0$ – vector of generalized coordinates of the link i coincides with the vectors of generalized coordinates of the base (link b): $\mathbf{q}^{(i)} = \mathbf{q}^{(b)}$ and $\tilde{\mathbf{q}}^{(b,i)} = \varnothing$,

$\tilde{n}_{b,j} = 0$ – vector of generalized coordinates of the link j coincides with the vectors of generalized coordinates of the base (link b): $\mathbf{q}^{(j)} = \mathbf{q}^{(b)}$ and $\tilde{\mathbf{q}}^{(b,j)} = \varnothing$.

Appropriate derivatives of the potential energy of elastic deformation $V_{i,j}^s$ are given by:

$$\frac{\partial V_{i,j}^s}{\partial q_k} = c_{i,j} \Delta d_{i,j} \frac{\partial}{\partial q_k} \Delta d_{i,j}, \tag{6.6}$$

where $q_k \in \left\{q_1^{(b)}, \dots, q_{n_b}^{(b)}, \tilde{q}_1^{(b,i)}, \dots, \tilde{q}_{\tilde{n}_{b,i}}^{(b,i)}, \tilde{q}_1^{(b,j)}, \dots, \tilde{q}_{\tilde{n}_{b,j}}^{(b,j)}\right\}$.

From (6.1) it follows that:

$$\frac{\partial}{\partial q_k} \Delta d_{i,j} = \frac{\partial |\mathbf{r}_{AB}|}{\partial q_k}. \tag{6.7}$$

The length of the vector $\mathbf{r}_{AB}$ is determined by:

$$|\mathbf{r}_{AB}| = \sqrt{(x_B - x_A)^2 + (y_B - y_A)^2 + (z_B - z_A)^2}, \tag{6.8}$$

additionally:

$$|\mathbf{r}_{AB}|^2 = (x_B - x_A)^2 + (y_B - y_A)^2 + (z_B - z_A)^2 = [\mathbf{r}_B - \mathbf{r}_A]^T [\mathbf{r}_B - \mathbf{r}_A]. \tag{6.9}$$

Leveraging the general identity:

$$\frac{\partial f}{\partial q_k} = \frac{1}{2f} \frac{\partial f^2}{\partial q_k}, \tag{6.10}$$

where f denotes an arbitrary function of the variable q, and taking (6.4) into account we obtain:

$$\frac{\partial |\mathbf{r}_{AB}|}{\partial q_k} = \frac{1}{|\mathbf{r}_{AB}|} [\mathbf{r}_B - \mathbf{r}_A]^T \left[\mathbf{T}_k^{(j)}\, \tilde{\mathbf{r}}_B^{(j)} - \mathbf{T}_k^{(i)}\, \tilde{\mathbf{r}}_A^{(i)}\right]. \tag{6.11}$$

The derivatives (6.6) of the potential energy of elastic deformation of the connecting element may be written as:

$$\frac{\partial V_{i,j}^s}{\partial q_k} = \frac{c_{i,j} \Delta d_{i,j}}{|\mathbf{r}_{AB}|} [\mathbf{r}_B - \mathbf{r}_A]^T \begin{cases} \mathbf{T}_k^{(j)}\, \tilde{\mathbf{r}}_B^{(j)} - \mathbf{T}_k^{(i)}\, \tilde{\mathbf{r}}_A^{(i)}, \text{ for } q_k \in \left\{q_1^{(b)}, \ldots, q_{n_b}^{(b)}\right\}, \\ -\mathbf{T}_k^{(i)}\, \tilde{\mathbf{r}}_A^{(i)}, \quad \text{for } q_k \in \left\{\tilde{q}_1^{(b,i)}, \ldots, \tilde{q}_{\tilde{n}_{b,i}}^{(b,i)}\right\}, \\ \mathbf{T}_k^{(j)}\, \tilde{\mathbf{r}}_B^{(j)}, \quad \text{for } q_k \in \left\{\tilde{q}_1^{(b,j)}, \ldots, \tilde{q}_{\tilde{n}_{b,j}}^{(b,j)}\right\}. \end{cases} \tag{6.12}$$

Analogous derivations may be performed when the interconnection is by a damping element. The dissipation function is given by:

$$D_{i,j} = \frac{1}{2} b_{i,j} \left(\Delta \dot{d}_{i,j}\right)^2, \tag{6.13}$$

where $b_{i,j}$ – damping coefficient of the connecting element,

$\Delta \dot{d}_{i,j} = |\dot{\mathbf{r}}_{AB}|$ – change of the length vector $\mathbf{r}_{AB}$ in time.

Differentiating (6.13) by $\dot{q}_k$ yields:

$$\frac{\partial D_{i,j}}{\partial \dot{q}_k} = \frac{b_{i,j} \Delta \dot{d}_{i,j}}{|\mathbf{r}_{AB}|} [\mathbf{r}_B - \mathbf{r}_A]^T \begin{cases} \mathbf{T}_k^{(j)}\, \tilde{\mathbf{r}}_B^{(j)} - \mathbf{T}_k^{(i)}\, \tilde{\mathbf{r}}_A^{(i)} \text{ for } \dot{q}_k \in \left\{\dot{q}_1^{(b)}, \ldots, \dot{q}_{n_b}^{(b)}\right\}, \\ -\mathbf{T}_k^{(i)}\, \tilde{\mathbf{r}}_A^{(i)} \quad \text{for } \dot{q}_k \in \left\{\dot{\tilde{q}}_1^{(b,i)}, \ldots, \dot{\tilde{q}}_{\tilde{n}_{b,i}}^{(b,i)}\right\}, \\ \mathbf{T}_k^{(j)}\, \tilde{\mathbf{r}}_B^{(j)} \quad \text{for } \dot{q}_k \in \left\{\dot{\tilde{q}}_1^{(b,j)}, \ldots, \dot{\tilde{q}}_{\tilde{n}_{b,j}}^{(b,j)}\right\}. \end{cases} \tag{6.14}$$

It is noteworthy for the sake of numerical efficiency that the partial derivatives defined by (6.12) and (6.14) are equal zero for $q_k \in \left\{q_1^{(b)}, \ldots, q_{n_b}^{(b)}\right\}$ and $\dot{q}_k \in \left\{\dot{q}_1^{(b)}, \ldots, \dot{q}_{n_b}^{(b)}\right\}$. The lengthy process of their calculation may therefore be avoided. This fact has the following physical interpretation: in an element connecting two links, a change of the (elastic or damping) force can only occur with an alteration of relative position of the points *A* and *B*. The common lifting motion of the attachment points of the connecting element does not influence the force occurring in it. A proof can be found in [Szczotka M., 2004]. It requires the assumption that *b* is a rigid link.

As a final remark let us state that another element, which is often of paramount importance in the analysis of dynamics of cranes, may be modelled similarly. That element is the hoisting line. When considering cases in which the length of the hoisting line changes significantly, the stiffness (and damping) coefficient should be dependent on its current length.

6.2 Spring-Damping Connections with Clearance

The considerations of models of connections with clearance are limited to kinematic joints of the 5th class (Fig. 6.2) [Harlecki A., 2002]. Let the displacement in a connection be denoted as:

$$d_{E^{(R,L)}} = q^{(B)} - q^{(A)}. \tag{6.15}$$

The connections depicted in Fig. 6.2 are two-way ones. In technical applications one-way connections also occur often. Therefore, for the rest of this work, models of one-way connections with clearance will be presented. Two types thereof are distinguished: *L* and *R* (Fig. 6.2). A two-way connection is obtainable by combining models of those two types.

Assuming the elastic characteristic of the spring-damping element (SDE) $E^{(L,R)}$ to be linear, we may present interaction between the links *A* and *B* as shown in Fig. 6.3. A model of this type of a spring-damping connection with clearance can be easily implemented on a computer, as it may be defined in the following way:

- for an element of type *L*:

$$F_{S,E^{(L)}} = \begin{cases} 0 & \text{when} \quad d_{E^{(L)}} \geq -\Delta_{E^{(L)}} \\ c_{E^{(L)}}\left(d_{E^{(L)}} + \Delta_{E^{(L)}}\right) & \text{when} \quad d_{E^{(L)}} < -\Delta_{E^{(L)}} \end{cases}, \tag{6.16.1}$$

- for an element of type *R*:

$$F_{S,E^{(R)}} = \begin{cases} c_{E^{(R)}}\left(d_{E^{(R)}} - \Delta_{E^{(R)}}\right) & \text{when} \quad d_{E^{(R)}} > \Delta_{E^{(R)}} \\ 0 & \text{when} \quad d_{E^{(R)}} \leq \Delta_{E^{(R)}} \end{cases}, \tag{6.16.2}$$

where $c_{E^{(L)}}, c_{E^{(R)}}$ – appropriate stiffness coefficients (Fig. 6.2).

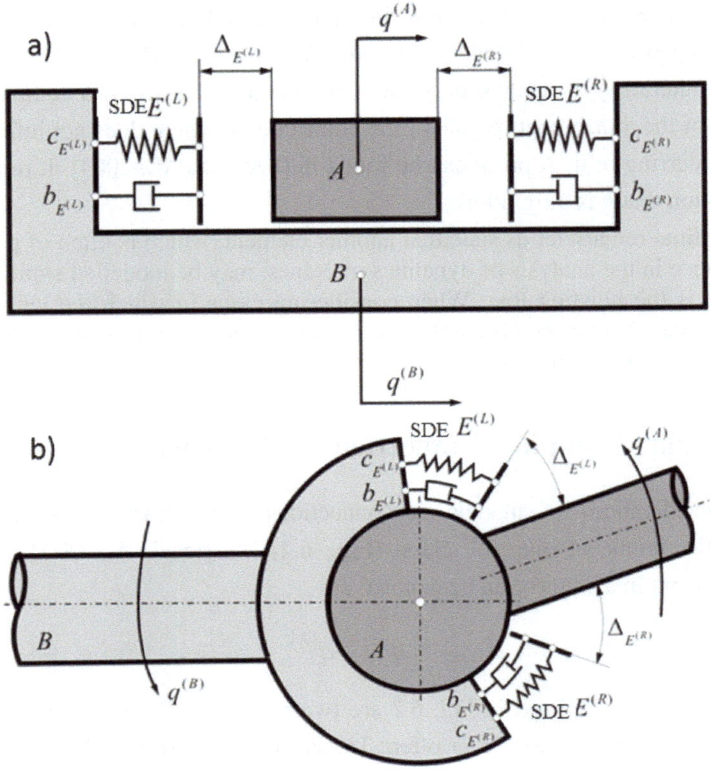

Fig. 6.2. A spring-damping connection with clearance: a) a slider, b) a revolute joint

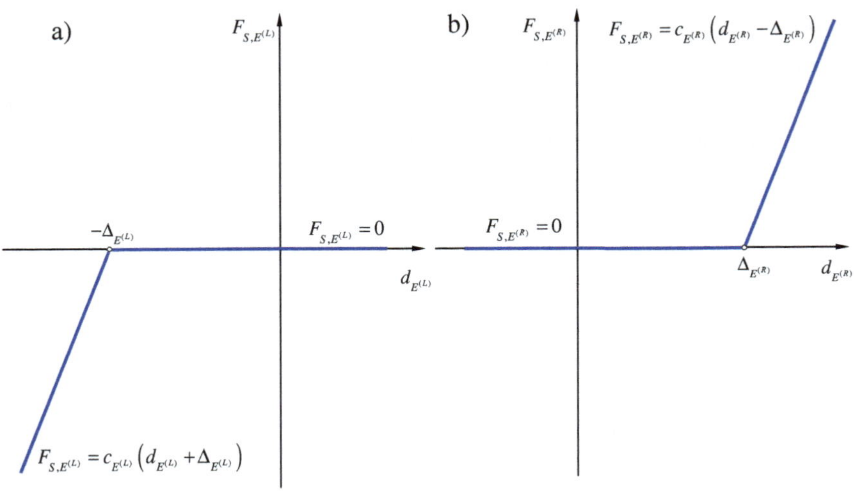

Fig. 6.3. The elastic force in a connection with clearance of type: a) L, b) R

Discontinuity of the derivative $F_{S,E^{(L)}}$ and $F_{S,E^{(R)}}$ at points $d_{E^{(L)}} = -\Delta_{E^{(L)}}$, $d_{E^{(R)}} = \Delta_{E^{(R)}}$ is a disadvantage of this model. To eliminate this inconvenience, a model SDE $E^{(L,\ R)}$ with a modified characteristic (Fig. 6.4) may be used. The functions describing the forces in the connection are then defined thus:

- for an element of type L:

$$F_{S,E^{(L)}} = \begin{cases} 0 & \text{when} \quad d_{E^{(L)}} > 0 \\ F^{l}_{S,E^{(L)}} & \text{when} \quad -a\Delta_{E^{(L)}} \le d_{E^{(L)}} \le 0\ , \\ c_{E^{(L)}}\left(d_{E^{(L)}} + \Delta_{E^{(L)}}\right) & \text{when} \quad d_{E^{(L)}} \le -a\Delta_{E^{(L)}} \end{cases} \tag{6.17.1}$$

- for an element of type R:

$$F_{S,E^{(R)}} = \begin{cases} c_{E^{(R)}}\left(d_{E^{(R)}} - \Delta_{E^{(L)}}\right) & \text{when} \quad d_{E^{(R)}} > a\Delta_{E^{(R)}} \\ F^{r}_{S,E^{(R)}} & \text{when} \quad 0 \le d_{E^{(R)}} \le a\Delta_{E^{(R)}}\ . \\ 0 & \text{when} \quad d_{E^{(R)}} \le 0 \end{cases} \tag{6.17.2}$$

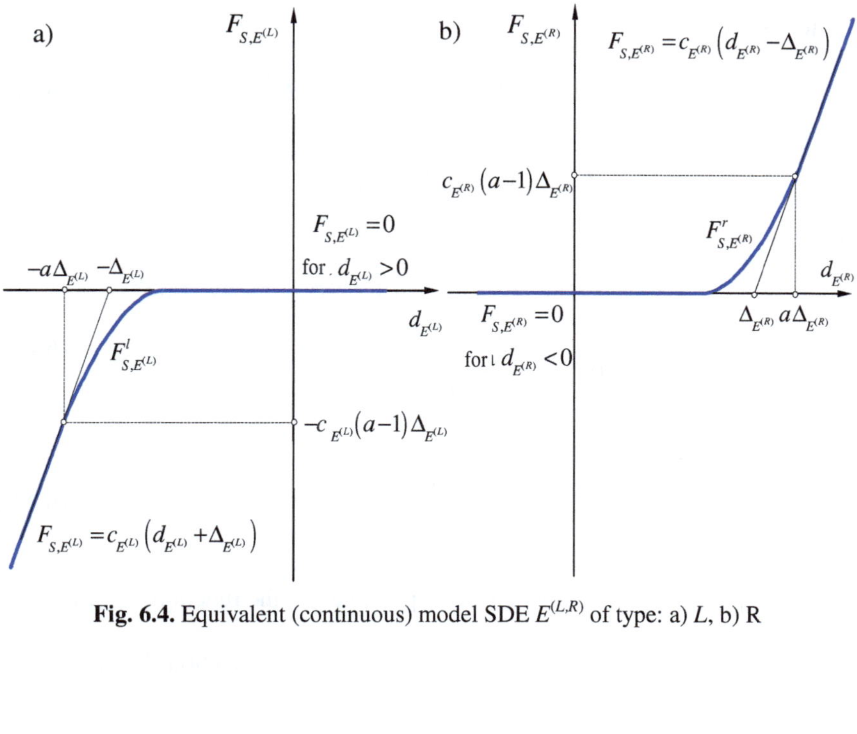

Fig. 6.4. Equivalent (continuous) model SDE $E^{(L,R)}$ of type: a) L, b) R

The functions $F^{l}_{S,E^{(L)}}$ and $F^{r}_{S,E^{(R)}}$ are assumed to take the form:

$$F^{(l,r)}_{S,E^{(L,R)}} = \chi^{(l,r)} d^{2}_{E^{(L,R)}} e^{\beta^{(l,r)} d_{E^{(L,R)}}}, \tag{6.18}$$

ensuring that these conditions are satisfied:

$$F^{(l,r)}_{S,E^{(L,R)}}(0) = 0, \tag{6.19.1}$$

$$F^{(l,r)'}_{S,E^{(L,R)}}(0) = 0. \tag{6.19.2}$$

The constants $\chi^{(l,r)}$, $\beta^{(l,r)}$ occurring in (6.18) may be obtained from the conditions, respectively:

- for an element of type L:

$$F^{l}_{S,E^{(L)}}\left(-a\Delta_{E^{(L)}}\right) = -c_{E^{(L)}}(a-1)\Delta_{E^{(L)}} = F^{l}_{S,a}, \tag{6.20.1}$$

$$F^{l}_{S,E^{(L)}}{}'\left(-a\Delta_{E^{(L)}}\right) = c_{E^{(L)}}, \tag{6.20.2}$$

- for an element of type R:

$$F^{r}_{S,E^{(R)}}\left(a\Delta_{E^{(R)}}\right) = c_{E^{(R)}}(a-1)\Delta_{E^{(R)}} = F^{r}_{S,a}, \tag{6.21.1}$$

$$F^{r}_{S,E^{(R)}}{}'\left(a\Delta_{E^{(R)}}\right) = c_{E^{(R)}}, \tag{6.21.2}$$

giving:

$$\beta^{(l,r)} = \frac{c_{E^{(L,R)}} - 2\dfrac{F^{(l,r)}_{S,a}}{d^{(l,r)}_{a}}}{F^{(l,r)}_{S,a}}, \tag{6.22.1}$$

$$\chi^{(l,r)} = \frac{F^{(l,r)}_{A,a}}{\left(d^{(l,r)}_{a}\right)^{2} e^{\beta^{(l,r)} d^{(l,r)}_{a}}}, \tag{6.22.2}$$

where $d^{l}_{0} = -a\Delta_{E^{(L)}}$ for elements of type L, i.e. for the function $F^{l}_{S,E^{(L)}}$,

$d^{r}_{0} = a\Delta_{E^{(R)}}$ for elements of type R, i.e. for the function $F^{r}_{S,E^{(R)}}$.

The functions $F_{S,E^{(R)}}\left(d_{E^{(L,R)}}\right)$ depend on a parameter a. Samples of their graphs for $c_{E^{(L,R)}} = 5\cdot 10^6 \dfrac{\mathrm{N}}{\mathrm{m}}$, $\Delta_{E^{(L,R)}} = 30\,\mathrm{mm}$ and different values of a are plotted in Fig. 6.5.

The damping forces occurring in connections with clearance may be written as:

- for an element of type L:

$$F_{D,E^{(L)}} = \begin{cases} b_{E^{(L)}}\dot{d}_{E^{(L)}} & \text{when} \quad d_{E^{(L)}} < -\Delta_{E^{(L)}} \\ 0 & \text{when} \quad d_{E^{(L)}} \geq -\Delta_{E^{(L)}} \end{cases}, \tag{6.23.1}$$

- for an element of type R:

$$F_{D,E^{(R)}} = \begin{cases} b_{E^{(R)}}\dot{d}_{E^{(R)}} & \text{when} \quad d_{E^{(R)}} > \Delta_{E^{(R)}} \\ 0 & \text{when} \quad d_{E^{(R)}} \leq \Delta_{E^{(R)}} \end{cases}, \tag{6.23.2}$$

where $b_{E^{(L)}}, b_{E^{(R)}}$ – appropriate damping coefficients (Fig. 6.2).

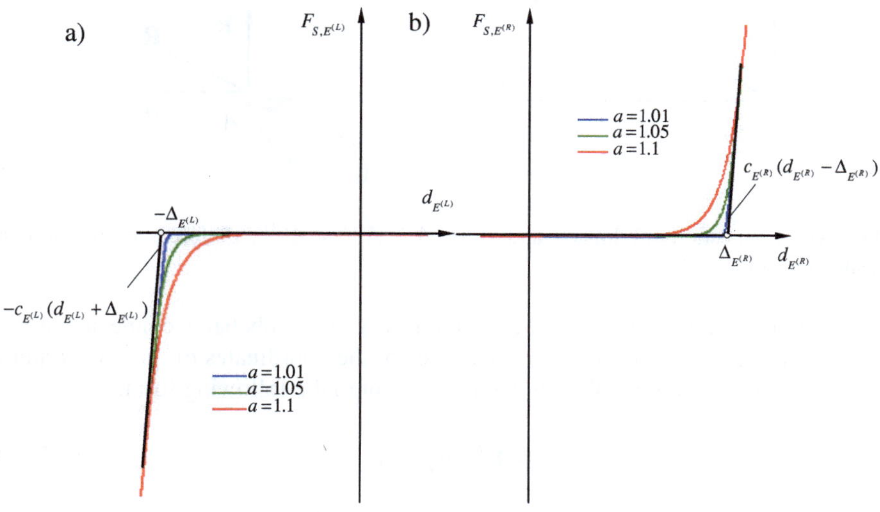

Fig. 6.5. Graphs of continuous characteristics SDE $E^{(L,R)}$ of type: a) L, b) R

6.3 Constraint Equations

When analysing closed kinematic chains or open ones which contain closed subchains (Fig. 6.6), it is necessary to externalize the reactions in the connection (which are internal forces) and to formulate proper equations relating the coordinates of the subchains.

In technical applications, the connections of links are usually revolute or spherical joints, rarely sliders. A detailed discussion of models of revolute joints and sliders is offered in [Harlecki A., 2002], including models of friction in such connections. The present book limits the considerations to spherical joints.

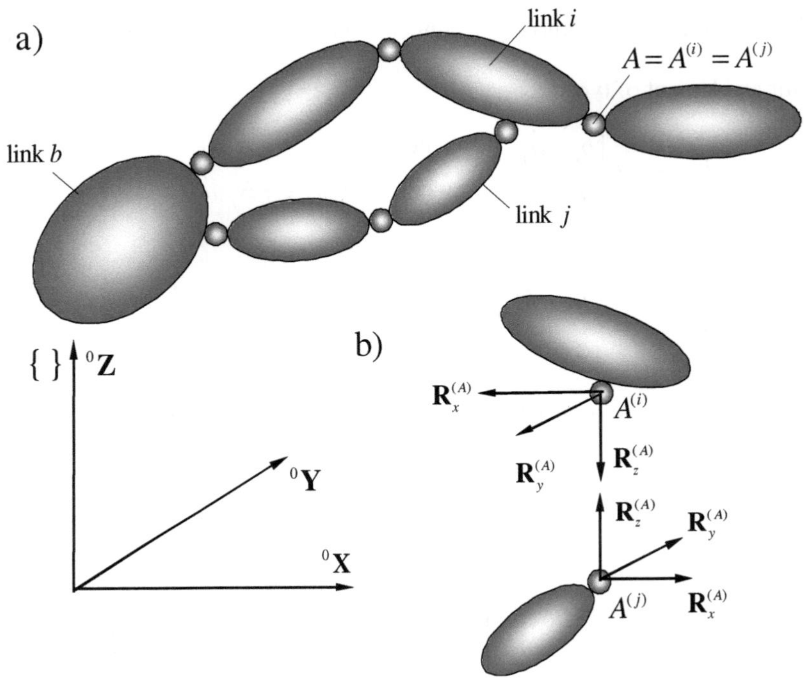

Fig. 6.6. Connections of links i and j: a) $A = A^{(i)} = A^{(j)}$, b) forces of reaction in a spherical joint

Assuming the link b, being a common base of two subchains connected at the point A, to be rigid, transformation matrices of the coordinates of the links i and j to the coordinate system of the link b may be put in the following form:

$$\tilde{\mathbf{T}}^{(b,i)} = \tilde{\mathbf{T}}^{(b,i)}\left(\tilde{\mathbf{q}}^{(b,i)}\right), \tag{6.24.1}$$

$$\tilde{\mathbf{T}}^{(b,j)} = \tilde{\mathbf{T}}^{(b,j)}\left(\tilde{\mathbf{q}}^{(b,j)}\right), \tag{6.24.2}$$

where $\tilde{\mathbf{q}}^{(b,i)}$, $\tilde{\mathbf{q}}^{(b,j)}$ are defined in (6.5).

Let the coordinates of the point A in the coordinate systems attached to the links i and j be given by the vectors:

$$\tilde{\mathbf{r}}_A^{(i)} = \begin{bmatrix} \tilde{x}_{i,A} & \tilde{y}_{i,A} & \tilde{z}_{i,A} & 1 \end{bmatrix}^T, \tag{6.25.1}$$

$$\tilde{\mathbf{r}}_A^{(j)} = \begin{bmatrix} \tilde{x}_{j,A} & \tilde{y}_{j,A} & \tilde{z}_{j,A} & 1 \end{bmatrix}^T . \tag{6.25.2}$$

Coordinates of the point A in the coordinate system of the body b may then be determined from the formulas:

$$\bar{\mathbf{r}}_A^{(b,i)} = \tilde{\mathbf{T}}^{(b,i)} \tilde{\mathbf{r}}_A^{(i)} , \tag{6.26.1}$$

$$\bar{\mathbf{r}}_A^{(b,j)} = \tilde{\mathbf{T}}^{(b,j)} \tilde{\mathbf{r}}_A^{(j)} . \tag{6.26.1}$$

Hence the constraint equation takes the form:

$$\bar{\mathbf{r}}_A^{(b,i)} = \bar{\mathbf{r}}_A^{(b,j)} . \tag{6.27}$$

The above may be decomposed into three equations resulting from comparisons of the components x, y, z of the vectors (6.27) and an identity due to their fourth components:

$$1^{\bar{\mathbf{r}}_A^{(b,i)}} = 1^{\bar{\mathbf{r}}_A^{(b,j)}} . \tag{6.28}$$

To eliminate the identity (6.28) an operator may be introduced which reduces vectors with four components to ones with three of them. The operator has a matrix form:

$$\boldsymbol{\theta} = \begin{bmatrix} 1 & 0 & 0 & 0 \\ 0 & 1 & 0 & 0 \\ 0 & 0 & 1 & 0 \end{bmatrix} . \tag{6.29}$$

The constraint equation (6.27) may then be rewritten as:

$$\bar{\mathbf{r}}_A^{(\theta,b,i)} = \bar{\mathbf{r}}_A^{(\theta,b,j)} , \tag{6.30}$$

where $\bar{\mathbf{r}}_A^{(\theta,b,i)} = \boldsymbol{\theta} \bar{\mathbf{r}}_A^{(b,i)}$,

$\bar{\mathbf{r}}_A^{(\theta,b,j)} = \boldsymbol{\theta} \bar{\mathbf{r}}_A^{(b,j)}$,

in which vectors have three components.

This reduction is implicit henceforth. The identity (6.28) will be omitted and equalities of type (6.27) will be deemed equivalent to three scalar equations. The equation (6.27) may therefore be rewritten in the following form:

$$\tilde{\mathbf{T}}^{(b,j)} \tilde{\mathbf{r}}_A^{(j)} - \tilde{\mathbf{T}}^{(b,i)} \tilde{\mathbf{r}}_A^{(i)} = \mathbf{0} . \tag{6.31}$$

Taking into account the facts:

$$\dot{\bar{\mathbf{r}}}_A^{(b,j)} = \dot{\tilde{\mathbf{T}}}^{(b,j)} \tilde{\mathbf{r}}_A^{(j)} = \sum_{k=1}^{\tilde{n}_{b,j}} \tilde{\mathbf{T}}_k^{(b,j)} \dot{\tilde{q}}_k^{(b,j)} \tilde{\mathbf{r}}_A^{(j)} , \tag{6.32.1}$$

$$\dot{\tilde{\mathbf{r}}}_A^{(b,i)} = \dot{\tilde{\mathbf{T}}}^{(b,i)} \tilde{\mathbf{r}}_A^{(i)} = \sum_{k=1}^{\tilde{n}_{b,i}} \tilde{\mathbf{T}}_k^{(b,i)} \dot{\tilde{q}}_k^{(b,i)} \tilde{\mathbf{r}}_A^{(i)}, \tag{6.32.2}$$

and:

$$\ddot{\tilde{\mathbf{r}}}_A^{(b,j)} = \sum_{k=1}^{\tilde{n}_{b,j}} \tilde{\mathbf{T}}_k^{(b,j)} \ddot{\tilde{q}}_k^{(b,j)} \tilde{\mathbf{r}}_A^{(j)} + \sum_{k=1}^{\tilde{n}_{b,j}} \sum_{l=1}^{\tilde{n}_{b,j}} \tilde{\mathbf{T}}_{k,l}^{(b,j)} \dot{\tilde{q}}_k^{(b,j)} \dot{\tilde{q}}_l^{(b,j)} \tilde{\mathbf{r}}_A^{(j)}, \tag{6.33.1}$$

$$\ddot{\tilde{\mathbf{r}}}_A^{(b,i)} = \sum_{k=1}^{\tilde{n}_{b,i}} \tilde{\mathbf{T}}_k^{(b,i)} \ddot{\tilde{q}}_k^{(b,i)} \tilde{\mathbf{r}}_A^{(i)} + \sum_{k=1}^{\tilde{n}_{b,i}} \sum_{l=1}^{\tilde{n}_{b,i}} \tilde{\mathbf{T}}_{k,l}^{(b,i)} \dot{\tilde{q}}_k^{(b,i)} \dot{\tilde{q}}_l^{(b,i)} \tilde{\mathbf{r}}_A^{(i)}, \tag{6.33.2}$$

one can replace the constraint equations (6.31) with their other form, called accelerative:

$$\mathbf{D}^{(b,j)T} \ddot{\tilde{\mathbf{q}}}^{(b,j)} - \mathbf{D}^{(b,i)T} \ddot{\tilde{\mathbf{q}}}^{(b,i)} = \mathbf{E}^{(i,j)}, \tag{6.34}$$

where $\mathbf{D}^{(b,j)} = \left(d_{k,s}^{(b,j)}\right)_{\substack{k=1,\dots,\tilde{n}_{b,j} \\ s=1,2,3}}$,

$$d_{s,k}^{(b,j)} = \sum_{\alpha=1}^{4} \left(\tilde{\mathbf{T}}_k^{(b,j)}\right)_{s,\alpha} \left(\tilde{\mathbf{r}}_A^{(j)}\right)_\alpha,$$

$$\mathbf{D}^{(b,i)} = \left(d_{k,s}^{(b,i)}\right)_{\substack{k=1,\dots,\tilde{n}_{b,i} \\ s=1,2,3}},$$

$$d_{s,k}^{(b,i)} = \sum_{\alpha=1}^{4} \left(\tilde{\mathbf{T}}_k^{(b,i)}\right)_{s,\alpha} \left(\tilde{\mathbf{r}}_A^{(j)}\right)_\alpha,$$

$$\mathbf{E}^{(i,j)} = \left(e^{(i,j)}\right)_{s=1,2,3},$$

$$\begin{aligned} e_s^{(i,j)} = & \sum_{k=1}^{\tilde{n}_{b,j}} \sum_{l=1}^{\tilde{n}_{b,j}} \dot{\tilde{q}}_k^{(b,j)} \dot{\tilde{q}}_l^{(b,j)} \sum_{\alpha=1}^{4} \left(\tilde{\mathbf{T}}_{k,l}^{(k,j)}\right)_{s,\alpha} \left(\tilde{\mathbf{r}}_A^{(j)}\right)_\alpha + \\ & - \sum_{k=1}^{\tilde{n}_{b,i}} \sum_{l=1}^{\tilde{n}_{b,i}} \dot{\tilde{q}}_k^{(b,i)} \dot{\tilde{q}}_l^{(b,i)} \sum_{\alpha=1}^{4} \left(\tilde{\mathbf{T}}_{k,l}^{(k,i)}\right)_{s,\alpha} \left(\tilde{\mathbf{r}}_A^{(i)}\right)_\alpha \end{aligned},$$

$$\tilde{\mathbf{T}}_k^{(b,\alpha)} = \frac{\partial \tilde{\mathbf{T}}^{(b,\alpha)}}{\partial \tilde{q}_k^{(b,\alpha)}},$$

$$\tilde{\mathbf{T}}_{k,l}^{(b,\alpha)} = \frac{\partial \tilde{\mathbf{T}}_k^{(b,\alpha)}}{\partial \tilde{q}_l^{(b,\alpha)}},$$

$$\alpha \in \{i, j\}.$$

Care must be taken to reflect in the system's equations of motion the occurrence of connections between links. When formulating the equations of motion of links *i* and *j*, the reaction forces of the constraints in the connection (Fig. 6.6b) must be taken into account. Assuming the vector of forces in the connection to be defined in the inertial system {} and its components to be:

$$\mathbf{R}^{(A)} = \begin{bmatrix} R_x^{(A)} & R_y^{(A)} & R_z^{(A)} & 0 \end{bmatrix}^T, \tag{6.35}$$

we can easily prove that additional terms (generalized forces) will appear in the equations of motion:

$$\mathbf{Q}_{\tilde{\mathbf{q}}^{(b,j)}}^{(b,j)}\left(\mathbf{R}^{(A)}\right) = \mathbf{D}^{(b,j)}\,\mathbf{R}^{(A)}, \tag{6.36.1}$$

$$\mathbf{Q}_{\tilde{\mathbf{q}}^{(b,i)}}^{(b,i)}\left(\mathbf{R}^{(A)}\right) = -\mathbf{D}^{(b,i)}\,\mathbf{R}^{(A)}. \tag{6.36.2}$$

Components of the vector $\mathbf{R}^{(A)}$ are additional unknowns and their number is equal to that of the constraint equations.

The modelling approach just described which consists in introducing the reactions of the constraints to the equations of motion and formulating the constraint equations as additional algebraic equations of the form (6.31) or differential equations of the form (6.34) raises the number of unknowns and the dimension of the problem. This can be avoided by introducing a spring-damping element with large stiffness and damping coefficients at the point *A* (Fig. 6.6). In this way the points $A^{(i)}$ and $A^{(j)}$ are kept close to each other during the motion. The values of the coefficients should be determined empirically. A formal description of this method is presented below.

Let:

$$\Delta\overline{\mathbf{r}}_A = \overline{\mathbf{r}}_A^{(j)} - \overline{\mathbf{r}}_A^{(i)}, \tag{6.37}$$

and:

$$\mathbf{C}^{(A)} = \begin{bmatrix} c_x^{(A)} & 0 & 0 \\ 0 & c_y^{(A)} & 0 \\ 0 & 0 & c_z^{(A)} \end{bmatrix}, \tag{6.38.1}$$

$$\mathbf{B}^{(A)} = \begin{bmatrix} b_x^{(A)} & 0 & 0 \\ 0 & b_y^{(A)} & 0 \\ 0 & 0 & b_z^{(A)} \end{bmatrix}, \tag{6.38.2}$$

where $c_x^{(A)}, c_y^{(A)}, c_z^{(A)}$ – stiffness coefficients,

$b_x^{(A)}, b_y^{(A)}, b_z^{(A)}$ – damping coefficients,

are the matrices of stiffness and damping of the SDE, respectively (Fig. 6.7).

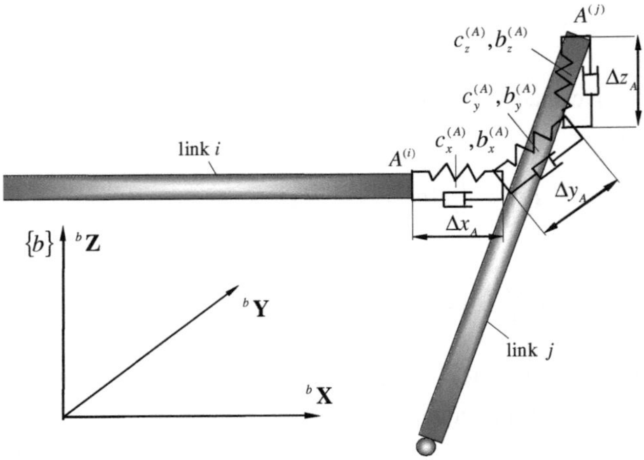

Fig. 6.7. A model of a spherical joint as a spring-damping element

The energy of elastic deformation and the dissipation of SDE's energy are expressed thus:

$$V_A^s = \frac{1}{2} \Delta\bar{\mathbf{r}}_A^{\ T} \mathbf{C}^{(A)} \Delta\bar{\mathbf{r}}_A, \tag{6.39.1}$$

$$D_A = \frac{1}{2} \Delta\dot{\bar{\mathbf{r}}}_A^{\ T} \mathbf{B}^{(A)} \Delta\dot{\bar{\mathbf{r}}}_A. \tag{6.39.2}$$

Taking into consideration the relations stated in [Wittbrodt E., et al., 2006] leads to:

$$\frac{\partial V_A^s}{\partial \tilde{\mathbf{q}}^{(b,j)}} + \frac{\partial D_A}{\partial \dot{\tilde{\mathbf{q}}}^{(b,j)}} = \mathbf{D}^{(b,j)} \left[\mathbf{C}^{(A)} \Delta\bar{\mathbf{r}}_A + \mathbf{D}^{(A)} \Delta\dot{\bar{\mathbf{r}}}_A \right], \tag{6.40.1}$$

$$\frac{\partial V_A^s}{\partial \tilde{\mathbf{q}}^{(b,i)}} + \frac{\partial D_A}{\partial \dot{\tilde{\mathbf{q}}}^{(b,i)}} = -\mathbf{D}^{(b,i)} \left[\mathbf{C}^{(A)} \Delta\mathbf{r}_A + \mathbf{D}^{(A)} \Delta\dot{\bar{\mathbf{r}}}_A \right]. \tag{6.40.2}$$

These values are to be subtracted from right-hand sides of the equations of motion corresponding to $\tilde{\mathbf{q}}^{(b,j)}$ and $\tilde{\mathbf{q}}^{(b,i)}$.

As can be easily seen, the presented method eliminates the need of separating the chain at spherical joints and increasing the dimension of the system of equation describing the system's dynamics. However, its drawbacks lie in the requirement of assuming large values of stiffness and damping coefficients in (6.38). Introducing such coefficients into the system causes high frequency oscillations to appear. It is therefore necessary to integrate the equations of motion of the system with a very small step or employ integration methods specialized for rigid systems [Press W. H., et al., 2002], [Wittbrodt E., et al., 2006]. A precaution is also due to the fact that the procedure described above consisting in the introduction of the constraints' reactions and formulation of appropriate constraint equations leads additionally to certain numerical complications [Frączek J., 2002].

7 Nonlinear Models of Materials

In numerous technical applications the supporting structure of a device is assumed to be subjected to stresses within the limits of proportionality, i.e. where the Hooke's law is applicable. It is also the case with offshore cranes. In the installation process of underwater pipelines with the reel method, however, the pipes are commonly deformed plastically when they are wound onto the reel. Furthermore, material exposed to prolonged deformation may show a tendency to creep. Hence, the present chapter which briefly introduces these models of construction materials: elasto-plastic and visco-elastic.

7.1 Basic Laws of Elasto-plastic Materials

A basic property of most materials working in their ranges of elasticity is the linear dependency of the deformation on the loads applied. The applicability of the Hooke's law describing the stresses σ within the range of elasticity is determined by the deformations ε which must not exceed certain values ε_L (Fig. 7.1a). For materials whose yield point ε_L is sharp it may be taken from a stress–strain curve. For other materials, a conventional value is assumed, which corresponds to 0.2% of permanent deformation of the object (Fig. 7.1b). Characteristic points on a stress–strain curve are (Fig. 7.1): 1 – proportionality limit, 2 – elastic limit, 3 and 4 – upper and lower plasticity limit, 4–5 – ideal plasticity region, 5–6 – strain hardening region, 7 – rupture.

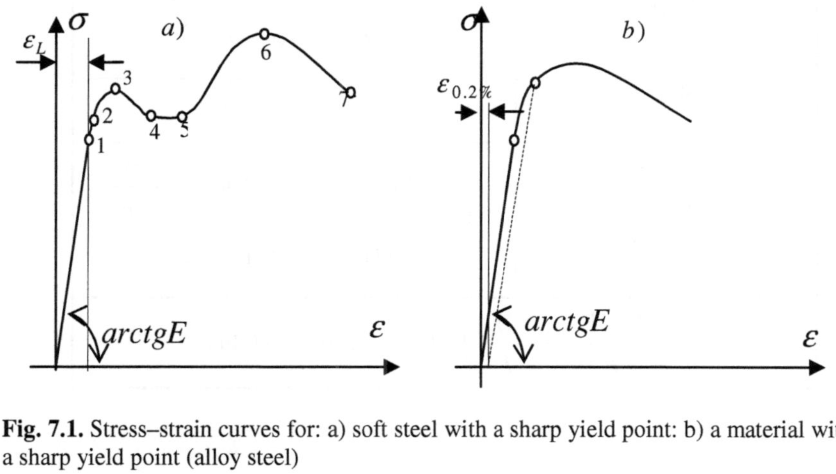

Fig. 7.1. Stress–strain curves for: a) soft steel with a sharp yield point: b) a material without a sharp yield point (alloy steel)

E. Wittbrodt et al.: Rigid Finite Element Method, OEO 1, pp. 91–97.
springerlink.com

Typical linear characteristics of stretching for various elasto-plastic materials are shown in Fig. 7.2 [Skrzypek J., 1986], [Ottosen N. S., Ristinmaa M., 2005]. The corresponding constitutive equations are summarized in Table 7.1.

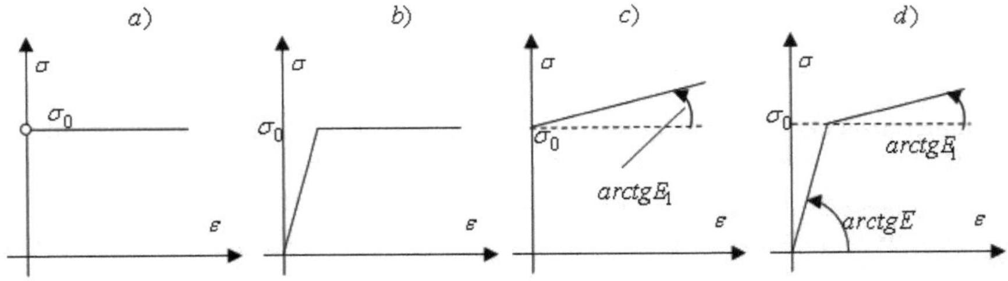

Fig. 7.2. Linear characteristics of stretching for ideal materials: a) rigid-plastic, b) elasto-plastic, c) rigid-plastic with reinforcement, d) elasto-plastic with reinforcement

Properties of many materials cannot be described with sufficient accuracy using linear constitutive dependencies. Various nonlinear models of materials are therefore formulated. Sample power characteristics of elasto-plastic materials with reinforcement are shown in Fig. 7.3 and their corresponding constitutive dependencies in Table 7.2.

Table 7.1. Dependencies describing linear characteristics of models of elasto-plastic materials

Model of a material	Equation $\sigma = f(\varepsilon)$ (linear)
rigid-plastic	$\sigma = \sigma_0$
elasto-plastic	$\sigma = \begin{cases} E\varepsilon & \text{for} \quad \varepsilon \le \frac{\sigma_0}{E} \\ \sigma_0 & \text{for} \quad \varepsilon > \frac{\sigma_0}{E} \end{cases}$
rigid-plastic with reinforcement	$\sigma = \sigma_0 + E_1\varepsilon$
elasto-plastic with reinforcement	$\sigma = \begin{cases} E\varepsilon & \text{for} \quad \varepsilon \le \frac{\sigma_0}{E} \\ E\varepsilon(1-\omega(\varepsilon)) & \text{for} \quad \varepsilon > \frac{\sigma_0}{E} \end{cases}$
where $\omega(\varepsilon) = \frac{E-E_1}{E}\left(1-\frac{\sigma_0}{E\varepsilon}\right)$, E_1 – elastic modulus	

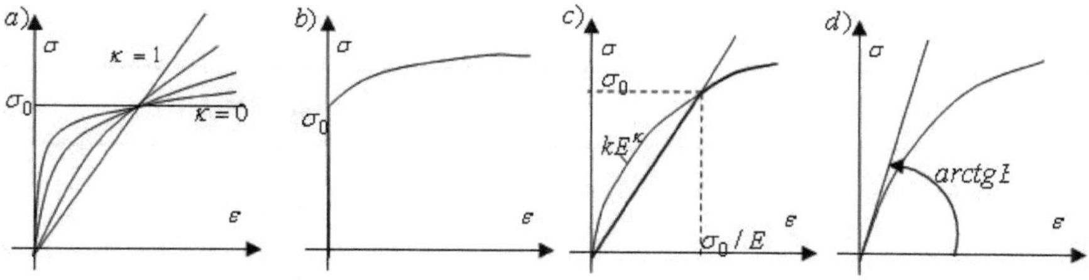

Fig. 7.3. Power characteristics of elasto-plastic materials: a) power elastic reinforcement, b) rigid-plastic reinforcement, c) a elasto-plastic model with reinforcement, d) a material according to Ramberg-Osgood [Ramberg W., Osgood W. R., 1943]

Table 7.2. Dependencies of power relations of models of elasto-plastic materials

Model of a material	Equation $\sigma = f(\varepsilon)$
power elastic reinforcement	$\frac{\sigma}{\sigma_0} = \left(\frac{\varepsilon}{\varepsilon_0}\right)^{\kappa}$, $0 \le \kappa \le 1$
power rigid-plastic reinforcement	$\sigma = \sigma_0 + k\varepsilon^{\kappa}$
elasto-plastic	$\sigma = \begin{cases} E\varepsilon & \text{for} \quad \varepsilon \le \frac{\sigma_0}{E} \\ k\varepsilon^{\kappa} & \text{for} \quad \varepsilon > \frac{\sigma_0}{E} \end{cases}$
Ramberg-Osgood	$\varepsilon = \frac{\sigma}{E} + k\left(\frac{\sigma}{E}\right)^{b}$
where k – a constant dependent on the material, b – exponent ≥ 1	

In Fig. 7.4, characteristics of ideally plastic materials are shown [Szuwalski K., Życzkowski M., 1973] and the dependencies corresponding to them are in Table 7.3.

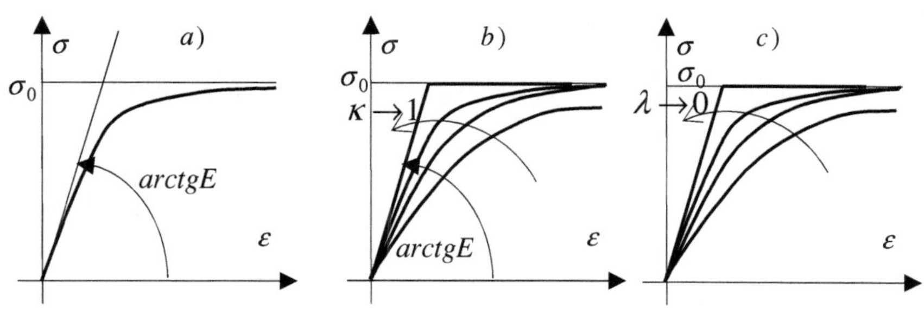

Fig. 7.4. An ideally plastic material: a) approximation with the hyperbolic tangent function, b) Ylinen approximation, c) approximation according to Szuwalski-Życzkowski

Table 7.3. Dependencies for models of ideally plastic materials

Model of a material	Equation $\sigma = f(\varepsilon)$
according to Prager [Prager W., 1938] (Fig.7.4a)	$\varepsilon = \frac{\sigma_0}{E} arctgh\left(\frac{\sigma}{\sigma_0}\right)$
according to Ylinen [Ylinen A., 1956] (Fig.7.4b)	$\varepsilon = \frac{1}{E}\left[\kappa\sigma - (1-\kappa)\sigma_0 \ln\left(1 - \frac{\sigma}{\sigma_0}\right)\right],$ $0 \le \kappa \le 1$
according to Szuwalski-Życzkowski [Szuwalski K., Życzkowski M., 1973] (Fig.7.4c)	$\varepsilon = \frac{\sigma}{E\left(1 - \frac{\sigma}{\sigma_0}\right)^{\lambda}}, \quad \lambda \ge 0$

The formulas presented in Tables 7.1 – 7.3 may be used to determine the bending moment in the analysis of large elasto-plastic deflections. An important problem is to determine the deformation under which the material changes from being elastic to plastic. The criterion may be defined based on various hypotheses. The most often used ones are: the Huber-Mises-Hencky (HMH) hypothesis and the Tresca-Guest hypothesis [Nowacki W., 1970], [Skrzypek J., 1986], [Ottosen N. S., Ristinmaa M., 2005]. Stresses causing transition to the plastic state under the HMH hypothesis form in the space of principal stresses a cylinder whose axis satisfies $\sigma_x = \sigma_y = \sigma_z$. Transition of the material to the plastic state will occur when the following equation is satisfied:

$$\sqrt{3J_2} - \sigma_0 = 0, \tag{7.1}$$

where σ_0 – initial plasticity limit of the material,

$J_2 = \frac{1}{2}\left[s_{xx}^2 + s_{yy}^2 + s_{zz}^2 + 2\left(s_{xy}^2 + s_{yz}^2 + s_{zx}^2\right)\right]$ – stress deviator invariant,

$$s_{ij} = \sigma_{ij} - \frac{1}{3}\sigma_{kk}\delta_{ij} = \begin{bmatrix} \sigma_x - \hat{\sigma} & \tau_{yx} & \tau_{zx} \\ \tau_{xy} & \sigma_y - \hat{\sigma} & \tau_{zy} \\ \tau_{xz} & \tau_{yz} & \sigma_z - \hat{\sigma} \end{bmatrix}$$ – stress deviator,

$\hat{\sigma} = \frac{1}{3}\left(\sigma_x + \sigma_y + \sigma_z\right)$ – mean stress,

$$\sigma_{ij} = \begin{bmatrix} \sigma_x & \tau_{yx} & \tau_{zx} \\ \tau_{xy} & \sigma_y & \tau_{zy} \\ \tau_{xz} & \tau_{yz} & \sigma_z \end{bmatrix} \quad - \text{ stress tensor.}$$

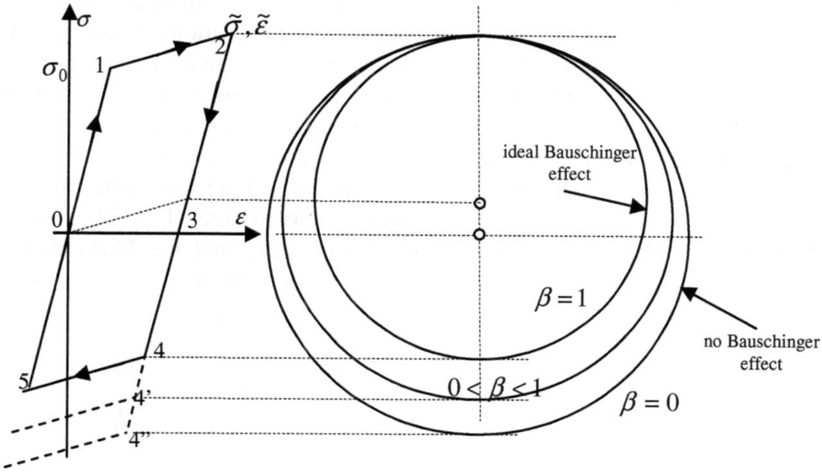

Fig. 7.5. Illustration of the Bauschinger effect

A result of plastic deformations is alteration of the position and the shape of the surface corresponding to the limit elastic state. The phenomenon is related to the yield point being shifted on the sides of both stretching and compression. It is described by various models of elastic reinforcement: isotropic, kinematic, combined, anisotropic. There exist multiple theories defining the shape of the stress-strain curve in the plastic region as a linear or nonlinear function of multiple parameters characterizing the material [Mróz Z., 1967], [Skrzypek J., 1986], [Ottosen N. S., Ristinmaa M., 2005]. In response to the plasticity surface being changed, the Bauschinger effect occurs which gives the material anisotropic properties (Fig. 7.5). After the reinforcement phase on the segment 1–2 follows relaxation 2–3 and transition to opposite stresses 3–4. The curves 2–4 and 4–5 are assumed to be parallel to 0–1 and 1–2 (initial stress and reinforcement phase), respectively. For actual materials, one can assume the coefficient of the Bauschinger effect $\beta \in \langle 0,\ldots,1 \rangle$, which relates isotropic reinforcement ($\beta = 0$) to kinematic reinforcement ($\beta = 1$).

7.2 A Model of Visco-elastic Material

In the linear theory of elasticity it is assumed that deformation of a body depends only on stress and material. The dependency between those values may in reality be also influenced by: temperature, time, generalized coordinates and velocities, etc. In order to take such influences into account, visco-elastic models of materials are introduced. Below selected models dealing with time are presented.

In visco-elastic problems [Nowacki W., 1963] two functions are of particular importance: creep function which increases the deformations under prolonged stress and relaxation function which describes how stress due to permanent deformation subsumes. Models of visco-elastic material are represented as systems of massless springs and dampers. Stiffness of the springs describes the elastic properties of the material and damping coefficients reflect the viscous traits. The simplest model taking into consideration both functions (creep and relaxation), which is often used, is the standard linear model (Fig. 7.6c). Also the Kelvin-Voigt model (Fig. 7.6a) only describing creep and the Maxwell model which deals just with relaxation (Fig. 7.6b) are very common.

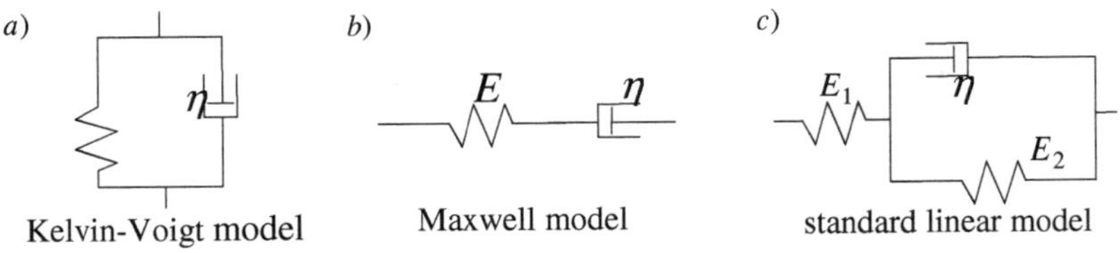

Fig. 7.6. Basic models of a visco-elastic material: a) Kelvin-Voigt, b) Maxwell, c) standard linear

The constitutive equation of the standard model (Fig. 7.6c) may be written in the form:

$$E_1\dot{\varepsilon} = \dot{\sigma} + \frac{1}{\eta}\left(\sigma(E_1 + E_2) - \varepsilon E_1 E_2\right). \tag{7.2}$$

In Fig. 7.7a, a graph $\varepsilon(t)$ is shown which corresponds to the solution of the equation (7.2) assuming load to be given by the function $\sigma(t) = \sigma_0[H(t) - H(t-T)]$ where $H(t)$ is the Heaviside function.

Later in this volume, methods of introducing nonlinear physical dependencies into systems' equations of motion will be presented.

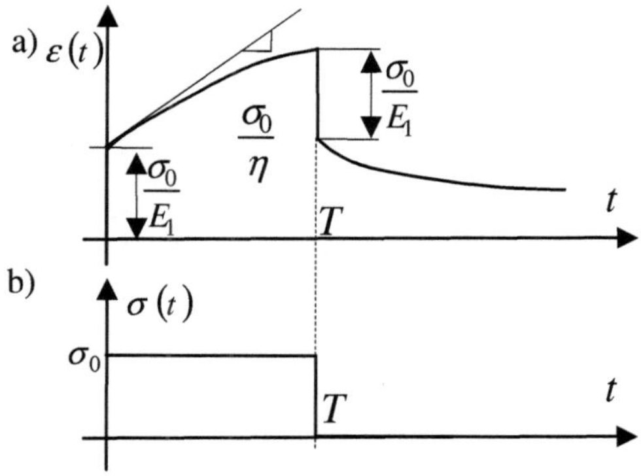

Fig. 7.7. Response of the standard model: a) material deformation, b) stress

8 The Rigid Finite Element Method

Actual kinematic chains commonly contain links whose flexibility greatly exceeds that of other links. It may then be necessary to take that flexibility into account. Booms of cranes and certain links of manipulators count among those. A large number of approaches in analysis of multibody systems can be found in literature with with one and more flexible links [Zienkiewicz O. C., 1972], [Wittbrodt E., 1983], [Wojciech S., 1984], [Huston R. L., Wanga Y., 1994], [Arteaga M. A., 1998], [Zienkiewicz O. C., Taylor R. L., 2000], [Berzeri M., et al., 2001], [Adamiec-Wójcik I., 2003], [Wittbrodt E., et al., 2006]. Chapter 9 introduces models of offshore cranes (a column one and an A-frame) which enable taking into account the flexibility of the supporting structure.

Let us consider a flexible link numbered p of a sample mechanism depicted in Fig. 8.1. Let $\{p,0\}$ be the coordinate system attached to the link p as if it were rigid. Its position relative to the preceding link s is given by the coordinates of the following vector:

$$\tilde{\mathbf{q}}^{(p,0)} = \left[\tilde{q}_1^{(p,0)} \quad \cdots \quad \tilde{q}_{\tilde{n}_{p,0}}^{(p,0)}\right]^T . \tag{8.1}$$

The number $\tilde{n}_{p,0}$ of coordinates of the vector $\tilde{\mathbf{q}}^{(p,0)}$ is less than 6 and depends on the class of the kinematic joint connecting the links s and p. These coordinates will henceforth be called rigid (configuration) coordinates of the link p.

In order to fully describe the relative motion of a flexible link, the vector (8.1) needs to be supplemented with a vector whose elements are called elastic coordinates. Their choice depends on the discretisation method used for the flexible link. Regardless of the method, the vector of generalized coordinates of the flexible link p describing its motion in the kinematic chain may be written as:

$$\tilde{\mathbf{q}}^{(p)} = \begin{bmatrix} \tilde{\mathbf{q}}^{(p,0)} \\ \tilde{\mathbf{q}}^{(p,f)} \end{bmatrix}, \tag{8.2}$$

where $\tilde{\mathbf{q}}^{(p,0)}$ – vector of generalized configuration (rigid) coordinates of the link p,

$\tilde{\mathbf{q}}^{(p,f)} = \left[\tilde{q}_1^{(p,f)} \quad \cdots \quad \tilde{q}_{\tilde{n}_{p,f}}^{(p,f)}\right]^T$ – vector of generalized elastic (flexible) coordinates of the link p,

$\tilde{n}_{p,f}$ – number of elastic coordinates of the link p.

E. Wittbrodt et al.: Rigid Finite Element Method, OEO 1, pp. 99–137.
springerlink.com

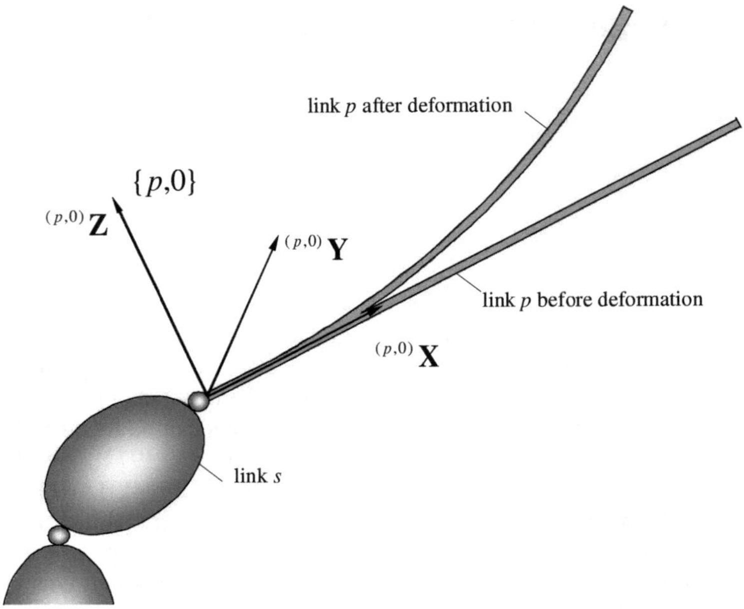

Fig. 8.1. A flexible link p

Let us also assume that the transformation of coordinates from the local coordinate system $\{p,0\}$ to the preceding coordinate system (with index s) is given by the matrix:

$$ {}_{(p,0)}^{\ s}\mathbf{T} = {}_{(p,0)}^{\ s}\mathbf{T}\left(\tilde{\mathbf{q}}^{(p,0)}\right). \tag{8.3} $$

One of many discretisation methods of flexible links will be presented below. This is the rigid finite element (RFE) method. It has two variants: classical and modified.

8.1 The RFE Method: Classical Formulation

The rigid finite element method has for many years been applied at the Gdańsk University of Technology, initially by Prof. Kruszewski, then by Prof. Wittbrodt, and their co-workers, to model multibody systems. The formulation of the method presented in [Kruszewski J., et al., 1975], in which each finite element is assumed to possess six degrees of freedom in its relative motion, is called classical. The description of the method expounded herein deviates from that which is found in papers by professor Kruszewski and his co-authors. Namely, joint coordinates and homogeneous transformations are used to derive the equations of motion, following [Adamiec-Wójcik I., 2003] and [Wittbrodt E., et al., 2006].

8.1.1 Generalized Coordinates: Transformation Matrices

Let p be a flexible beam link in a kinematic chain. That link is replaced with a series of rigid finite elements connected with spring-damping elements using discretisation which is detailed by Kruszewski and co-authors in [Kruszewski J., et al., 1975], [Kruszewski J., et al., 1999]. In the case of a beam with constant section, the procedure is as follows: first, this is the so-called primary division, the beam of length L_p is divided into m_p equally long segments (Fig. 8.2a).

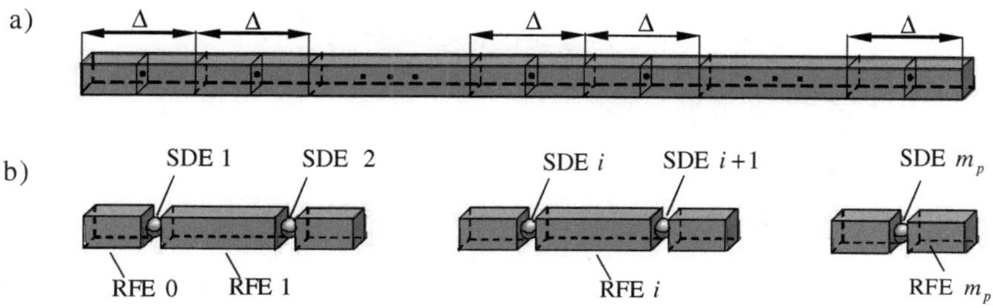

Fig. 8.2. Division of a flexible link: a) primary division, b) secondary division

Flexibility traits of the elements are inherited by the spring-damping elements (SDE) placed at the centre of each segment of length Δ. In this way, one obtains a secondary division of the flexible link into m_p+1 rigid finite elements (RFEs) connected by m_p massless and dimensionless spring-damping elements (Fig. 8.2b).

Division of beam links with variable sections and a method of determining characteristic parameters of RFEs and SDE are expounded, among other things, in the work [Wittbrodt E., et al., 2006]. Since each RFE (except RFE 0) has a coordinate system attached with origin in its centre of mass and axes coinciding with the principal axes of inertia (Fig. 8.3), the position of the element in undeformed state can be determined unambiguously relative to the system $\{p,0\}$ of RFE 0, provided that the transformation matrices are known:

$$\tilde{\mathbf{T}}^{(p,i')} = \text{const} . \tag{8.4}$$

In the general case, the transformation matrices with constant coefficientstake the form:

$$\tilde{\mathbf{T}}^{(p,i')} = \begin{bmatrix} \tilde{\mathbf{R}}^{(p,i')} & \tilde{\mathbf{r}}^{(p,i')} \\ \mathbf{0} & 1 \end{bmatrix}, \tag{8.5}$$

where $\tilde{\mathbf{R}}^{(p,i')}$ – direction cosine matrix of the axes of the system $\{p,i'\}$ relative to the system $\{p,0\}$,

$\tilde{\mathbf{r}}^{(p,i')}$ – vector of coordinates of the origin system of the system $\{p,i'\}$ in $\{p,0\}$.

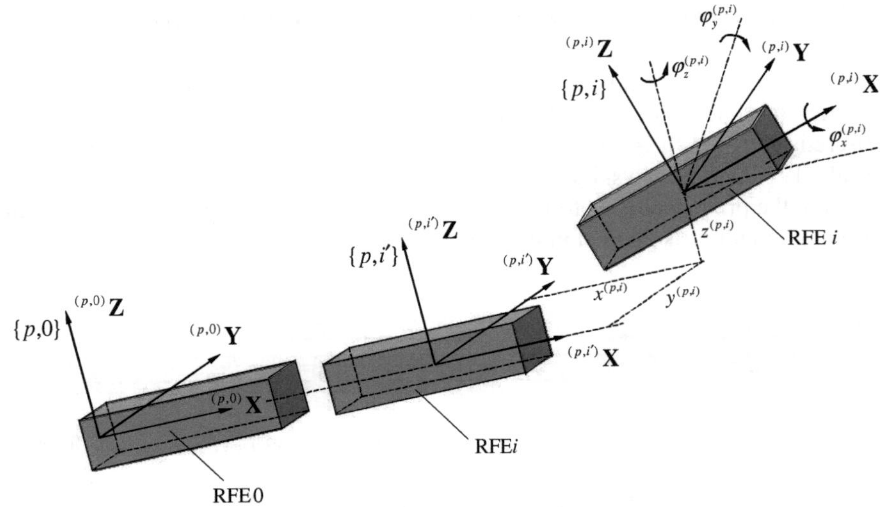

Fig. 8.3. Coordinate systems related to a flexible link: {} – the inertial system, {p,0} – the system attached to RFE 0, {p,i'} – the system attached to RFE i in undeformed state of the beam, {p,i} – the system attached in a fixed way to RFE i whose axes coincide with the principal central axes of inertia of the element, $x^{(p,i)}, y^{(p,i)}, z^{(p,i)}$ – coordinates of the origin of the coordinate system {p,i} in {p,i'}, $\varphi_x^{(p,i)}, \varphi_y^{(p,i)}, \varphi_z^{(p,i)}$ – ZYX Euler angles described in chapter 4

If the system {p,i'} has axes parallel to the axes of the system {p,0}, the rotation matrix $\widetilde{\mathbf{R}}^{(p,i')}$ is the identity matrix. Due to the lifting motion and external loads, individual RFEs are subjected to displacements. The generalized coordinates being the components of the vector:

$$\widetilde{\mathbf{q}}^{(p,i)} = \begin{bmatrix} x^{(p,i)} & y^{(p,i)} & z^{(p,i)} & \varphi_x^{(p,i)} & \varphi_y^{(p,i)} & \varphi_z^{(p,i)} \end{bmatrix}^T, \tag{8.6}$$

decribe the motion of the i-th RFE ($i = 1,...,m_p$) of the link p relative to the system {p,i'} attached to the RFE i in undeformed state. The transformation matrix ${}_{i}^{i'}\widetilde{\mathbf{T}}^{(p)}$ from the system {p,i} to the system {p,i'} in the nonlinear model, allowing the rotation angles $\varphi_x^{(p,i)}, \varphi_y^{(p,i)}, \varphi_z^{(p,i)}$ to be large, takes the following form:

$${}_{i}^{i'}\widetilde{\mathbf{T}}^{(p)} = \begin{bmatrix} c_z^{(p,i)}c_y^{(p,i)} & c_z^{(p,i)}s_y^{(p,i)}s_x^{(p,i)} - s_z^{(p,i)}c_x^{(p,i)} & c_z^{(p,i)}s_y^{(p,i)}c_x^{(p,i)} + s_z^{(p,i)}s_x^{(p,i)} & x^{(p,i)} \\ s_z^{(p,i)}c_y^{(p,i)} & s_z^{(p,i)}s_y^{(p,i)}s_x^{(p,i)} + c_z^{(p,i)}c_x^{(p,i)} & s_z^{(p,i)}s_y^{(p,i)}c_x^{(p,i)} - c_z^{(p,i)}s_x^{(p,i)} & y^{(p,i)} \\ -s_y^{(p,i)} & c_y^{(p,i)}s_x^{(p,i)} & c_y^{(p,i)}c_x^{(p,i)} & z^{(p,i)} \\ 0 & 0 & 0 & 1 \end{bmatrix}, \tag{8.7}$$

where $c_\alpha^{(p,i)} = \cos\varphi_\alpha^{(p,i)}, \quad s_\alpha^{(p,i)} = \sin\varphi_\alpha^{(p,i)} \quad$ for $\alpha \in \{x, y, z\}$..

When small rotation angles of RFEs are assumed, leading to omission of higher rank small terms from the approximations of trigonometric functions of the angles $\varphi_x^{(p,i)}$, $\varphi_y^{(p,i)}$, $\varphi_z^{(p,i)}$ (the linear model), the matrix ${}_{i}^{i'}\tilde{\mathbf{T}}^{(p)}$ may be written [Adamiec-Wójcik I., 2003] as:

$$
{}_{i}^{i'}\tilde{\mathbf{T}}^{(p)} = \begin{bmatrix} 1 & -\varphi_z^{(p,i)} & \varphi_y^{(p,i)} & x^{(p,i)} \\ \varphi_z^{(p,i)} & 1 & -\varphi_x^{(p,i)} & y^{(p,i)} \\ -\varphi_y^{(p,i)} & \varphi_x^{(p,i)} & 1 & z^{(p,i)} \\ 0 & 0 & 0 & 1 \end{bmatrix}. \tag{8.8}
$$

The transformation matrix from the system $\{p,i\}$ to the system $\{p,0\}$, whether the model is linear or nonlinear, has this form:

$$
\tilde{\mathbf{T}}^{(p,i)} = \tilde{\mathbf{T}}^{(p,i)}\left(\tilde{\mathbf{q}}^{(p,i)}\right) = \tilde{\mathbf{T}}^{(p,i')}\,{}_{i}^{i'}\tilde{\mathbf{T}}^{(p)}. \tag{8.9}
$$

8.1.2 Kinetic Energy of a Flexible Link

Let us assume, as in chapter 5, that the concerned multibody system is situated on a movable base $\{A\}$ (Fig. 5.1) whose motion relative to the inertial (global) system $\{0\} = \{\}$ is known.

Rigid finite elements of the link p may be treated as m_p+1 consecutive bodies appended to the link s of the kinematic chain. In further considerations, the first rigid finite element in the chain (RFE 0) is treated separately, because the generalized coordinates describing the relative motion of this RFE depend on the type of the kinematic joint connecting the link p with its preceding link s and their number is less than 6. The coordinate system $\{p,0\}$ plays the role of the configuration system of the link p.

Let the vector of generalized coordinates $\overline{\mathbf{q}}^{(p)}$ contain the coordinates of RFE 0 of the link p and the coordinates of the link s which precedes the link p. Let also the transformation matrix $\mathbf{T}^{(p,0)}$ define the transformation from the system $\{p,0\}$ attached to RFE 0 of the flexible link p to the inertial system. The following notation is introduced:

$$
\overline{\mathbf{q}}^{(p)} = \begin{bmatrix} \mathbf{q}^{(s)} \\ \tilde{\mathbf{q}}^{(p,0)} \end{bmatrix}, \tag{8.10.1}
$$

$$
\mathbf{T}^{(p,0)} = {}_{A}^{0}\mathbf{T}(t)\,{}_{s}^{A}\mathbf{T}\left(\mathbf{q}^{(s)}\right)\,{}_{(p,0)}^{s}\mathbf{T}\left(\tilde{\mathbf{q}}^{(p,0)}\right) = {}_{A}^{0}\mathbf{T}(t)\,\overline{\mathbf{T}}^{(p,0)}\left(\overline{\mathbf{q}}^{(p)}\right), \tag{8.10.2}
$$

where $\overline{\mathbf{T}}^{(p,0)} = {}_{s}^{A}\mathbf{T}\left(\mathbf{q}^{(s)}\right)\,{}_{(p,0)}^{s}\mathbf{T}\left(\tilde{\mathbf{q}}^{(p,0)}\right)$.

The kinetic energy of RFE 0 of the link p is given by the expression:

$$E_{p,0}=\frac{1}{2}\mathrm{tr}\left\{\dot{\mathbf{T}}^{(p,0)}\,\mathbf{H}^{(p,0)}\,\dot{\mathbf{T}}^{(p,0)T}\right\}, \tag{8.11}$$

where $\mathbf{H}^{(p,0)}$ – matrix of inertia of RFE 0 of the link p.

A derivation similar to that in chapter 5 yields:

$$\boldsymbol{\varepsilon}_{\overline{q}_k^{(p)}}(E_{p,0})=\sum_{i=1}^{n_{p,0}}a_{k,i}^{(p,0)}\ddot{q}_i^{(p,0)}+e_k^{(p,0)}, \tag{8.12}$$

where $a_{k,i}^{(p,0)}=\mathrm{tr}\left\{\mathbf{T}_k^{(p,0)}\,\mathbf{H}^{(p,0)}\,\mathbf{T}_i^{(p,0)T}\right\}$,

$$e_k^{(p,0)}=\sum_{i=1}^{n_{p,0}}\sum_{j=1}^{n_{p,0}}\mathrm{tr}\left\{\mathbf{T}_k^{(p,0)}\mathbf{H}^{(p,0)}\,\mathbf{T}_{i,j}^{(p,0)}\right\}\dot{q}_i^{(p,0)}\,\dot{q}_j^{(p,0)}+ \\ +\mathrm{tr}\left\{\mathbf{T}_k^{(p,0)}\,\mathbf{H}^{(p,0)}\left[{}_A^0\ddot{\mathbf{T}}\widetilde{\mathbf{T}}^{(p,0)}+2\,{}_A^0\dot{\mathbf{T}}\dot{\widetilde{\mathbf{T}}}^{(p,0)}\right]^T\right\},$$

$n_{p,0}=n_s+\widetilde{n}_{p,0}$.

The equation (8.12) may be put in a matrix form:

$$\boldsymbol{\varepsilon}_{\overline{q}_k^{(p)}}(E_{p,0})=\begin{bmatrix}\mathbf{A}_{s,s}^{(p,0)} & \mathbf{A}_{s,0}^{(p,0)}\\ \mathbf{A}_{0,s}^{(p,0)} & \mathbf{A}_{0,0}^{(p,0)}\end{bmatrix}\begin{bmatrix}\ddot{\mathbf{q}}^{(s)}\\ \ddot{\widetilde{\mathbf{q}}}^{(p,0)}\end{bmatrix}+\begin{bmatrix}\mathbf{e}_s^{(p,0)}\\ \mathbf{e}_0^{(p,0)}\end{bmatrix}. \tag{8.13}$$

The remaining RFEs of the flexible link are treated as elements of the kinematic chain appended to RFE 0. Hence, the coordinates of an arbitrary point in the local system $\{p,i\}$ of RFE i of the link p ($i = 1,...,m_p$) may be transformed, following the procedure presented in chapter 5, to the inertial system. The following equality is used:

$$\mathbf{r}^{(p,i)}=\mathbf{T}^{(p,i)}\,\widetilde{\mathbf{r}}^{(p,i)}, \tag{8.14}$$

where $\mathbf{T}^{(p,i)}=\mathbf{T}^{(p,0)}\,\widetilde{\mathbf{T}}^{(p,i)}$

$\mathbf{r}^{(p,i)}$ – vector of coordinates in the inertial system { },

$\widetilde{\mathbf{r}}^{(p,i)}$ – vector of local coordinates in the system $\{p,i\}$.

The kinetic energy of REF i of the link p equals:

$$E_{p,i}=\frac{1}{2}\mathrm{tr}\left\{\dot{\mathbf{T}}^{(p,i)}\,\mathbf{H}^{(p,i)}\,\dot{\mathbf{T}}^{(p,i)T}\right\}, \tag{8.15}$$

where $\mathbf{H}^{(p,i)}$ – matrix of inertia of RFE i of the link p.

Defining a vector with $n_{p,i}=n_s+\widetilde{n}_{p,0}+6=n_{p,0}+6$ components:

$$\mathbf{q}^{(p,i)}=\begin{bmatrix}\overline{\mathbf{q}}^{(p)}\\ \widetilde{\mathbf{q}}^{(p,i)}\end{bmatrix}, \tag{8.16}$$

the following may be written:

$$\boldsymbol{\varepsilon}_{\mathbf{q}^{(p,i)}}(E_{p,i}) = \mathbf{A}^{(p,i)}\ddot{\mathbf{q}}^{(p,i)} + \mathbf{e}^{(p,i)}, \tag{8.17}$$

where $\mathbf{A}^{(p,i)} = \left(\tilde{a}_{l,s}^{(p,i)}\right)_{l,s=1,\dots,n_{p,i}} = \mathrm{tr}\left\{\mathbf{T}_l^{(p,i)}\,\mathbf{H}^{(p,i)}\,\mathbf{T}_s^{(p,i)T}\right\},$

$$\mathbf{e}^{(p,i)} = \left(\tilde{e}_l^{(p,i)}\right)_{l=1,\dots,n_{p,i}} = \sum_{s=1}^{n_{p,i}}\sum_{j=1}^{n_{p,i}} \mathrm{tr}\left\{\mathbf{T}_l^{(p,i)}\,\mathbf{H}^{(p,i)}\,\mathbf{T}_{s,j}^{(p,i)T}\right\}\dot{q}_s^{(p,i)}\,\dot{q}_j^{(p,i)} + $$
$$+ \mathrm{tr}\left\{\mathbf{T}_l^{(p,i)}\,\mathbf{H}^{(p,i)}\left[{}_A^0\ddot{\mathbf{T}}\overline{\mathbf{T}}^{(p,i)} + {}_A^0\dot{\mathbf{T}}\dot{\overline{\mathbf{T}}}^{(p,i)}\right]\right\}.$$

The same may be expressed in the block form, thus:

$$\boldsymbol{\varepsilon}_{\mathbf{q}^{(p,i)}}\left(E_{p,i}\right) = \begin{bmatrix} \mathbf{A}_{s,s}^{(p,i)} & \mathbf{A}_{s,0}^{(p,i)} & \mathbf{A}_{s,i}^{(p,i)} \\ \mathbf{A}_{0,s}^{(p,i)} & \mathbf{A}_{0,0}^{(p,i)} & \mathbf{A}_{0,i}^{(p,i)} \\ \mathbf{A}_{i,s}^{(p,i)} & \mathbf{A}_{i,0}^{(p,i)} & \mathbf{A}_{i,i}^{(p,i)} \end{bmatrix} \begin{bmatrix} \ddot{\mathbf{q}}^{(s)} \\ \ddot{\tilde{\mathbf{q}}}^{(p,0)} \\ \ddot{\tilde{\mathbf{q}}}^{(p,i)} \end{bmatrix} + \begin{bmatrix} \mathbf{e}_s^{(p,i)} \\ \mathbf{e}_0^{(p,i)} \\ \mathbf{e}_i^{(p,i)} \end{bmatrix}. \tag{8.18}$$

8.1.3 Potential Energy of Gravity Forces and Deformations of a Flexible Link p

The potential energy of gravity forces of the RFE i is given by:

$$V_{p,i}^{g} = m^{(p,i)}\, g\, \boldsymbol{\theta}_3 \mathbf{T}^{(p,i)}\tilde{\mathbf{r}}_C^{(p,i)}, \tag{8.19}$$

where $\tilde{\mathbf{r}}_C^{(p,i)}$ – vector determining the position of the centre of mass of the RFE i in the local coordinate system $\{p,i\}$,

$m^{(p,i)}$ – mass of REF i.

Hence, after ironing out the differences in the definitions of matrices $\mathbf{T}^{(p,0)}$ and $\mathbf{T}^{(p,i)}$, for $i = 1,\dots,m_p$ the following holds:

$$\mathbf{G}^{(p,0)} = \frac{\partial V_{p,0}^{g}}{\partial \overline{\mathbf{q}}^{(p)}} = \begin{bmatrix} \mathbf{G}_s^{(p,0)} \\ \mathbf{G}_0^{(p,0)} \end{bmatrix}, \tag{8.20.1}$$

$$\mathbf{G}^{(p,i)} = \frac{\partial V_{p,i}^{g}}{\partial \mathbf{q}^{(p,i)}} = \begin{bmatrix} \mathbf{G}_s^{(p,i)} \\ \mathbf{G}_0^{(p,i)} \\ \mathbf{G}_i^{(p,i)} \end{bmatrix}, \tag{8.20.2}$$

where $\mathbf{G}^{(p,i)} = \left(g_k^{(p,i)} \right)_{k=1,\ldots,n_{p,i}}$, $g_k^{(p,i)} = m^{(p,i)} g \, \mathbf{T}_k^{(p,i)} \tilde{\mathbf{r}}_C^{(p,i)}$ for $i = 0,1,\ldots,m_p$,

$\mathbf{G}_s^{(p,0)}, \mathbf{G}_0^{(p,0)}, \mathbf{G}_s^{(p,i)}, \mathbf{G}_0^{(p,i)}, \mathbf{G}_i^{(p,i)}$ – appropriate blocks of vectors $\mathbf{G}^{(p,0)}$ and $\mathbf{G}^{(p,i)}$ corresponding to the coordinates $\mathbf{q}^{(s)}, \tilde{\mathbf{q}}^{(p,0)}, \tilde{\mathbf{q}}^{(p,i)}$.

Since the considered link is flexible, before formulating its equations of motion the expressions resulting from the energy of elastic deformation of SDE must be determined. Their derivations in the case of linear physical dependencies describing the properties of the material are presented below. The way with nonlinear physical dependencies will be discussed later. In the considerations pertaining to the deformation of spring-damping elements the reference coordinate system is assumed to be $\{p,0\}$, which is attached to RFE 0, and the matrices $\tilde{\mathbf{R}}^{(p,i')}$, which occur in (8.5), to be identity matrices. A consequence of this is the proposition that in the undeformed state of the link p the axes of all the coordinate systems attached to RFEs from 0 to m_p are parallel. A general algorithm omitting this assumption is presented in [Wittbrodt E., et al., 2006]. A numerically efficient modification of the algorithm will also be described later in this chapter.

Let SDE e connect the RFEs l and r of a flexible link p (Fig. 8.4).

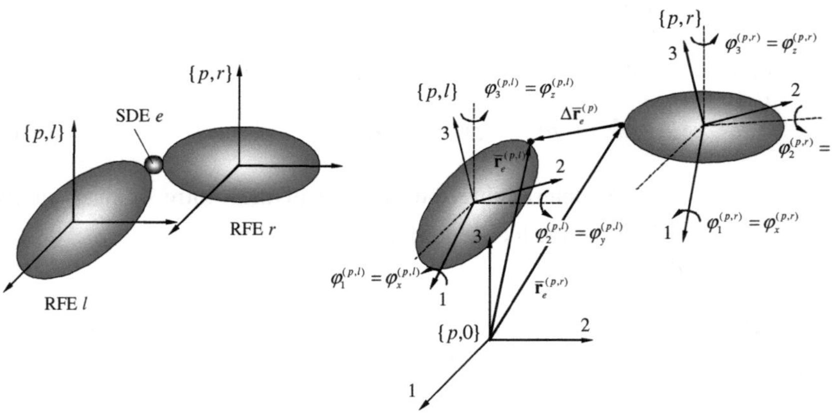

Fig. 8.4. A model of a spring-damping element: a) connection of RFEs l and r by SDE e, b) notation assumed

The energy of elastic deformation of this element is given by the formula:

$$V_{p,e}^{s} = \frac{1}{2}\sum_{j=1}^{3} c_{e,j}^{(p)} \left[\bar{x}_{r,j} - \bar{x}_{l,j} \right]^2 + \frac{1}{2}\sum_{j=1}^{3} c_{e,j+3}^{(p)} \left[\varphi_j^{(p,r)} - \varphi_j^{(p,l)} \right]^2, \tag{8.21}$$

where $c_{e,j}^{(p)}$ for j=1,2,3 – coefficients of translational stiffness of the SDE e of the link p,

$c_{e,j}^{(p)}$ for j=4,5,6 – coefficients of rotational stiffness of the SDE e of the link p,

$\overline{\mathbf{r}}_e^{(p,l)} = \begin{bmatrix}\overline{x}_{l,1} & \overline{x}_{l,2} & \overline{x}_{l,3} & 1\end{bmatrix}^T$, $\overline{\mathbf{r}}_e^{(p,r)} = \begin{bmatrix}\overline{x}_{r,1} & \overline{x}_{r,2} & \overline{x}_{r,3} & 1\end{bmatrix}^T$ – vectors of coordinates of the SDE e (treated first as a point of the RFE l, and next as a point of the RFE r) expressed in the system $\{p,0\}$,

$\varphi_j^{(p,r)}, \varphi_j^{(p,l)}$ – rotation angles of the RFEs r and l of the link p.

In Fig. 8.4 and formula (8.21), the axes of the coordinate system are denoted with (1, 2, 3) instead of (**X** ,**Y**, **Z**) used hitherto. This shortens the formulas considerably.

The coordinates of the SDE e in the systems attached to the RFEs l and r are assumed to be represented by vectors $\tilde{\mathbf{r}}_e^{(p,l)}$ and $\tilde{\mathbf{r}}_e^{(p,r)}$ in Fig. 8.4, respectively. Consequently, the coordinates of this spring-damping element in the reference coordinate system $\{p,0\}$ are expressed by:

$$\overline{\mathbf{r}}_e^{(p,i)} = \tilde{\mathbf{T}}^{(p,i)}\,\tilde{\mathbf{r}}_e^{(p,i)}, \tag{8.22}$$

where $i \in \{r,l\}$.

The vector $\Delta\overline{\mathbf{r}}_e^{(p)}$ (Fig. 8.4b) is given as:

$$\Delta\overline{\mathbf{r}}_e^{(p)} = \overline{\mathbf{r}}_e^{(p,r)} - \overline{\mathbf{r}}_e^{(p,l)} = \tilde{\mathbf{T}}^{(p,r)}\tilde{\mathbf{r}}_e^{(p,r)} - \tilde{\mathbf{T}}^{(p,l)}\tilde{\mathbf{r}}_e^{(p,l)}, \tag{8.23}$$

and the potential energy of elastic deformation of the SDE e may be put in the following form:

$$V_{p,e}^{s} = \frac{1}{2}\Delta\overline{\mathbf{r}}_e^{(p)T}\mathbf{C}_T^{(p,e)}\Delta\overline{\mathbf{r}}_e^{(p)} + \frac{1}{2}\left[\tilde{\mathbf{q}}^{(p,r)} - \tilde{\mathbf{q}}^{(p,l)}\right]^T \mathbf{C}_R^{(p,e)}\left[\tilde{\mathbf{q}}^{(p,r)} - \tilde{\mathbf{q}}^{(p,l)}\right], \tag{8.24}$$

where $\mathbf{C}_T^{(p,e)} = \begin{bmatrix} c_{e,1}^{(p)} & 0 & 0 & 0 \\ 0 & c_{e,2}^{(p)} & 0 & 0 \\ 0 & 0 & c_{e,3}^{(p)} & 0 \\ 0 & 0 & 0 & 0 \end{bmatrix}$, $\mathbf{C}_R^{(p,e)} = \begin{bmatrix} 0 & 0 & 0 & 0 & 0 & 0 \\ 0 & 0 & 0 & 0 & 0 & 0 \\ 0 & 0 & 0 & 0 & 0 & 0 \\ 0 & 0 & 0 & c_{e,4}^{(p)} & 0 & 0 \\ 0 & 0 & 0 & 0 & c_{e,5}^{(p)} & 0 \\ 0 & 0 & 0 & 0 & 0 & c_{e,6}^{(p)} \end{bmatrix}$.

The above considerations pertain to the general case in which the transformation matrices $\tilde{\mathbf{T}}^{(p,i)}$ dla $i \in \{l,r\}$ are nonlinear. When small oscillations are considered, i.e. when the transformation matrices ${}_{i}^{i'}\tilde{\mathbf{T}}^{(p)}$ conform to the formula (8.8), the transformation formula taking the system $\{p,i\}$ to the system $\{p,0\}$ may be represented thusly:

$$\overline{\mathbf{r}}_e^{(p,i)} = \mathbf{r}_e'^{(p,i)} + \mathbf{D}_e'^{(p,i)}\widetilde{\mathbf{q}}^{(p,i)}, \tag{8.25}$$

where $\mathbf{r}_e'^{(p,i)} = \begin{bmatrix} \widetilde{x}_{i,1} + a_1^{(p,i')} \\ \widetilde{x}_{i,2} + a_2^{(p,i')} \\ \widetilde{x}_{i,3} + a_3^{(p,i')} \\ 1 \end{bmatrix}$ – vector with constant coefficients,

$\mathbf{D}_e'^{(p,i)} = \begin{bmatrix} 1 & 0 & 0 & 0 & \widetilde{x}_{i,3} & -\widetilde{x}_{i,2} \\ 0 & 1 & 0 & -\widetilde{x}_{i,3} & 0 & \widetilde{x}_{i,1} \\ 0 & 0 & 1 & \widetilde{x}_{i,2} & -\widetilde{x}_{i,1} & 0 \\ 0 & 0 & 0 & 0 & 0 & 0 \end{bmatrix}$ – matrix with constant coefficients,

$a_j^{(p,i')}$ – components of the vector $\widetilde{\mathbf{r}}^{(p,i')}$ from the formula (8.5) for $j = 1, 2, 3$,

$\widetilde{x}_{i,1}, \widetilde{x}_{i,2}, \widetilde{x}_{i,3}$ – coordinates of considered point in $\{p,i\}$.

The formula (8.21) for small deformations takes the form:

$$\begin{aligned} V_{p,e}^s = &\frac{1}{2}\left[\Delta\mathbf{r}_e'^{(p)} + \mathbf{D}_e'^{(p,r)}\tilde{\mathbf{q}}^{(p,r)} - \mathbf{D}_e'^{(p,l)}\tilde{\mathbf{q}}^{(p,l)}\right]^T \mathbf{C}_T^{(p,e)}\left[\Delta\mathbf{r}_e'^{(p)} + \mathbf{D}_e'^{(p,r)}\tilde{\mathbf{q}}^{(p,r)} - \mathbf{D}_e'^{(p,l)}\tilde{\mathbf{q}}^{(p,l)}\right] + \\ &+\frac{1}{2}\left[\tilde{\mathbf{q}}^{(p,r)} - \tilde{\mathbf{q}}^{(p,l)}\right]^T \mathbf{C}_R^{(p,e)}\left[\tilde{\mathbf{q}}^{(p,r)} - \tilde{\mathbf{q}}^{(p,l)}\right] \end{aligned}, \tag{8.26}$$

where $\Delta\mathbf{r}_e'^{(p)} = \widetilde{\mathbf{r}}_e'^{(p,r)} - \widetilde{\mathbf{r}}_e'^{(p,l)}$.

In the case of beam links, the SDE i connects the RFE i–1 with the RFE i, therefore $\widetilde{\mathbf{q}}^{(p,l)} = \widetilde{\mathbf{q}}^{(p,i-1)}$ and $\widetilde{\mathbf{q}}^{(p,r)} = \widetilde{\mathbf{q}}^{(p,i)}$.

The potential energy of elastic deformation of the link p equals the sum of energies of all the SDE:

$$V_p^s = \sum_{e=1}^{m_p} V_{p,e}^s. \tag{8.27}$$

One should take into account that the formula expressing the elastic energy of SDE 1 of the link p is a variant of the formulas (8.21) and (8.24), and it takes the form:

$$V_{(p,1)}^s = \frac{1}{2}\Delta\overline{\mathbf{r}}_1^{(p)T}\mathbf{C}_T^{(p,e)}\Delta\overline{\mathbf{r}}_1^{(p)} + \frac{1}{2}\widetilde{\mathbf{q}}^{(p,1)T}\mathbf{C}_R^{(p,e)}\widetilde{\mathbf{q}}^{(p,1)}, \tag{8.28}$$

where $\Delta\overline{\mathbf{r}}_1^{(p)} = \widetilde{\mathbf{T}}^{(p,1)}\overline{\mathbf{r}}_1^{(p,1)} - \widetilde{\mathbf{r}}_1^{(p,0)}$.

Taking (8.27) into account leads to:

$$\frac{\partial V_p^s}{\partial \tilde{\mathbf{q}}^{(p,i)}} = -\mathbf{C}_R^{(p,i)}\tilde{\mathbf{q}}^{(p,i-1)} + \left(\mathbf{C}_R^{(p,i)} + \mathbf{C}_R^{(p,i+1)}\right)\tilde{\mathbf{q}}^{(p,i)} - \mathbf{C}_R^{(p,i+1)}\tilde{\mathbf{q}}^{(p,i+1)} + \tilde{\mathbf{S}}^{(p,i)}, \quad (8.29)$$

where $\tilde{\mathbf{S}}^{(p,i)} = \frac{\partial \Delta\overline{\mathbf{r}}_i^{(p)T}}{\partial \tilde{\mathbf{q}}^{(p,i)}}\mathbf{C}_T^{(p,i)}\Delta\overline{\mathbf{r}}_i^{(p)} + \frac{\partial \Delta\overline{\mathbf{r}}_{i+1}^{(p)T}}{\partial \tilde{\mathbf{q}}^{(p,i)}}\mathbf{C}_T^{(p,i+1)}\Delta\overline{\mathbf{r}}_{i+1}^{(p)}$.

Let us remark that for $i = 0$ and $i = m_p$ the following should be assumed, respectively: $\mathbf{C}_R^{(p,0)} = \mathbf{0}$, $\mathbf{C}_R^{(p,i_p+1)} = \mathbf{C}_T^{(p,i_p+1)} = \mathbf{0}$. The formula (8.29) is valid both for linear and nonlinear oscillations. The form of the vectors $\tilde{\mathbf{S}}^{(p,i)}$ in the linear case may be determined easily by means of the formula (8.26). Problems related to the choice of stiffness coefficients when analysing large deflections are discussed in the following papers: [Adamiec-Wójcik I., 1992], [Wojciech S., Adamiec-Wójcik I., 1993], [Wojciech S., Adamiec-Wójcik I., 1994] and [Wittbrodt E., et al., 2006].

8.1.4 Generalized Forces: Equations of Motion

Let us assume that the following act upon the RFE *i*: a force $\tilde{\mathbf{F}}^{(p,i)}$ and a pair of forces whose moment $\tilde{\mathbf{M}}^{(p,i)}$ has the components:

$$\tilde{\mathbf{F}}^{(p,i)} = \begin{bmatrix}\tilde{F}_x^{(p,i)} & \tilde{F}_y^{(p,i)} & \tilde{F}_z^{(p,i)} & 0\end{bmatrix}^T, \quad (8.30.1)$$

$$\tilde{\mathbf{M}}^{(p,i)} = \begin{bmatrix}\tilde{M}_x^{(p,i)} & \tilde{M}_y^{(p,i)} & \tilde{M}_z^{(p,i)} & 0\end{bmatrix}^T. \quad (8.30.2)$$

Applying the formulas (5.40) and (5.42) along with the procedure presented in [Adamiec-Wójcik I., et al., 2008] yields these forms of generalized forces due to their presence:

$$\begin{aligned} Q_k^{(p,i)}\left(\tilde{\mathbf{F}}^{(p,i)},\tilde{\mathbf{M}}^{(p,i)}\right) &= \tilde{\mathbf{F}}^{(p,i)T}\mathbf{T}^{(p,i)T}\mathbf{T}_k^{(p,i)}\tilde{\mathbf{r}}^{(p,i)} + \tilde{M}_1^{(p,i)}\sum_{j=1}^{3}\left(\mathbf{T}^{(p,i)}\right)_{j,3}\left(\mathbf{T}_k^{(p,i)}\right)_{j,2} + \\ &+ \tilde{M}_2^{(p,i)}\sum_{j=1}^{3}\left(\mathbf{T}^{(p,i)}\right)_{j,1}\left(\mathbf{T}_k^{(p,i)}\right)_{j,3} + \tilde{M}_3^{(p,i)}\sum_{j=1}^{3}\left(\mathbf{T}^{(p,i)}\right)_{j,2}\left(\mathbf{T}_k^{(p,i)}\right)_{j,1}. \end{aligned} \quad (8.31)$$

When forces acting on RFE 0 are considered, it may be written:

$$\mathbf{Q}^{(p,0)} = \begin{bmatrix}\mathbf{Q}_s^{(p,0)} \\ \mathbf{Q}_0^{(p,0)}\end{bmatrix}, \quad (8.32.1)$$

whereas for a force $\tilde{\mathbf{F}}^{(p,i)}$ and a pair of forces with moment $\tilde{\mathbf{M}}^{(p,i)}$ acting on RFEs from 1 to m_p the following holds:

$$\mathbf{Q}^{(p,i)} = \begin{bmatrix} \mathbf{Q}_s^{(p,i)} \\ \mathbf{Q}_0^{(p,i)} \\ \mathbf{Q}_i^{(p,i)} \end{bmatrix}. \tag{8.32.2}$$

In the case of a flexible link decomposed into m_p+1 rigid finite elements, the following vector of generalized coordinates of the link and expressions giving the kinetic energy and the potential energy of the gravity forces may be defined:

$$\mathbf{q}^{(p)} = \begin{bmatrix} \mathbf{q}^{(s)} \\ \tilde{\mathbf{q}}^{(p,0)} \\ \tilde{\mathbf{q}}^{(p,1)} \\ \vdots \\ \tilde{\mathbf{q}}^{(p,m_p)} \end{bmatrix}, \tag{8.33.1}$$

$$E_p = \sum_{i=1}^{m_p} E_{p,i}\,, \tag{8.33.2}$$

$$V_p^g = \sum_{i=1}^{m_p} V_{p,i}^g\,. \tag{8.33.3}$$

From the equations (8.13), (8.18), (8.20), (8.29) and (8.32) it follows that the equations of motion of the link p, including the term due to the energy of elastic deformation, take the form:

$$\mathbf{A}^{(p)}\ddot{\mathbf{q}}^{(p)} + \mathbf{K}_R^{(p)}\mathbf{q}^{(p)} = -\mathbf{e}^{(p)} - \mathbf{G}^{(p)} - \mathbf{S}^{(p)} + \mathbf{Q}^{(p)}, \tag{8.34}$$

where
$$\mathbf{A}^{(p)} = \begin{bmatrix} \sum_{i=0}^{m_p} \mathbf{A}_{s,s}^{(p,i)} & \sum_{i=0}^{m_p} \mathbf{A}_{s,0}^{(p,i)} & \mathbf{A}_{s,1}^{(p,1)} & \cdots & \mathbf{A}_{s,m_p}^{(p,m_p)} \\ \sum_{i=0}^{m_p} \mathbf{A}_{0,s}^{(p,i)} & \sum_{i=0}^{m_p} \mathbf{A}_{0,0}^{(p,i)} & \sum_{i=0}^{m_p} \mathbf{A}_{0,1}^{(p,i)} & \cdots & \mathbf{A}_{0,m_p}^{(p,m_p)} \\ \mathbf{A}_{1,s}^{(p,1)} & \mathbf{A}_{1,0}^{(p,1)} & \mathbf{A}_{1,1}^{(p,1)} & \cdots & 0 \\ \vdots & \vdots & \vdots & \ddots & \vdots \\ \mathbf{A}_{m_p,s}^{(p,m_p)} & \mathbf{A}_{m_p,0}^{(p,m_p)} & 0 & \cdots & \mathbf{A}_{m_p,m_p}^{(p,m_p)} \end{bmatrix},$$

$$\mathbf{K}_R^{(p)} = \begin{bmatrix} \mathbf{0} & \mathbf{0} & \mathbf{0} & \cdots & \mathbf{0} \\ \mathbf{0} & \mathbf{C}_R^{(p,1)} & -\mathbf{C}_R^{(p,1)} & \cdots & \mathbf{0} \\ \mathbf{0} & -\mathbf{C}_R^{(p,1)} & \mathbf{C}_R^{(p,1)} + \mathbf{C}_R^{(p,2)} & \cdots & \mathbf{0} \\ \vdots & \vdots & \vdots & \ddots & \vdots \\ \mathbf{0} & \mathbf{0} & \mathbf{0} & \cdots & \mathbf{C}_R^{(p,m_p)} \end{bmatrix},$$

$$\mathbf{G}^{(p)} = \begin{bmatrix} \sum_{i=0}^{m_p} \mathbf{G}_s^{(p,i)} \\ \sum_{i=0}^{m_p} \mathbf{G}_0^{(p,i)} \\ \mathbf{G}_1^{(p,1)} \\ \vdots \\ \mathbf{G}_{m_p}^{(p,m_p)} \end{bmatrix}, \mathbf{e}^{(p)} = \begin{bmatrix} \sum_{i=0}^{m_p} \mathbf{e}_s^{(p,i)} \\ \sum_{i=0}^{m_p} \mathbf{e}_0^{(p,i)} \\ \mathbf{e}_1^{(p,1)} \\ \vdots \\ \mathbf{e}_{m_p}^{(p,m_p)} \end{bmatrix}, \mathbf{S}^{(p)} = \begin{bmatrix} \mathbf{0} \\ \tilde{\mathbf{S}}^{(p,0)} \\ \tilde{\mathbf{S}}^{(p,1)} \\ \vdots \\ \tilde{\mathbf{S}}^{(p,m_p)} \end{bmatrix}, \mathbf{Q}^{(p)} = \begin{bmatrix} \sum_{i=0}^{m_p} \mathbf{Q}_s^{(p,i)} \\ \sum_{i=0}^{m_p} \mathbf{Q}_0^{(p,i)} \\ \mathbf{Q}_1^{(p,1)} \\ \vdots \\ \mathbf{Q}_i^{(p,m_p)} \end{bmatrix}.$$

A remark is due that the matrices $\mathbf{A}^{(p)}$ and $\mathbf{K}_R^{(p)}$ contain many zeroes. This fact may be leveraged in an implementation of the algorithm on a computer. The equations of motion of the system's links from 1 to p, forming a kinematic chain, may be generated in the way described in section 5.4. The equations for a rigid link may be obtained as a special case of a flexible link taking $m_p = 0$. The rigid link may then be treated as RFE 0.

When a link p follows a flexible link in a kinematic chain, the model includes a connection between the last RFE of the flexible link s and the next link (namely, with RFE 0 of the next link). If linear oscillations are considered, i.e. the transformation matrix for the RFE i of the flexible link takes the form (8.18), the matrix of masses $\mathbf{A}^{(p)}$ is a diagonal matrix in the fragment from RFE 1 to RFE m_p of the link p. Calculations are considerably simples when this fact is used in the integration of the equations (8.34). Additionally, the stiffness matrix $\mathbf{K}_R^{(p)}$ is a block-tridiagonal matrix, which is also helpful in solving the equations of motion. A product of matrices with constant coefficients may be distinguished in the vector $\mathbf{S}^{(p,f)} = \left[\tilde{\mathbf{S}}^{(p,1)} \quad \ldots \quad \tilde{\mathbf{S}}^{(p,m_p)}\right]^T$ in the linear case [Wojnarowski J., Adamiec-Wójcik I., 2005], thus assuming:

$$\mathbf{S}^{(p,f)} = \mathbf{K}_T^{(p,f)} \tilde{\mathbf{q}}^{(p,f)} + \mathbf{S}_c^{(p,f)}, \tag{8.35}$$

where $\mathbf{K}_T^{(p,f)}, \mathbf{S}_c^{(p,f)}$ – a matrix and a vector with constant coefficients,

$$\tilde{\mathbf{q}}^{(p,f)} = \left[\tilde{\mathbf{q}}^{(p,1)T} \quad \cdots \quad \tilde{\mathbf{q}}^{(p,\tilde{n}_p)T}\right]^T.$$

The presented model includes all possible displacements of the RFEs into which a flexible link is divided. If just one type of flexibility (e.g. to bending in one plane or torsion) is dominant in the link, models with fewer degrees of freedom of the RFEs may be easily obtained as a special case of the given formulas by appropriately fixing the vector of generalized coordinates of the rigid element.

8.2 Modification of the Rigid Finite Element Method

The classical rigid finite element method enables taking into account arbitrary displacements of finite elements and therefore analysis of the following deformations: lateral, longitudinal, rotational and shear. The displacements of each element are considered relative to the reference coordinate system attached to RFE 0. In this section, a modification of the rigid finite element method is presented which also has applications to discretisation of flexible beam links. In the modification only lateral and rotational deformations are assumed, and displacements of each RFE are defined relative to its preceding RFE. The method is presented in [Wojciech S., 1984] for planar systems and in the papers [Wojciech S., 1990], [Adamiec-Wójcik I., 1992], [Adamiec-Wójcik I., 1993] and [Adamiec-Wójcik I., 2003] as well as in [Wittbrodt E., et al., 2006] for spatial systems. The modification allows large deflections of flexible links to be analysed.

8.2.1 Generalized Coordinates: Transformation Matrices

Discretisation of a flexible beam link is performed in the same way as in the classical rigid finite element method, i.e. with primary and secondary divisions (Fig. 8.4). To each rigid finite element, a coordinate system is attached whose origin is located in its preceding spring-damping element (Fig. 8.5).

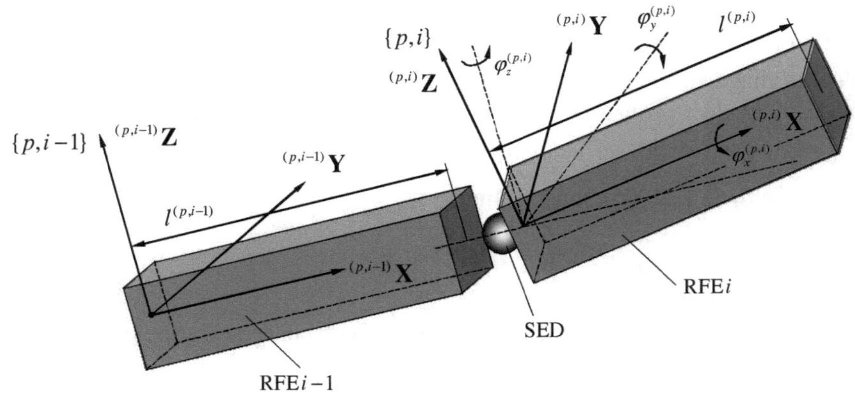

Fig. 8.5. Generalized coordinates of the *i*-th RFE and local coordinate systems

The generalized coordinates describing the position of the *i*-th RFE relative to the preceding *i*-1-th RFE of the flexible link *p* are the angles $\varphi_x^{(p,i)}$, $\varphi_y^{(p,i)}$, $\varphi_z^{(p,i)}$, of which the latter two correspond to bending and the first one to torsion of the element. Upon discretisation, the flexible link may be viewed as a system of rigid links connected by joints of the 3rd class. Similarly to the model formed using classical finite elements, a rigid link is a special case of a flexible link ($m_p = 0$). The transformation matrix $\tilde{\mathbf{T}}^{(p,i)}$ from the system $\{p,i\}$ attached to RFE i ($i = 1,\ldots,m_p$) to the system $\{p,i-1\}$ in the nonlinear model, i.e. allowing the angles $\varphi_\alpha^{(p,i)}$ for $\alpha \in \{x, y, z\}$ to be large, takes the form:

$$\tilde{\mathbf{T}}^{(p,i)} = \begin{bmatrix} c_z^{(p,i)}c_y^{(p,i)} & c_z^{(p,i)}s_y^{(p,i)}s_x^{(p,i)} - s_z^{(p,i)}c_x^{(p,i)} & c_z^{(p,i)}s_y^{(p,i)}c_x^{(p,i)} + s_z^{(p,i)}s_x^{(p,i)} & l^{(p,i-1)} \\ s_z^{(p,i)}c_y^{(p,i)} & s_z^{(p,i)}s_y^{(p,i)}s_x^{(p,i)} + c_z^{(p,i)}c_x^{(p,i)} & s_z^{(p,i)}s_y^{(p,i)}c_x^{(p,i)} - c_z^{(p,i)}s_x^{(p,i)} & 0 \\ -s_y^{(p,i)} & c_y^{(p,i)}s_x^{(p,i)} & c_y^{(p,i)}c_x^{(p,i)} & 0 \\ 0 & 0 & 0 & 1 \end{bmatrix}, \quad (8.36.1)$$

where $c_\alpha^{(p,i)} = \cos\varphi_\alpha^{(p,i)}$, $s_\alpha^{(p,i)} = \sin\varphi_\alpha^{(p,i)}$ for $\alpha \in \{x, y, z\}$,

$l^{(p,i-1)}$ – length of RFE i–1 of the link p.

When the angles $\varphi_\alpha^{(p,i)}$ are small, the following may be assumed:

$$\tilde{\mathbf{T}}^{(p,i)} = \begin{bmatrix} 1 & -\varphi_z^{(p,i)} & \varphi_y^{(p,i)} & l^{(p,i-1)} \\ \varphi_z^{(p,i)} & 1 & -\varphi_x^{(p,i)} & 0 \\ -\varphi_y^{(p,i)} & \varphi_x^{(p,i)} & 1 & 0 \\ 0 & 0 & 0 & 1 \end{bmatrix}. \quad (8.36.2)$$

When all three types of oscillations are considered (rotational and lateral in two planes), the generalized coordinates describing the motion of the i-th RFE of the link p relative to its predecessor may be written as components of the following vector:

$$\tilde{\mathbf{q}}^{(p,i)} = \begin{bmatrix} \phi_x^{(p,i)} & \phi_y^{(p,i)} & \phi_z^{(p,i)} \end{bmatrix}^T \quad \text{for } i = 1,\ldots,m_p. \quad (8.37)$$

RFE 0 is treated like in the classical rigid finite element method, its generalized coordinates being given by the vector $\tilde{\mathbf{q}}^{(p,0)}$. A series of intermediate transformations yields the transformation matrix from the local system $\{p,i\}$ (i=1,...,m_p) to the global system:

$$\mathbf{T}^{(p,i)} = \mathbf{T}^{(p,i-1)}\tilde{\mathbf{T}}^{(p,i)}, \quad (8.38)$$

where $\mathbf{T}^{(p,i-1)} = \mathbf{T}^{(p,0)}\tilde{\mathbf{T}}^{(p,1)}\ldots\tilde{\mathbf{T}}^{(p,i-1)}$

$\mathbf{T}^{(p,0)}$ – matrix given by (8.10.2),

$\tilde{\mathbf{T}}^{(p,i)}$ – matrix defined by the formula (8.36) for i=1,...,m_p.

The kinetic energy, the potential energy of gravity forces and the generalized forces caused by external forces and moments thereof acting on the flexible link are calculated as in section 5.3.

An important property of formula (8.38) is that the matrix $\mathbf{T}^{(p,i)}$ depends not only on the vector $\mathbf{q}^{(s)}$ of generalized coordinates of the link which precedes the flexible link, but also on all the RFEs preceding the RFE i. Defining the vectors:

$$\mathbf{q}^{(p,i)} = \begin{bmatrix} \mathbf{q}^{(s)} \\ \tilde{\mathbf{q}}^{(p,0)} \\ \tilde{\mathbf{q}}^{(p,1)} \\ \vdots \\ \tilde{\mathbf{q}}^{(p,i-1)} \\ \tilde{\mathbf{q}}^{(p,i)} \end{bmatrix}, \tag{8.39}$$

and taking (8.2) into account allows us to write:

$$\mathbf{q}^{(p)} = \begin{bmatrix} \mathbf{q}^{(s)} \\ \tilde{\mathbf{q}}^{(p,0)} \\ \tilde{\mathbf{q}}^{(p,f)} \end{bmatrix}, \tag{8.40}$$

where $\tilde{\mathbf{q}}^{(p,f)} = \begin{bmatrix} \tilde{\mathbf{q}}^{(p,1)T} & \cdots & \tilde{\mathbf{q}}^{(p,m_p)T} \end{bmatrix}^T$.

8.2.2 Kinetic Energy: Lagrange Operators

From (8.38) it follows:

$$\mathbf{T}^{(p,i)} = \mathbf{T}^{(p,0)}\,\hat{\mathbf{T}}^{(p,i)}\left(\tilde{\mathbf{q}}^{(p,1)},\ldots,\tilde{\mathbf{q}}^{(p,i)}\right), \tag{8.41}$$

where $\hat{\mathbf{T}}^{(p,i)} = \prod_{j=1}^{i} \tilde{\mathbf{T}}^{(p,j)}\left(\tilde{\mathbf{q}}^{(p,j)}\right)$.

Since the kinetic energy of the link p may be written as:

$$E_p = \sum_{i=0}^{m_p} E_{p,i}\,, \tag{8.42}$$

where $E_{p,i} = \mathrm{tr}\left\{\dot{\mathbf{T}}^{(p,i)}\mathbf{H}^{(p,i)}\dot{\mathbf{T}}^{(p,i)T}\right\}$, calculations analogous to those presented in chapter 5 give:

$$\boldsymbol{\varepsilon}_{\mathbf{q}^{(p,i)}}\left(E_{p,i}\right)=\begin{bmatrix}\mathbf{A}_{s,s}^{(p,i)} & \mathbf{A}_{s,0}^{(p,i)} & \mathbf{A}_{s,1}^{(p,i)} & \cdots & \mathbf{A}_{s,j}^{(p,i)} & \cdots & \mathbf{A}_{s,i}^{(p,i)}\\ \mathbf{A}_{0,s}^{(p,i)} & \mathbf{A}_{0,0}^{(p,i)} & \mathbf{A}_{0,1}^{(p,i)} & \cdots & \mathbf{A}_{0,j}^{(p,i)} & \cdots & \mathbf{A}_{0,i}^{(p,i)}\\ \mathbf{A}_{1,s}^{(p,i)} & \mathbf{A}_{1,0}^{(p,i)} & \mathbf{A}_{1,1}^{(p,i)} & \cdots & \mathbf{A}_{1,j}^{(p,i)} & \cdots & \mathbf{A}_{1,i}^{(p,i)}\\ \vdots & \vdots & \vdots & \cdots & \vdots & \cdots & \vdots\\ \mathbf{A}_{i,s}^{(p,i)} & \mathbf{A}_{i,0}^{(p,i)} & \mathbf{A}_{i,1}^{(p,i)} & \cdots & \mathbf{A}_{i,j}^{(p,i)} & \cdots & \mathbf{A}_{i,i}^{(p,i)}\end{bmatrix}\begin{bmatrix}\ddot{\mathbf{q}}^{(s)}\\ \ddot{\tilde{\mathbf{q}}}^{(p,0)}\\ \ddot{\tilde{\mathbf{q}}}^{(p,1)}\\ \vdots\\ \ddot{\tilde{\mathbf{q}}}^{(p,j)}\\ \vdots\\ \ddot{\tilde{\mathbf{q}}}^{(p,i)}\end{bmatrix}+\begin{bmatrix}\mathbf{e}_s^{(p,i)}\\ \mathbf{e}_0^{(p,i)}\\ \mathbf{e}_1^{(p,i)}\\ \vdots\\ \mathbf{e}_j^{(p,i)}\\ \vdots\\ \mathbf{e}_i^{(p,i)}\end{bmatrix}= \tag{8.43}$$

$$=\mathbf{A}^{(p,i)}\ddot{\mathbf{q}}^{(p,i)}+\mathbf{e}^{(p,i)},$$

where $\mathbf{A}_{\alpha,\beta}^{(p,i)}$ – appropriate blocks of the matrix $\mathbf{A}^{(p,i)}$,

$$\mathbf{A}^{(p,i)}=\left(a_{k,j}^{(p,i)}\right)_{k,j=1,\ldots,n_{p,i}}=\mathrm{tr}\left\{\mathbf{T}_k^{(p,i)}\,\mathbf{H}^{(p,i)}\,\mathbf{T}_j^{(p,i)T}\right\}$$

$$\mathbf{e}^{(p,i)}=\left(e_k^{(p,i)}\right)_{k=1,\ldots,n_{p,i}}=\sum_{j=1}^{n_{p,i}}\sum_{l=1}^{n_{p,i}}\mathrm{tr}\left\{\mathbf{T}_k^{(p,i)}\,\mathbf{H}^{(p,i)}\mathbf{T}_{j,l}^{(p,i)}\right\}\dot{q}_j^{(p,i)}\,\dot{q}_l^{(p,i)}+$$

$$+\mathrm{tr}\left\{\mathbf{T}_k^{(p,i)}\mathbf{H}^{(p,i)}\left[{}_A^0\ddot{\mathbf{T}}\overline{\mathbf{T}}^{(p,i)}+2{}_A^0\dot{\mathbf{T}}\dot{\overline{\mathbf{T}}}^{(p,i)}\right]^T\right\},$$

$$\overline{\mathbf{T}}^{(p,i)}={}_s^A\mathbf{T}\left(\mathbf{q}^{(s)}\right)\prod_{j=0}^{i}\tilde{\mathbf{T}}^{(p,j)},$$

$$n_{p,i}=n_s+\tilde{n}_{p,0}+3i\,.$$

As before, the gravity forces of the RFEs and their derivatives may be put in the form:

$$V_{p,i}^g=m^{(p,i)}g\,\boldsymbol{\theta}_3\mathbf{T}^{(p,i)}\tilde{\mathbf{r}}_C^{(p,i)}, \tag{8.44}$$

and further:

$$\frac{\partial V_{p,i}^g}{\partial\mathbf{q}^{(p,i)}}=\begin{bmatrix}\mathbf{G}_s^{(p,i)}\\ \mathbf{G}_0^{(p,i)}\\ \vdots\\ \mathbf{G}_i^{(p,i)}\end{bmatrix}, \tag{8.45}$$

where $\mathbf{G}_\alpha^{(p,i)}$ – appropriate blocks of the vector $\mathbf{G}^{(p,i)}$,

$$\mathbf{G}^{(p,i)}=\left(g_k^{(p,i)}\right)_{k=1,\ldots,n_{p,i}},$$

$$g_k^{(p,i)}=m^{(p,i)}g\,\boldsymbol{\theta}_3\mathbf{T}_k^{(p,i)}\tilde{\mathbf{r}}_C^{(p,k)}.$$

The generalized forces may be similarly presented. If $\widetilde{\mathbf{F}}^{(p,i)}$ and $\widetilde{\mathbf{M}}^{(p,i)}$ specified in (8.30) act on RFE i of the link, then:

$$\mathbf{Q}^{(p,i)} = \begin{bmatrix} \mathbf{Q}_s^{(p,i)} \\ \mathbf{Q}_0^{(p,i)} \\ \vdots \\ \mathbf{Q}_i^{(p,i)} \end{bmatrix}, \tag{8.46}$$

where $\mathbf{Q}_\alpha^{(p,i)}$ – appropriate blocks of the vector $\mathbf{Q}^{(p,i)}$,

$$\mathbf{Q}^{(p,i)} = \left(\mathbf{Q}_k^{(p,i)}\right)_{k=1,\dots,n_{p,i}} = \widetilde{\mathbf{F}}^{(p,i)T}\mathbf{T}^{(p,i)T}\,\mathbf{T}_k^{(p,i)}\,\widetilde{\mathbf{r}}^{(p,i)} + \widetilde{M}_1^{(p,i)}\sum_{j=1}^{3}\left(\mathbf{T}^{(p,i)}\right)_{j,3}\left(\mathbf{T}_k^{(p,i)}\right)_{j,2} +$$
$$+\widetilde{M}_2^{(p,i)}\sum_{j=1}^{3}\left(\mathbf{T}^{(p,i)}\right)_{j,1}\left(\mathbf{T}_k^{(p,i)}\right)_{j,3} + \widetilde{M}_3^{(p,i)}\sum_{j=1}^{3}\left(\mathbf{T}^{(p,i)}\right)_{j,2}\left(\mathbf{T}_k^{(p,i)}\right)_{j,1},$$

$\widetilde{\mathbf{r}}^{(p,i)}$ – vector giving the coordinates of the point to which the force in the system $\{p,i\}$.

Formulation of the equations of motion further requires the determination of the elastic energy and its derivatives. The reasoning below pertains to linear physical dependencies.

8.2.3 Energy of Elastic Deformation

The potential energy of elastic deformation of an SDE of a flexible link is calculated based on the fact that the generalized coordinates specify relative angles. For the spring-damping element connecting the RFEs i–1 and i it is given by the formula:

$$V_{p,i}^s = \frac{1}{2}\sum_{j=1}^{3} c_{i,3+j}^{(p)}\left[\varphi_j^{(p,i)}\right]^2, \tag{8.47}$$

where $c_{i,3+j}^{(p)}$ are the appropriate coefficients of rotational stiffness defined in (8.21).

The formula (8.47) may be rewritten as:

$$V_{p,i}^s = \frac{1}{2}\widetilde{\mathbf{q}}^{(p,i)T}\mathbf{C}^{(p,i)}\,\widetilde{\mathbf{q}}^{(p,i)}, \tag{8.48}$$

where $\mathbf{C}^{(p,i)} = \begin{bmatrix} c_{i,4}^{(p)} & 0 & 0 \\ 0 & c_{i,5}^{(p)} & 0 \\ 0 & 0 & c_{i,6}^{(p)} \end{bmatrix}$.

The derivatives of the potential energy of elastic deformation relative to the generalized coordinates have the form:

$$\frac{\partial V^{s}_{(p,i)}}{\partial \tilde{\mathbf{q}}^{(p,i)}} = \mathbf{C}^{(p,i)}\tilde{\mathbf{q}}^{(p,i)} . \tag{8.49}$$

8.2.4 Equations of Motion

Whereas the kinetic energy and the potential energy of gravity forces of the link p are given by the formulas:

$$E_p = \sum_{i=0}^{m_p} E_{p,i} , \tag{8.50.1}$$

$$V_p^g = \sum_{i=0}^{m_p} V_{p,i}^g , \tag{8.50.2}$$

and taking (8.43), (8.45), (8.46) and (8.49) into account, equations of motion of the link p may be written as:

$$\mathbf{A}^{(p)}\ddot{\mathbf{q}}^{(p)} = \mathbf{f}^{(p)}, \tag{8.51.1}$$

or decomposed with blocks:

$$\begin{bmatrix} \mathbf{A}_{s,s}^{(p)} & \mathbf{A}_{s,0}^{(p)} & \mathbf{A}_{s,1}^{(p)} & \cdots & \mathbf{A}_{s,j}^{(p)} & \cdots & \mathbf{A}_{s,m_p}^{(p)} \\ \mathbf{A}_{0,s}^{(p)} & \mathbf{A}_{0,0}^{(p)} & \mathbf{A}_{0,1}^{(p)} & \cdots & \mathbf{A}_{0,j}^{(p)} & \cdots & \mathbf{A}_{0,m_p}^{(p)} \\ \vdots & \vdots & \vdots & \cdots & \vdots & \cdots & \vdots \\ \mathbf{A}_{i,s}^{(p)} & \mathbf{A}_{i,0}^{(p)} & \mathbf{A}_{i,1}^{(p)} & \cdots & \mathbf{A}_{i,j}^{(p)} & \cdots & \mathbf{A}_{i,m_p}^{(p)} \\ \vdots & \vdots & \vdots & \cdots & \vdots & \cdots & \vdots \\ \mathbf{A}_{m_p,s}^{(p)} & \mathbf{A}_{m_p,0}^{(p)} & \mathbf{A}_{m_p,1}^{(p)} & \cdots & \mathbf{A}_{m_p,j}^{(p)} & \cdots & \mathbf{A}_{m_p,m_p}^{(p)} \end{bmatrix} \begin{bmatrix} \ddot{\mathbf{q}}^{(s)} \\ \ddot{\tilde{\mathbf{q}}}^{(p,0)} \\ \vdots \\ \ddot{\tilde{\mathbf{q}}}^{(p,j)} \\ \vdots \\ \ddot{\tilde{\mathbf{q}}}^{(p,m_p)} \end{bmatrix} = \begin{bmatrix} \mathbf{f}_s^{(p)} \\ \mathbf{f}_0^{(p)} \\ \vdots \\ \mathbf{f}_i^{(p)} \\ \vdots \\ \mathbf{f}_{m_p}^{(p)} \end{bmatrix}, \tag{8.51.2}$$

where $\mathbf{A}_{s,s}^{(p)} = \sum_{i=0}^{m_p} \mathbf{A}_{s,s}^{(p,i)}$, $\quad \mathbf{A}_{s,j}^{(p)} = \sum_{i=j}^{m_p} \mathbf{A}_{s,j}^{(p,i)}$, $\quad \mathbf{A}_{i,s}^{(p)} = \sum_{j=i}^{m_p} \mathbf{A}_{i,s}^{(p,j)}$,

$$\mathbf{A}_{i,j}^{(p)} = \sum_{l=\max\{i,j\}}^{m_p} \mathbf{A}_{i,j}^{(p,l)} \qquad \text{for } i, j = 0,1,\ldots,m_p ,$$

$$\mathbf{f}_s^{(p)} = \sum_{i=0}^{m_p} \left[-\mathbf{e}_s^{(p,i)} - \mathbf{G}_s^{(p,i)} + \mathbf{Q}_s^{(p,i)} \right],$$

$$\mathbf{f}_0^{(p)} = \sum_{i=0}^{m_p} \left[-\mathbf{e}_0^{(p,i)} - \mathbf{G}_0^{(p,i)} + \mathbf{Q}_0^{(p,i)} \right],$$

$$\mathbf{f}_i^{(p)} = \sum_{j=i}^{m_p} \left[-\mathbf{e}_i^{(p,j)} - \mathbf{G}_i^{(p,j)} + \mathbf{Q}_i^{(p,j)} \right] - \mathbf{C}^{(p,i)} \tilde{\mathbf{q}}^{(p,i)} \qquad \text{for } i = 1,\ldots,m_p .$$

8.3 Modelling of Planar System

By means of the rigid finite element method, an arbitrary description of the geometry of a system may be given. The traditional approach may be used instead of homogeneous transformations and joint coordinates proposed in earlier chapters. In the present chapter an example is given of modelling a planar system using the rigid finite element method in its modified form and a classical description of the system's geometry.

8.3.1 Determination of Generalized Coordinates

In Fig. 8.6, a sample decomposition of a k-th flexible links into n_k+1 rigid finite elements connected at points $A_1^{(k)},\ldots,A_{n_k}^{(k)}$ by n_k massless spring-damping elements is presented. Since the problem considered is plane, in the relative motion each RFE enjoys one degree of freedom which is the inclination angle of the axis ${}^{(k,i)}\mathbf{X}$ of the RFE i to the axis $\mathbf{X}$ of the global system (Fig. 8.7). Further analysis assumes the angles to be measured relative to the global system.

The position of the link k being discredited is therefore described by n_k+3 coordinates. Two of them, (x_k, y_k), are the coordinates of the point $A^{(k)}$ which equals the point $A_0^{(k)}$ of the first RFE (usually being one of the nodes of the whole mechanism). The remaining coordinates are the angles already mentioned which will be denoted $\varphi^{(k,0)},\ldots,\varphi^{(k,n_k)}$. As a noteworthy observation, these angles correspond to those from (8.37) – $\varphi_y^{(p,i)}$. Thus, the vector of coordinates of the link k may be defined:

$$\mathbf{q}^{(k)} = \left[x_k, y_k, \varphi^{(k,0)}, \varphi^{(k,1)}, \ldots, \varphi^{(k,n_k)} \right]^T . \tag{8.52}$$

Following [Wojciech S., 1984], [Szczotka M., 2011b], when introducing denotations for coordinates of the point $A_i^{(k)}\left(a_{i,i}^{(k)}, b_{i,i}^{(k)}\right)$ and $A_{i+1}^{(k)}\left(a_{i,i+1}^{(k)}, b_{i,i+1}^{(k)}\right)$ in the local coordinate system $0_i^{(k)}\xi_i^{(k)}\eta_i^{(k)}$ attached to the centre of mass of the

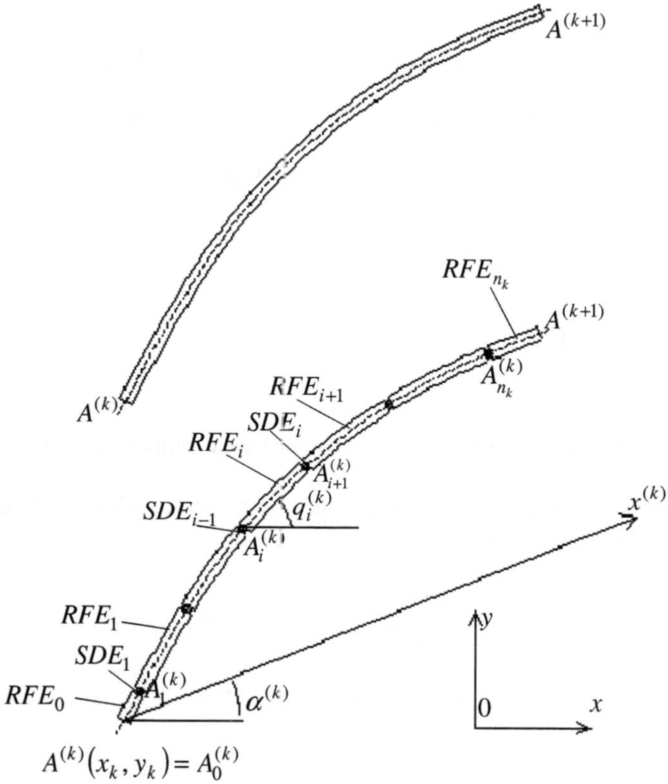

Fig. 8.6. Decomposition of a flexible link into rigid finite elements

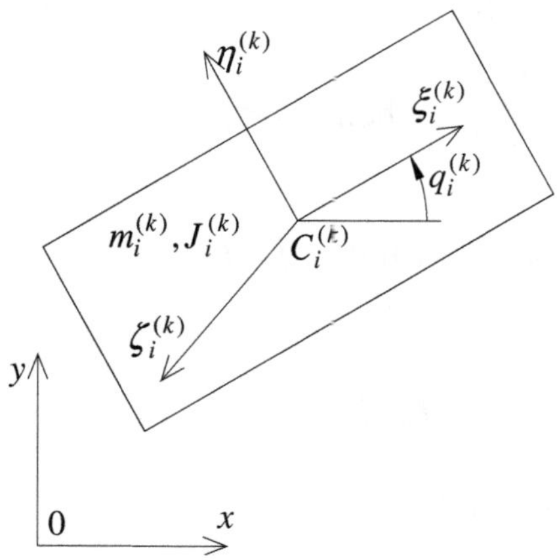

Fig. 8.7. Inclination angles of an RFE to the axes of a stationary coordinate system

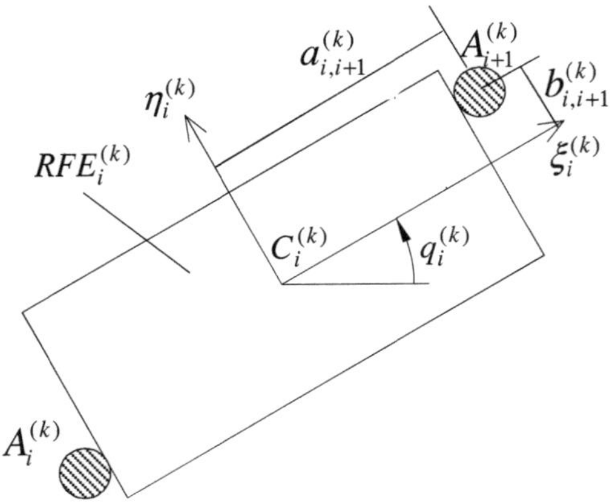

Fig. 8.8. Coordinates of a point in the local coordinate system

RFE i (Fig. 8.8) we may write the coordinates of the centre of mass of the RFE i in the coordinate system $\{A\}$ as follows:

$$x_{C_i}^{(k)} = x_k + \sum_{j=0}^{i} h_{i,j}^{(k)} \cos\left(\varphi^{(k,j)} + \varphi_{i,j}^{(k)}\right)$$
$$y_{C_i}^{(k)} = y_k + \sum_{j=0}^{i} h_{i,j}^{(k)} \sin\left(\varphi^{(k,j)} + \varphi_{i,j}^{(k)}\right), \tag{8.53}$$

where $h_{i,j}^{(k)} = \begin{cases} \sqrt{\left(-a_{j,j}^{(k)} + a_{j,j+1}^{(k)}\right)^2 + \left(-b_{j,j}^{(k)} + b_{j,j+1}^{(k)}\right)^2} & \text{when} \quad j<i \\ \sqrt{\left(a_{i,i}^{(k)}\right)^2 + \left(b_{i,i}^{(k)}\right)^2} & \text{when} \quad j=i \end{cases},$

$$\varphi_{i,j}^{(k)} = \begin{cases} \operatorname{arctg} \dfrac{b_{j,j+1}^{(k)} - b_{j,j}^{(k)}}{a_{j,j+1}^{(k)} - a_{j,j}^{(k)}} & \text{when} \quad j<i \\ \operatorname{arctg} \dfrac{b_{i,i}^{(k)}}{a_{i,i}^{(k)}} + \pi & \text{when} \quad j=i \end{cases}.$$

8.3.2 Equations of Motion of a Link

The kinetic energy of the i-th RFE equals:

$$E_{k,i} = \frac{1}{2} m_i^{(k)} \left\{ \left[\dot{x}_{C_i}^{(k)}\right]^2 + \left[\dot{y}_{C_i}^{(k)}\right]^2 \right\} + \frac{1}{2} J_i^{(k)} \left[\varphi^{(k,j)}\right]^2, \tag{8.54}$$

where $m_i^{(k)}$ – mass of the i-th RFE,

$J_i^{(k)}$ – moment of inertia of the i-th RFE relative to the central axis perpendicular to the plane **XY**,

and then the following sum gives the energy of the entire link k:

$$E_k = \sum_{i=0}^{n_k} E_{k,i} \,. \tag{8.55}$$

Using (8.53), (8.54) and the following identities, which may be proved by induction:

$$\sum_{i=0}^{n_k} m_i^{(k)} \sum_{j=0}^{i} \dot{\varphi}^{(k,j)} b_{i,j}^{(k)} = \sum_{i=0}^{n_k} \dot{\varphi}^{(k,i)} \sum_{j=0}^{n_k} m_j^{(k)} b_{j,i}^{(k)} \,, \tag{8.56.1}$$

$$\sum_{i=0}^{n_k} m_i^{(k)} \left[\sum_{j=0}^{i} \dot{\varphi}^{(k,j)} b_{i,j}^{(k)} \right]^2 = \sum_{i=0}^{n_k} \dot{\varphi}^{(k,i)} \sum_{j=0}^{n_k} \dot{\varphi}^{(k,j)} \sum_{l=\max\{i,j\}}^{n_k} m_l^{(k)} b_{l,i}^{(k)} b_{l,j}^{(k)} \,, \tag{8.56.2}$$

we may express the kinetic energy of the link k as:

$$E_k = \frac{1}{2} \left[\dot{\mathbf{q}}^{(k)}\right]^T \mathbf{A}^{(k)} \dot{\mathbf{q}}^{(k)} , \tag{8.57}$$

where $\mathbf{A}^{(k)} = \begin{bmatrix} \mathbf{M}_1^{(k)} & \mathbf{M}_2^{(k)} \\ \left[\mathbf{M}_2^{(k)}\right]^T & \mathbf{M}^{(k)} \end{bmatrix}$,

$$\mathbf{M}_1^{(k)} = \begin{bmatrix} m^{(k)} & 0 \\ 0 & m^{(k)} \end{bmatrix},$$

$$\mathbf{M}_2^{(k)} = \begin{bmatrix} -\overline{A}_0^{(k)} \sin\left(\varphi^{(k,0)} + \alpha_0^{(k)}\right) \ldots - \overline{A}_i^{(k)} \sin\left(\varphi^{(k,i)} + \alpha_i^{(k)}\right) \ldots - \overline{A}_{n_k}^{(k)} \sin\left(\varphi^{(k,n_k)} + \alpha_{n_k}^{(k)}\right) \\ \overline{A}_0^{(k)} \cos\left(\varphi^{(k,0)} + \alpha_0^{(k)}\right) \ldots + \overline{A}_i^{(k)} \cos\left(\varphi^{(k,i)} + \alpha_i^{(k)}\right) \ldots + \overline{A}_{n_k}^{(k)} \cos\left(\varphi^{(k,n_k)} + \alpha_{n_k}^{(k)}\right) \end{bmatrix},$$

$$\mathbf{M}^{(k)} = \left(m_{i,j}^{(k)}\right)_{i,j=0}^{n_k}, \qquad m_{i,j}^{(k)} = \overline{A}_{i,j}^{(k)} \cos\left(\varphi^{(k,i)} - \varphi^{(k,j)} - \alpha_{i,j}^{(k)}\right),$$

$$\overline{A}_i^{(k)} = \sqrt{\left[\overline{a}_i^{(k)}\right]^2 + \left[\overline{b}_i^{(k)}\right]^2} \,, \qquad \alpha_i^{(k)} = \operatorname{arctg} \frac{\overline{b}_i^{(k)}}{\overline{a}_i^{(k)}} \,,$$

$$\overline{a}_i^{(k)} = \sum_{j=0}^{n_k} m_j^{(k)} h_{j,i} \cos \varphi_{j,i}^{(k)} \,, \qquad \overline{b}_i^{(k)} = \sum_{j=0}^{n_k} m_j^{(k)} h_{j,i} \sin \varphi_{j,i}^{(k)} \,,$$

$$\overline{A}_{i,j}^{(k)} = \sqrt{\left[\overline{a}_{i,j}^{(k)}\right]^2 + \left[\overline{b}_{i,j}^{(k)}\right]^2} \,, \qquad \alpha_{i,j}^{(k)} = \operatorname{arctg} \frac{\overline{b}_{i,j}^{(k)}}{\overline{a}_{i,j}^{(k)}} \,,$$

$$\overline{a}_{i,j}^{(k)} = \sum_{l=\max\{i,j\}}^{n_k} m_l^{(k)} h_{l,i} h_{l,j} \cos\left(\varphi_{l,i}^{(k)} - \varphi_{l,j}^{(k)}\right) + \delta_{i,j} J_i^{(k)},$$

$$\overline{b}_{i,j}^{(k)} = \sum_{l=\max\{i,j\}}^{n_k} m_l^{(k)} h_{l,i} h_{l,j} \sin\left(\varphi_{l,i}^{(k)} - \varphi_{l,j}^{(k)}\right),$$

$\delta_{i,j}$ – Kronecker delta,

$$m^{(k)} = \sum_{i=0}^{n_k} m_i^{(k)}.$$

This enables transforming the Lagrange equation of the link k to:

$$\boldsymbol{\varepsilon}_{\mathbf{q}^{(k)}} + \frac{\partial D^{(k)}}{\partial \dot{\mathbf{q}}^{(k)}} + \frac{\partial V_k^g}{\partial \mathbf{q}^{(k)}} + \frac{\partial V_k^s}{\partial \mathbf{q}^{(k)}} = \mathbf{Q}^{(k)}, \tag{8.58}$$

where $\boldsymbol{\varepsilon}_{\mathbf{q}^{(k)}} = \begin{bmatrix} \varepsilon_{x_k} & \varepsilon_{y_k} & \varepsilon_{\varphi^{(k,1)}} & \ldots & \varepsilon_{\varphi^{(k,n_k)}} \end{bmatrix}^T$,

$D^{(k)}$ – dissipation function of the k link's energy,

V_k^g, V_k^s – potential energy of deformation and gravity forces of the link k,

$\mathbf{Q}^{(k)}$ – vector of generalized forces.

By taking into account (8.57), the following is obtained:

$$\boldsymbol{\varepsilon}_{\mathbf{q}^{(k)}} = \mathbf{A}^{(k)} \ddot{\mathbf{q}}^{(k)} + \overline{\mathbf{B}}^{(k)} \dot{\mathbf{q}}^{(k)}, \tag{8.59}$$

where $\mathbf{A}^{(k)}$ – defined in (8.57),

$\overline{\mathbf{B}}^{(k)}$ – matrix with the following elements:

$$\overline{b}_{1,i+3}^{(k)} = -\dot{\varphi}^{(k,i)} \overline{A}_i^{(k)} \cos\left(\varphi^{(k,i)} + \alpha_i^{(k)}\right), \qquad \overline{b}_{i+3,1}^{(k)} = 0,$$

$$\overline{b}_{2,i+3}^{(k)} = -\dot{\varphi}^{(k,i)} \overline{A}_i^{(k)} \sin\left(\varphi^{(k,i)} + \alpha_i^{(k)}\right), \qquad \overline{b}_{i+3,2}^{(k)} = 0,$$

$$\overline{b}_{3+j,i+3}^{(k)} = -\dot{\varphi}^{(k,i)} \overline{A}_{i,j}^{(k)} \sin\left(\varphi^{(k,i)} - \varphi^{(k,j)} - \alpha_{i,j}^{(k)}\right) \qquad i, j = 0, 1, \ldots, n_k,$$

$$\overline{b}_{1,1}^{(k)} = \overline{b}_{1,2}^{(k)} = \overline{b}_{2,1}^{(k)} = \overline{b}_{2,2}^{(k)} = 0.$$

The following formula gives the potential energy of gravity forces:

$$V_k^g = \sum_{i=0}^{n_k} m_i^{(k)} \left[y_{c_i}^{(k)} - y_{c_i,0}^{(k)} \right], \tag{8.60}$$

whereby $y_{c_i,0}^{(k)} = \text{const}$.

It then follows from (8.53):

$$\frac{\partial V_k^g}{\partial \mathbf{q}^{(k)}} = \begin{bmatrix} 0 \\ m^{(k)} \\ m_0^{(k)} \sum_{i=0}^{n_k} h_{i,0}^{(k)} \cos\left(\varphi^{(k,0)} + \varphi_{i,0}^{(k)}\right) \\ \vdots \\ m_j^{(k)} \sum_{i=j}^{n_k} h_{i,1}^{(k)} \cos\left(\varphi^{(k,j)} + \varphi_{i,j}^{(k)}\right) \\ \vdots \\ m_{n_k}^{(k)} h_{n_k,n_k}^{(k)} \cos\left(\varphi^{(k,n_k)} + \varphi_{n_k,n_k}^{(k)}\right) \end{bmatrix}. \tag{8.61}$$

The expression giving the energy V_k^s of elastic deformation and the dissipation function $D^{(k)}$ of the k link's energy depend on the form of assumed physical dependencies between the deformations and stresses characteristic to the spring-damping elements. In the case of linear Kelvin-Voigt model, counting V_k^s and $D^{(k)}$ as components due to deformation of the spring-damping elements (Fig. 8.6), it may be written:

$$\frac{\partial V_k^s}{\partial \mathbf{q}^{(k)}} = \mathbf{C}^{(k)} \mathbf{q}^{(k)}, \tag{8.62}$$

$$\frac{\partial D^{(k)}}{\partial \mathbf{q}^{(k)}} = \mathbf{D}^{(k)} \dot{\mathbf{q}}^{(k)}, \tag{8.63}$$

where $\mathbf{C}^{(k)}$ and $\mathbf{D}^{(k)}$ are the stiffness and damping matrix, respectively, whose coefficients are constant and dependent on the geometry of the link and the constants determining the stiffness and the damping of the SDE.

The vector of generalized forces $\mathbf{Q}^{(k)}$ is formed by the values of forces caused by external loads and reactions in the joints. If load shown in Fig. 8.9 is applied to the i-th RFE, the components of the generalized forces due to the loads take the form:

$$\begin{aligned} &Q_{i,1}^{(k)} = P_{ix}^{(k)} \\ &Q_{i,2}^{(k)} = P_{iy}^{(k)} \\ &Q_{i,j+2}^{(k)} = -\left[h_{i,j}^{(k)} \sin\left(\phi^{(k,j)} + \phi_{i,j}^{(k)}\right) + \delta_{i,j} r_i^{(k)} \sin\left(\phi^{(k,j)} + \gamma_i^{(k)}\right) \right] P_{ix}^{(k)} + \\ &\qquad + \left[h_{i,j}^{(k)} \cos\left(\phi^{(k,j)} + \phi_{i,j}^{(k)}\right) + \delta_{i,j} r_i^{(k)} \cos\left(\phi^{(k,j)} + \gamma_i^{(k)}\right) \right] P_{iy}^{(k)} + \delta_{i,j} M_i^{(k)} \\ &\qquad\qquad\qquad\qquad \text{for } j = 0,1,\ldots i \\ &Q_{i,j+2}^{(k)} = 0 \qquad\qquad \text{for } j > i \end{aligned} \tag{8.64}$$

where $h_{i,j}^{(k)}, \varphi_{i,j}^{(k)}$ – defined as in formula (8.53),

$r_i^{(k)}, \gamma_i^{(k)}$ – polar coordinates (relative to the middle $C_i^{(k)}$ of the i-th RFE of the link) of the point to which the load is applied.

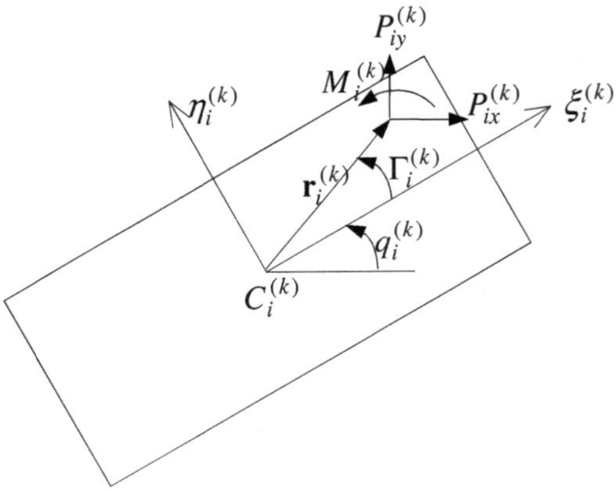

Fig. 8.9. Coordinates of the point of application of external load in the local coordinate system of the i-th RFE of the k-th link

Summing:

$$Q_i^{(k)} = \sum_{j=0}^{n_k} Q_{i,j}^{(k)}, \tag{8.65}$$

yields the components of the vector of generalized forces. Furthermore, with (8.58), (8.59), (8.61), (8.62), (8.63) the equations of motion may be rewritten:

$$\mathbf{A}^{(k)}\ddot{\mathbf{q}}^{(k)} + \mathbf{B}^{(k)}\dot{\mathbf{q}}^{(k)} + \mathbf{C}^{(k)}\mathbf{q}^{(k)} = \mathbf{Q}^{(k)} - \frac{\partial V_k^g}{\partial \mathbf{q}^{(k)}}, \tag{8.66}$$

where $\mathbf{B}^{(k)} = \overline{\mathbf{B}}^{(k)} + \mathbf{D}^{(k)}$.

In the vector of generalized forces $\mathbf{Q}^{(k)}$ both the reactions of constraints and known loads are included. For some operations, it is convenient to have this vector written as:

$$\mathbf{Q}^{(k)} = -\mathbf{K}^{(k)}\mathbf{R}^{(k)} + \mathbf{Q}_P^{(k)}, \tag{8.67}$$

where $\mathbf{K}^{(k)}$ – matrix with n_k+3 rows whose elements depend on $\mathbf{q}^{(k)}$,

$\mathbf{R}^{(k)}$ – vector of reaction (its elements are the components of reaction in revolute and translational connections and forces occurring therein as well as moments of undeveloped friction),

$\mathbf{Q}_P^{(k)}$ – vector of generalized forces due to known external loads, reactions in flexible connections and forces of developed dry friction and viscous friction.

Given the form of the vector (8.67), the equations of motion of the link k may be written as follows:

$$\mathbf{A}^{(k)}\ddot{\mathbf{q}}^{(k)} + \mathbf{B}^{(k)}\dot{\mathbf{q}}^{(k)} + \mathbf{C}^{(k)}\mathbf{q}^{(k)} + \mathbf{K}^{(k)}\mathbf{R}^{(k)} = \mathbf{F}^{(k)}, \tag{8.68}$$

where $\mathbf{F}^{(k)} = -\frac{\partial V_k^g}{\partial \mathbf{q}^{(k)}} + \mathbf{Q}_P^{(k)}$.

The elements of the matrices $\mathbf{A}^{(k)}$, $\mathbf{C}^{(k)}$, $\mathbf{K}^{(k)}$ depend on $\mathbf{q}^{(k)}$ and the elements of the matrix $\mathbf{B}^{(k)}$ and the vector $\mathbf{F}^{(k)}$ depend on $\mathbf{q}^{(k)}$ and $\dot{\mathbf{q}}^{(k)}$. In the special case of n_k=0, the concerned link is modelled as rigid.

Motion of the base $\{A\}$ may be taken into consideration by assuming it to be the RFE 0 whose motion is described by:

$$\begin{aligned} x_0 &= x_{org}^{(A)}(t) \\ y_0 &= y_{org}^{(A)}(t) \\ \varphi_{0,0}^{(0)} &= \psi^{(A)}(t) \end{aligned} \quad . \tag{8.69}$$

The vector of reaction in the connection is thence defined by the vector:

$$\mathbf{R}^{(0)} = \left[F_x^{(0)} \quad F_y^{(0)} \quad M_z^{(0)}\right]^T, \tag{8.70}$$

whose components describe the forces and moment which realize the excitation (8.69).

A detailed description of the algorithm of combining the equations of subsystems for revolute and translational connections is presented in [Wojciech S., 1984].

8.4 Modelling Large Deflections and Inclusion of Nonlinear Physical Dependencies

Most of the applications already discussed in which the RFE method is used pertain to systems containing beam links. Some of the considerations in this book are for pipelines which may be subjected to deflections much larger than typical beam systems. Although the RFE method enables analysis involving large deflections, the specific dynamic behaviour of offshore pipelines and cableswhen laid on the bottom of a sea, calls for considerable modifications in the formulation of the equations of motion according to this method [Szczotka M., 2011b]. They will be later applied in some of the examples presented.

When the deflections of the link are large, the length of the chord AB' may differ (be smaller) from its primary length $AB=l$ (Fig. 8.10). Let us remind that, according to (5.5), the motion of the base (the vessel's hull) is known to be given by the vector:

$$\mathbf{q}^{(A)} = \begin{bmatrix} x_{org}^{(A)} & y_{org}^{(A)} & z_{org}^{(A)} & \varphi_z^{(A)} & \varphi_y^{(A)} & \varphi_x^{(A)} \end{bmatrix}^T. \tag{8.71}$$

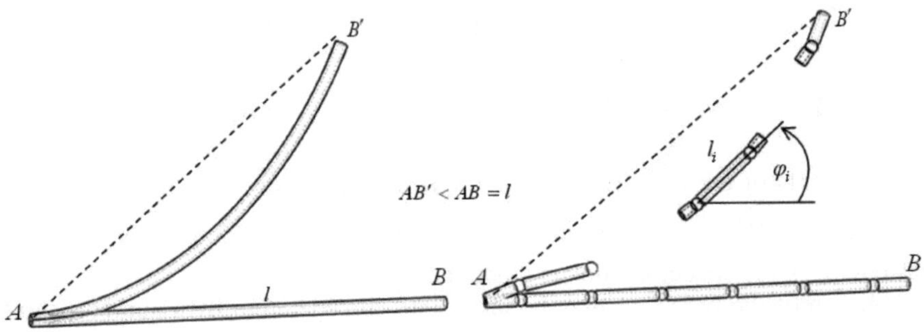

Fig. 8.10. Division of a beam with length l into RFEs and SDE: a) primary beam, b) equivalent system of RFEs and SDE

Let the components of the following vector determine the displacements and orientation of the RFE i in the system $\{A\}$:

$$\tilde{\mathbf{q}}^{(i)} = \begin{bmatrix} \tilde{\mathbf{r}}^{(i)T} & \tilde{\mathbf{\Phi}}^{(i)T} \end{bmatrix}^T, \tag{8.72}$$

where $\tilde{\mathbf{r}}^{(i)} = \begin{bmatrix} x^{(i)} & y^{(i)} & z^{(i)} \end{bmatrix}^T$ – coordinates of the origin of the system $\{i\}$ attached to the RFE i in the system $\{A\}$,

$\tilde{\mathbf{\Phi}}^{(i)} = \begin{bmatrix} \varphi_z^{(i)} & \varphi_y^{(i)} & \varphi_x^{(i)} \end{bmatrix}^T$ – ZYX Euler angles determining the orientation of the axes of the system $\{i\}$ relative to $\{A\}$.

Based on the information from previous chapters, let us write the transformation matrices from the system $\{i\}$ to the base system $\{A\}$ in the form:

$$\tilde{\mathbf{T}}^{(i)} = \begin{bmatrix} \tilde{\mathbf{R}}^{(i)} & \tilde{\mathbf{r}}^{(i)} \\ \mathbf{0} & 1 \end{bmatrix}, \tag{8.73}$$

where $$\tilde{\mathbf{R}}^{(i)} = \begin{bmatrix} c\varphi_z^{(i)} & -s\varphi_z^{(i)} & 0 \\ s\varphi_z^{(i)} & c\varphi_z^{(i)} & 0 \\ 0 & 0 & 1 \end{bmatrix} \begin{bmatrix} c\varphi_y^{(i)} & 0 & s\varphi_y^{(i)} \\ 0 & 1 & 0 \\ -s\varphi_y^{(i)} & 0 & c\varphi_y^{(i)} \end{bmatrix} \begin{bmatrix} 1 & 0 & 0 \\ 0 & c\varphi_x^{(i)} & -s\varphi_x^{(i)} \\ 0 & s\varphi_x^{(i)} & c\varphi_x^{(i)} \end{bmatrix},$$

and the transformation matrices from the system $\{i\}$ to the global system, according to (5.4.1) and (5.6), are as follows:

$$\mathbf{T}^{(i)} = \mathbf{T}^{(i)}\left(\tilde{\mathbf{q}}^{(i)}, t\right) = {}_A^0\mathbf{T}(t)\tilde{\mathbf{T}}^{(i)}\left(\tilde{\mathbf{q}}^{(i)}\right). \tag{8.74}$$

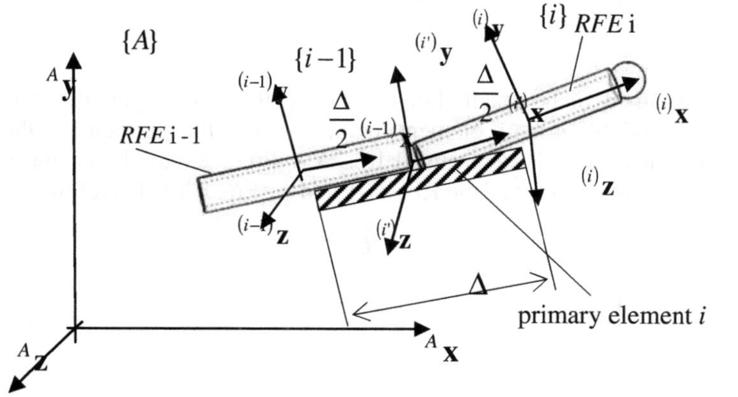

Fig. 8.11. The primary element i, RFEs i-1 and i having load applied to the beam

Large displacements of the links cause the primary element as well as the RFEs i and i-1 created in the secondary division to be in the configuration depicted in Fig. 8.11.

The coordinate systems $\{i\text{-}1\}$ and $\{i\}$ are attached to RFEs i-1 and i. On the other hand, to the primary element i the coordinate system $\{i'\}$ is attached. The coordinate system $\{A\}$ may in further considerations be the global system or one attached to the deck of a vessel or platform.

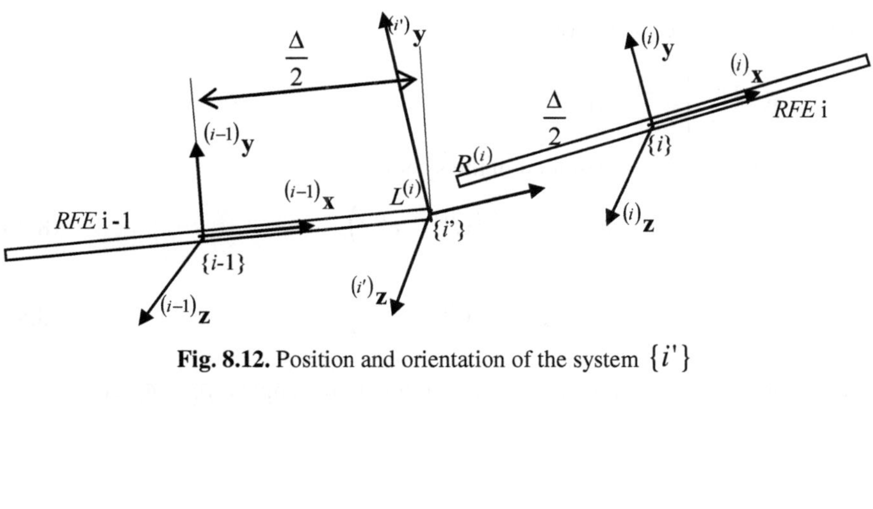

Fig. 8.12. Position and orientation of the system $\{i'\}$

If division of a beam into rigid finite elements is fine enough, differences between the angles $\varphi_z^{(i)} - \varphi_z^{(i-1)}$, $\varphi_y^{(i)} - \varphi_y^{(i-1)}$, $\varphi_x^{(i)} - \varphi_x^{(i-1)}$ which are components of the vector:

$$\boldsymbol{\Delta}\tilde{\boldsymbol{\Phi}}^{(i)} = \tilde{\boldsymbol{\Phi}}^{(i)} - \tilde{\boldsymbol{\Phi}}^{(i-)} = \begin{bmatrix} \varphi_z^{(i)} - \varphi_z^{(i-1)} \\ \varphi_y^{(i)} - \varphi_y^{(i-1)} \\ \varphi_x^{(i)} - \varphi_x^{(i-1)} \end{bmatrix} \tag{8.75}$$

may be assumed to be small. Let us assume that the origin of the coordinate system $\{i'\}$ (of the primary element) coincides with the right end of the RFE i-1 and its orientation to is determined by ZYX Euler angles being the arithmetic means of the Euler angles of the RFEs i-1 and i – Fig. 8.12. Therefore:

$$\mathbf{r}^{(i')} = \tilde{\mathbf{r}}^{(i-1)} + \tilde{\mathbf{R}}^{(i-1)} \tilde{\mathbf{r}}_R^{(i-1')}, \tag{8.76}$$

where $\tilde{\mathbf{r}}_R^{(i-1')} = [\Delta/2 \quad 0 \quad 0]^T$,

and:

$$\tilde{\boldsymbol{\Phi}}^{(i')} = \begin{bmatrix} \varphi_z^{(i')} \\ \varphi_y^{(i')} \\ \varphi_x^{(i')} \end{bmatrix} = \frac{1}{2}\left[\tilde{\boldsymbol{\Phi}}^{(i-1)} + \tilde{\boldsymbol{\Phi}}^{(i)}\right]. \tag{8.77}$$

The coordinates of the right end of the RFE i-1 (point $L^{(i)}$) and the left RFE i (point $R^{(i)}$) in the base system $\{A\}$ are determined thus:

$$\tilde{\mathbf{r}}_L^{(i)} = \tilde{\mathbf{r}}^{(i-1)} + \tilde{\mathbf{R}}^{(i-1)} \tilde{\mathbf{r}}_R^{(i-1')}, \tag{8.78}$$

$$\tilde{\mathbf{r}}_R^{(i)} = \tilde{\mathbf{r}}^{(i1)} + \tilde{\mathbf{R}}^{(i1)} \tilde{\mathbf{r}}_L^{(i-1')}, \tag{8.79}$$

where $\tilde{\mathbf{r}}_R^{(i-1')}$ – defined in (8.76),

$$\tilde{\mathbf{r}}_L^{(i-1')} = \begin{bmatrix} -\frac{\Delta}{2} & 0 & 0 \end{bmatrix}^T.$$

These vectors may be represented in the system $\{i'\}$:

$$\mathbf{r}_L^{(i')} = \mathbf{R}^{(i')T}\left(\tilde{\mathbf{r}}_L^{(i)} - \mathbf{r}^{(i')}\right), \tag{8.80}$$

$$\mathbf{r}_R^{(i')} = \mathbf{R}^{(i')T}\left(\tilde{\mathbf{r}}_R^{(i)} - \mathbf{r}^{(i')}\right), \tag{8.81}$$

where $\mathbf{R}^{(i')}$ – rotation matrix corresponding to the angles $\varphi_x^{(i')}, \varphi_y^{(i')}, \varphi_z^{(i')}$, whereas, considering (8.77), the vectors $\mathbf{\Phi}_L^{(i')}$ and $\mathbf{\Phi}_R^{(i')}$ are:

$$\mathbf{\Phi}_L^{(i')} = \tilde{\mathbf{\Phi}}^{(i-1)} - \tilde{\mathbf{\Phi}}^{(i')} = -\frac{1}{2}\Delta\tilde{\mathbf{\Phi}}^{(i)}, \tag{8.82}$$

$$\mathbf{\Phi}_R^{(i')} = \tilde{\mathbf{\Phi}}^{(i)} - \tilde{\mathbf{\Phi}}^{(i')} = \frac{1}{2}\Delta\tilde{\mathbf{\Phi}}^{(i)}. \tag{8.83}$$

The vector of deformation of the SDE i takes the form:

$$\Delta\mathbf{q}^{(i')} = \begin{bmatrix} \Delta\mathbf{r}^{(i')} \\ \Delta\mathbf{\Phi}^{(i')} \end{bmatrix}, \tag{8.84}$$

where $\Delta\mathbf{r}^{(i')} = \mathbf{r}_R^{(i')} - \mathbf{r}_L^{(i')}$,

$\Delta\mathbf{\Phi}^{(i')} = \mathbf{\Phi}_R^{(i')} - \mathbf{\Phi}_L^{(i')}$.

Taking (8.80) – (8.83) into account:

$$\Delta\mathbf{q}^{(i')} = \begin{bmatrix} \mathbf{R}^{(i')T}\left[\tilde{\mathbf{r}}_R^{(i)} - \mathbf{r}^{(i')} - \tilde{\mathbf{r}}_L^{(i)} + \mathbf{r}^{(i')}\right] \\ \frac{1}{2}\Delta\tilde{\mathbf{\Phi}} - \left(-\frac{1}{2}\Delta\tilde{\mathbf{\Phi}}\right) \end{bmatrix} = \begin{bmatrix} \mathbf{R}^{(i')T}\left[\tilde{\mathbf{r}}_R^{(i)} - \tilde{\mathbf{r}}_L^{(i)}\right] \\ \Delta\tilde{\mathbf{\Phi}} \end{bmatrix}. \tag{8.85}$$

The axes of the SDE are the principal deformation axes, hence the following formulas for the forces and moments caused by the deformation of the SDE i:

$$\tilde{\mathbf{F}}^{(i)} = \mathbf{C}_r^{(i)}\Delta\mathbf{r}^{(i')}, \tag{8.86}$$

$$\tilde{\mathbf{M}}^{(i)} = \mathbf{C}_\Phi^{(i)}\Delta\mathbf{\Phi}^{(i')}, \tag{8.87}$$

where $\tilde{\mathbf{F}}^{(i)} = \left[\tilde{F}_x^{(i)} \quad \tilde{F}_y^{(i)} \quad \tilde{F}_z^{(i)}\right]^T$,

$\tilde{\mathbf{M}}^{(i)} = \left[\tilde{M}_x^{(i)} \quad \tilde{M}_y^{(i)} \quad \tilde{M}_z^{(i)}\right]^T$,

$\mathbf{C}_r^{(i)} = diag\left\{c_x^{(i)}, c_y^{(i)}, c_z^{(i)}\right\}$,

$\mathbf{C}_\Phi^{(i)} = diag\left\{c_\psi^{(i)}, c_\theta^{(i)}, c_\varphi^{(i)}\right\}$,

$c_x^{(i)}, \ldots, c_\varphi^{(i)}$ – stiffness coefficients.

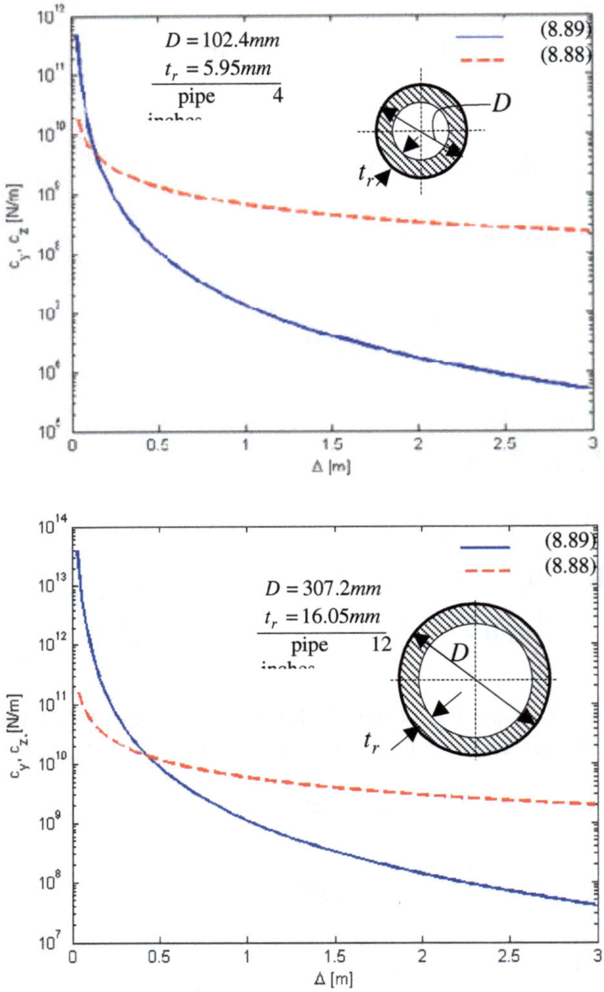

Fig. 8.13. Values of shear coefficients according to (8.88) and (8.89)

In [Kruszewski J., et al., 1999] the following formulas for stiffness coefficients of the elements are given:

$$c_x = \frac{EA}{\Delta}, \qquad c_y = \frac{GA}{\kappa\Delta}, \qquad c_z = \frac{GA}{\kappa\Delta},$$
$$c_\varphi = \frac{GJ_x}{\Delta}, \qquad c_\theta = \frac{GJ_y}{\Delta}, \qquad c_\psi = \frac{GJ_z}{\Delta}. \tag{8.88}$$

The work [Szczotka M., 2011b] takes another approach to defining the shear stiffness coefficients c_y and c_z by assuming:

$$c_y = \frac{12EJ_z}{\Delta^3}, \qquad c_z = \frac{12EJ_y}{\Delta^3}, \tag{8.89}$$

and maintaining the conformance of the values of remaining coefficients to (8.88). The above modification of the coefficients c_y and c_z enables the same expressions to give the potential energy of elastic deformation of the primary element obtained with the RFE method and the energy of elastic deformation of the deformable element considered in FEM.

The values of shear stiffness coefficients determined by formulas (8.88) and (8.89) are shown in Fig. 8.13 for different lengths of the element. Calculations were performed for two different sections of pipes which are analysed in later chapters of this volume. Appropriate division into finite elements enables both coefficients to share the same value. The stiffness coefficients c_y and c_z in the formulas (8.89) have smaller values when the elements resulting from the division are longer.

Forces $+\tilde{\mathbf{F}}_{est}^{(i)}$ applied to the point whose coordinates are given by (8.78) and pairs of forces $+\tilde{\mathbf{M}}_{est}^{(i)}$ act on the RFE i-1. Forces $-\tilde{\mathbf{F}}_{est}^{(i)}$ applied to the point whose coordinates are given by (8.79) and pairs of forces $-\tilde{\mathbf{M}}_{est}^{(i)}$ act on the RFE i. (Fig. 8.14).

The forces (8.86) and the moments (8.87) are given in the coordinate system $\{i'\}$. Their transformation to the global system is done as follows:

$$\mathbf{F}^{(i)} = \mathbf{R}^{(i')}\tilde{\mathbf{F}}^{(i)}, \tag{8.90.1}$$

$$\mathbf{M}^{(i)} = \mathbf{R}^{(i')}\tilde{\mathbf{M}}^{(i)}, \tag{8.90.2}$$

and in the following way to the coordinate systems of the RFEs i and i-1:

$$\tilde{\mathbf{F}}_{est}^{(i-1)} = \mathbf{R}^{(i-1)^T}\mathbf{F}^{(i)}, \tag{8.91.1}$$

$$\tilde{\mathbf{F}}_{est}^{(i)} = \mathbf{R}^{(i)^T}\mathbf{F}^{(i)}, \tag{8.91.2}$$

$$\tilde{\mathbf{M}}_{est}^{(i-1)} = \mathbf{R}^{(i-1)^T}\mathbf{M}^{(i)}, \tag{8.91.3}$$

$$\tilde{\mathbf{M}}_{est}^{(i)} = \mathbf{R}^{(i)^T}\mathbf{M}^{(i)}. \tag{8.91.4}$$

The presented discussion shows that the crucial change introduced with respect to the original formulation of the RFE method (section 8.1) is having the system

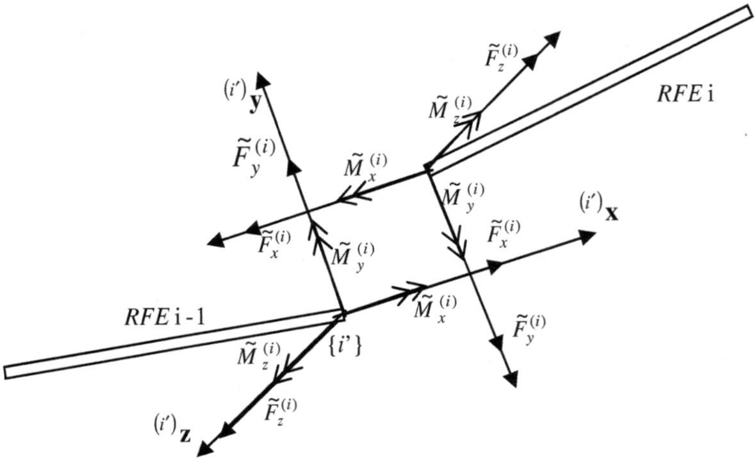

Fig. 8.14. Forces and moments acting on the RFEs *i*-1 and *i* caused by deformation of the SDE *i*

of principal deformation of the SDE {*i*'} "follow" large displacements of the finite elements. A similar approach to planar systems with variable configuration is presented in [Wittbrodt E., 1983]. Furthermore, the modification of shear stiffness coefficients enables the energy of an element's deformation to be expressed in the same form as in the method of deformable finite elements. This conclusion holds for linear physical dependencies. Also of importance is an observation that since the presented proposal assumes the vectors of generalized coordinates of the RFEs take forms described by (8.84), no distinction is made among the variables to configuration (describing the motion of the beam as a rigid body) and flexible ones, as in section 8.1.

8.4.1 Equations of Motion When Using the Classical RFE Method

The equations of motion of a system taking into account the dependencies from previous chapters may be put in the form:

$$\mathbf{A}\ddot{\mathbf{q}} = \mathbf{f}(t, \mathbf{q}, \dot{\mathbf{q}}), \tag{8.92}$$

where $\mathbf{A} = diag\left[\tilde{\mathbf{A}}_0, \ldots, \tilde{\mathbf{A}}_n\right]$,

$$\mathbf{q} = \left[\tilde{\mathbf{q}}_0^T, \ldots, \tilde{\mathbf{q}}_n^T\right]^T,$$

$$\mathbf{f} = \mathbf{Q} - \mathbf{e} - \mathbf{G} = \left[\tilde{\mathbf{f}}_0^T, \ldots, \tilde{\mathbf{f}}_n^T\right]^T,$$

$$\mathbf{Q} = \left[\left(\mathbf{Q}_0^{(est)} + \mathbf{Q}_i^{(S)}\right)^T, \ldots, \left(\mathbf{Q}_n^{(est)} + \mathbf{Q}_n^{(S)}\right)^T\right]^T,$$

$$\mathbf{e} = \left[\tilde{\mathbf{e}}_0^T, \dots, \tilde{\mathbf{e}}_n^T\right]^T,$$

$$\mathbf{G} = \left[\left(\frac{\partial V_0}{\partial \tilde{\mathbf{q}}_0}\right)^T, \dots, \left(\frac{\partial V_n}{\partial \tilde{\mathbf{q}}_n}\right)^T\right]^T,$$

n – number of RFEs in the concerned model.

Note that the matrix **A** is diagonal, which is of great importance when integrating the system's equations of motion. Such form is characteristic of systems modelled with the classical RFE method.

Let us assume that at the point whose coordinates are given by the vector $\tilde{\mathbf{r}}_i'$ in the local system of the RFE i there act: an external force and a pair of forces given by:

$$\begin{aligned} \tilde{\mathbf{F}}_i &= \left[\tilde{F}_i^{(x)} \quad \tilde{F}_i^{(y)} \quad \tilde{F}_i^{(z)} \quad 0\right]^T, \\ \tilde{\mathbf{M}}_i &= \left[\tilde{M}_i^{(x)} \quad \tilde{M}_i^{(y)} \quad \tilde{M}_i^{(z)} \quad 0\right]^T. \end{aligned} \tag{8.93}$$

Their corresponding generalized forces may then be determined from the formulas [Wittbrodt E., et al., 2006]:

$$\begin{aligned} \left(\mathbf{Q}_i\left(\tilde{\mathbf{F}}_i\right)\right)_{k=1,\dots,6} &= \tilde{\mathbf{F}}_i^T \mathbf{T}_i^T \mathbf{T}_{i,k} \tilde{\mathbf{r}}_i', \\ \left(\mathbf{Q}_i\left(\tilde{\mathbf{M}}_i\right)\right)_{k=1,\dots,6} &= \tilde{M}_i^{(x)} \sum_{j=1}^{3} \left(\mathbf{T}_i\right)_{j,3} \left(\mathbf{T}_{i,k}\right)_{j,2} + \tilde{M}_i^{(y)} \sum_{j=1}^{3} \left(\mathbf{T}_i\right)_{j,1} \left(\mathbf{T}_{i,k}\right)_{j,3} + \tilde{M}_i^{(z)} \sum_{j=1}^{3} \left(\mathbf{T}_i\right)_{j,2} \left(\mathbf{T}_{i,k}\right)_{j,1}. \end{aligned} \tag{8.94}$$

The relations (8.94) allow us to determine the generalized forces and moments pertaining to the impact of the sea environment:

$$\mathbf{Q}_i^{(S)} = \mathbf{Q}_i^{(h)}\left(\tilde{\mathbf{F}}_i^h, \tilde{\mathbf{M}}_i^h\right) + \mathbf{Q}_i^{(b)}\left(\tilde{\mathbf{F}}_i^b, \tilde{\mathbf{M}}_i^b\right) + \mathbf{Q}_i^{(w)}\left(\tilde{\mathbf{F}}_i^t, \tilde{\mathbf{M}}_i^t\right) + \mathbf{Q}_i^{(d)}\left(\tilde{\mathbf{F}}_i^d, \tilde{\mathbf{M}}_i^d\right) \tag{8.95}$$

where $\tilde{\mathbf{F}}_i^h, \tilde{\mathbf{M}}_i^h$ – vectors of forces and moments due to interaction of the element i with the liquid (including the influence of waves, sea currents and hydrodynamics),

$\tilde{\mathbf{F}}_i^b, \tilde{\mathbf{M}}_i^b$ – vectors of forces and hydrostatic buoyancy moments (hydrostatic buoyancy of the pipeline and additional buoyant modules),

$\tilde{\mathbf{F}}_i^t, \tilde{\mathbf{M}}_i^t$ – vectors of forces and moments of the action of guiding structures (e.g. reel, guiding ramp, mechanisms),

$\tilde{\mathbf{F}}_i^d, \tilde{\mathbf{M}}_i^d$ – vectors of forces and moments due to the action of the seabed.

The forces and moments caused by the deformation of the SDE may be included similarly. According to (8.86) and (8.87), forces and moments caused by the deformation of the SDE i (left end of the RFE i) and the SDE i+1 (right end of the RFE i) act upon the RFE i. Hence:

$$\mathbf{Q}_i^{(est)} = \mathbf{Q}_i^{(est)}\left(-\mathbf{F}_i^{(est)}, -\widetilde{\mathbf{M}}_i^{(est)}, \mathbf{F}_{i+1}^{(est)}, \widetilde{\mathbf{M}}_{i+1}^{(est)}\right), \tag{8.96}$$

where $\mathbf{F}^{(i)}, \widetilde{\mathbf{M}}^{(i)}$ – defined in (8.86) and (8.87),

$$\begin{aligned}\mathbf{Q}_{i,k}^{(est)} = \mathbf{F}_{i+1}^{(est)^T}\mathbf{T}_{i,k}\widetilde{\mathbf{r}}_{R,i}' - \mathbf{F}_i^{(est)^T}\mathbf{T}_{i,k}\widetilde{\mathbf{r}}_{L,i}' + \\ + \left[\widetilde{M}_{i+1,x}^{(est)} - \widetilde{M}_{i,x}^{(est)}\right]\cdot\sum_{j=1}^{3}(\mathbf{T}_i)_{j,3}(\mathbf{T}_{i,k})_{j,2} + \\ + \left[\widetilde{M}_{i+1,y}^{(est)} - \widetilde{M}_{i,y}^{(est)}\right]\cdot\sum_{j=1}^{3}(\mathbf{T}_i)_{j,1}(\mathbf{T}_{i,k})_{j,3} + \\ + \left[\widetilde{M}_{i+1,z}^{(est)} - \widetilde{M}_{i,z}^{(est)}\right]\cdot\sum_{j=1}^{3}(\mathbf{T}_i)_{j,2}(\mathbf{T}_{i,k})_{j,1}.\end{aligned}$$

8.4.2 Inclusion of Nonlinear Physical Dependencies

In the rigid finite element method, flexibility is described in an approximate manner (displacements are realized in the SDE only). Therefore, the tensor $\sigma_{jk}^{(i)}$ present in (7.1), defined for each SDE i in the plane normal to the beam's axis, is given as [Szczotka M., 2011b]:

$$\sigma_{jk}^{(i)} = \begin{bmatrix} \sigma_x^{(i)} & \tau_{yx}^{(i)} & \tau_{zx}^{(i)} \\ \tau_{xy}^{(i)} & 0 & 0 \\ \tau_{xz}^{(i)} & 0 & 0 \end{bmatrix}, \tag{8.97}$$

where $\sigma_x = \dfrac{\widetilde{F}_x^{(i)}}{A}$,

$\tau_{xy}^{(i)} = \dfrac{\widetilde{F}_y^{(y)}}{A} + \dfrac{1}{2}\tau_S^{(i)}$,

$\tau_{xz}^{(i)} = \dfrac{\widetilde{F}_z^{(y)}}{A} + \dfrac{1}{2}\tau_S^{(i)}$,

$\tau_{yx}^{(i)} = \tau_{xy}^{(i)}$, $\qquad \tau_{zx}^{(i)} = \tau_{xz}^{(i)}$,

$\tau_S^{(i)} = \dfrac{\widetilde{M}_x^{(i)}}{2At_{\min}^{(i)}}$ – stress tangent to the torque,

$t_{\min}^{(i)}$ – minimal thickness of the section's side.

The form of the stress tensor in a flexible beam modelled with the RFE method is similar to the tensor obtained in the Saint-Venant problem for torsion and bending of beams [Nowacki W., 1970]. The stresses $\sigma_{g,y}^{(i)}$, $\sigma_{g,z}^{(i)}$ due to the bending moments at which the transition from elasticity to plasticity occurs, may be determined from the following equation taking (7.1) into account:

$$\sqrt{3J_2^{(i)}+\left(\tilde{\sigma}_g^{(i)}\right)^2}-\sigma_0=0, \tag{8.98}$$

where $\tilde{\sigma}_g^{(i)}$ – equivalent bending stress,

$J_2^{(i)}$ – specified as in (7.1).

Some models of pipelines presented later use the given dependencies to construct a module allowing us to determine the forces and moments in the SDE when occurrence of elasto-plastic deformations is possible. Fig. 8.15 schematically shows a flowchart of actions comprising the procedure of determining the bending moment acting on the RFE on the assumption that $\vartheta_i \in \{\varphi_{yi}, \varphi_{zi}\}$.

The diagram uses the following notation:

F_i – mark specifying the state of the material,

X_s – maximal deformation which causes the phase to change from elastic to plastic,

$\tilde{\vartheta}_i^0$ – neutral value of displacement (at which $m_\vartheta^{(i)}=0$),

$C_e^{(i)}$ – stiffness coefficient of the SDE within the elastic region,

$C_p^{(i)}$ – stiffness coefficient of the SDE within the plastic region, $C_p^{(i)}=\mu C_e^{(i)}$, $\mu=0.01, 0.1, \ldots$,

$f_{mat}(\ )$ – function describing the shape of the characteristic $\sigma=f(\varepsilon_p)$ within the plastic region,

$M_B^{(i)}$ – value $m_\vartheta^{(i)}$ determined in the previous step $t-h$,

$\vartheta_B^{(i)}$ – plastic deformation in the previous step $t-h$.

A control procedure for the mark F_i and for calculating the values $\vartheta_B^{(i)}$ and $M_B^{(i)}$ (Fig. 8.16) is also necessary.

Approximation of the characteristic of a material may be performed for arbitrary data obtained e.g. from measurements. The linear segments (elastic region, linear reinforcement in the plastic region) may be interspersed with nonlinear ones, thus leading to significantly greater stability of the calculations. An example of such characteristic can be found in [Szczotka M., 2010].

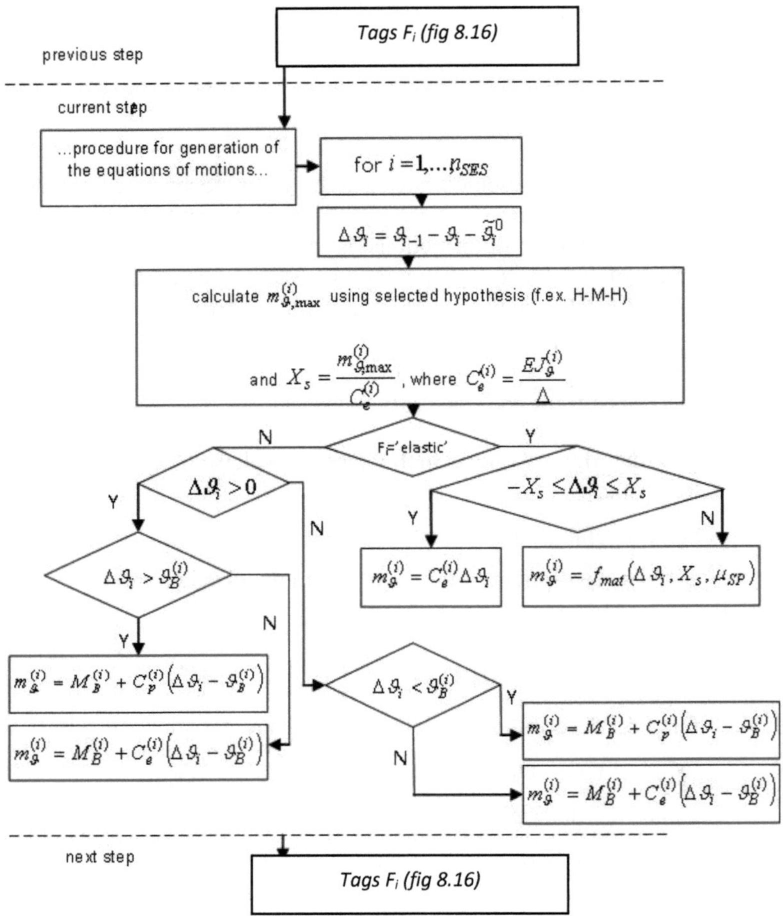

Fig. 8.15. Flow diagram of the algorithm determining the value $m_\vartheta^{(i)}$

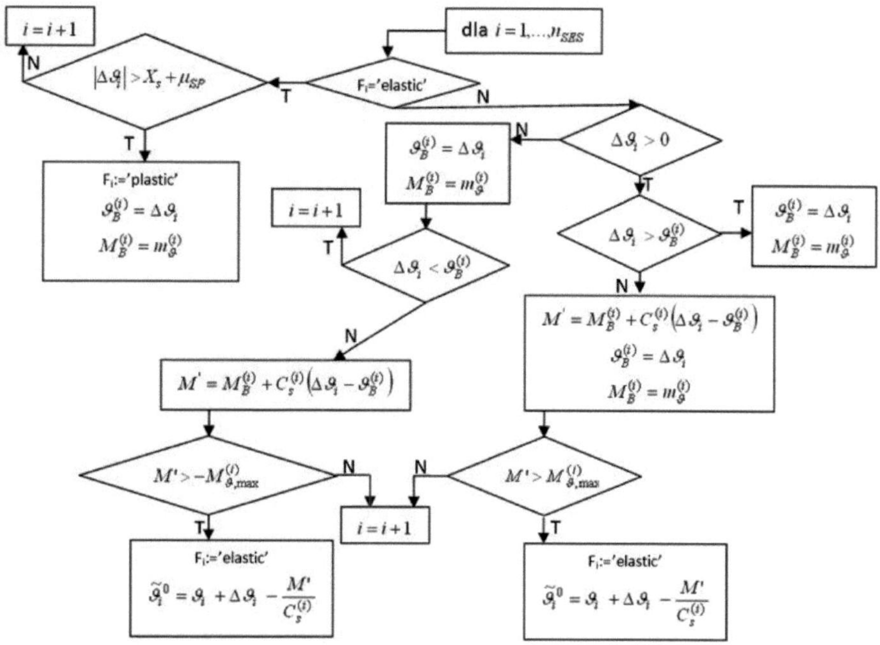

Fig. 8.16. Flow diagram of state markers control, F_i and calculations of the values $\vartheta_B^{(i)}$ and $M_B^{(i)}$

9 Applications of Models of Offshore Structures

Each offshore structure is unique in the sense that it is built only after a customer with a specific need actually places an order. Design companies and manufacturers of engineering systems of this type are often small and medium enterprises, which cannot afford purchasing costly computer software packages for numerical computation involved in dynamics of mechanical systems. Therefore, they often employ custom, in-house dynamic models of the structures designed. In the present chapter, dynamic models of the following are presented: a gantry suited for relocating sets of BOP valves on an extraction platform, a column crane and a device for laying pipes on the seabed. The formulation of models thereof leverages the methods described in earlier chapters.

9.1 BOP Transportation Gantry

One of the types of offshore cranes is a BOP crane. The construction of Protea from Gdańsk is presented in Fig. 9.1. It is a gantry crane installed on a drilling platform designed to transport a system of valves named BOP (Blowout Preventor). BOP is used to block an uncontrolled outflow of oil or natural gas from a wellbore at the seabed. After drilling the wellbore, the BOP is put inside it, and afterwards risers are being connected to the BOP. The risers drain off oil or gas into suitable tanks. In view of the plug task, weight of the BOP reaches hundreds of tons. During the transportation process (during the travel of a gantry crane) the BOP is protected by a system of guides presented in Fig. 9.2.

Clearance between the load and the guide system equals a few centimetres. Weight of the presented crane is 200 T, hoisting capacity 550 T and height about 30 m. The analysis of a travel system is an interesting and important problem concerning the dynamics of a BOP crane. The crane is supported on rails and its motion is realized by the means of a rack and a toothed wheel (Fig. 9.3). Maximum velocity of travel of the crane is equal to 3 m/min. Due to the movement of the platform's deck caused by sea weaving and wind forces, the protection systems are used. These systems limit the movement of the crane in vertical direction and horizontal one, perpendicular to the longitudinal axis of rails. This task is particularly realized by an anti-lift system presented in Fig. 9.4.

E. Wittbrodt et al.: Rigid Finite Element Method, OEO 1, pp. 139–201.
springerlink.com

Fig. 9.1. BOP crane

Fig. 9.2. Guide system

Fig. 9.3. Rack travel system

Fig. 9.4. Anti-lift system

9.1.1 Mathematical Model of the System

The schema of the model of the BOP crane together with more important coordinate systems is presented in Fig. 9.5. The following basic assumptions for modelling are established:

- movement of the base (system $\{A\}$) is known and described by functions:

$$
\begin{aligned}
&y_1 = x_{org}^{(A)}(t); \quad y_2 = x_{org}^{(A)}(t); \quad y_3 = x_{org}^{(A)}(t); \\
&y_6 = \psi_{org}^{(A)}(t); \quad y_5 = \theta_{org}^{(A)}(t); \quad y_4 = \varphi_{org}^{(A)}(t),
\end{aligned}
\tag{9.1}
$$

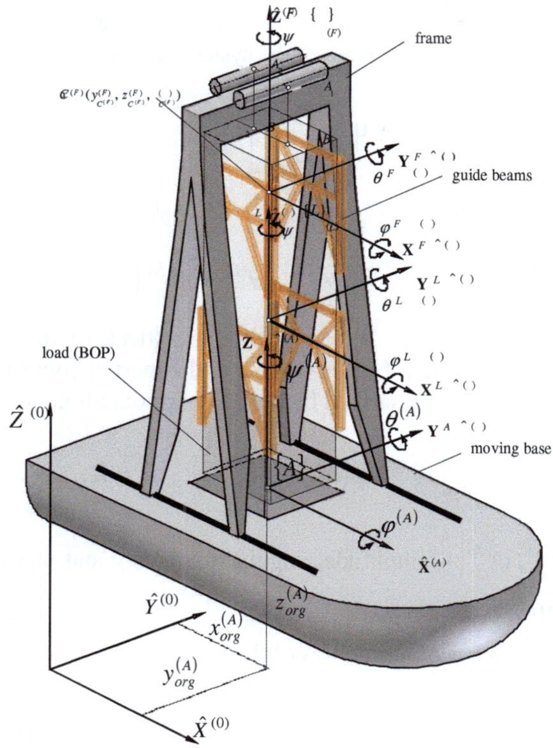

Fig. 9.5. A model of a BOP gantry with load

- structure of the crane (frame) is treated as a rigid body – it should be noticed that the construction of the BOP crane is a kind of combination of two A-frames; an A-frame has been a subject of many analyses presented in [Fałat P., 2004]; these analyses have shown that the influence of flexibility of the frame on dynamics of the whole system (on motion of the load) is slight,
- load is a rigid body of rectangular shape,
- load is suspended on two ropes – their flexibility and damping are taken into account,
- load can touch the guides only along its edges,
- clearance and flexibility between the load and guides are taken into consideration,
- frame is fixed flexibly to the deck and, additionally, in $\hat{\mathbf{Y}}^{(A)}$ direction clearance can occur,
- input in the drive system has been modelled in two ways: a kinematic input via a spring-damping element and a force input,
- wind force can be taken into consideration,
- homogenous transformations are used to describe the system's geometry.

Both the load (system $\{L\}$ in Fig. 9.5) and the frame (system $\{F\}$) have 6 degrees of freedom in respect to the deck (system $\{A\}$). So, the model has 12 degrees of freedom and the vector of generalized coordinates of the system has a following form:

$$\mathbf{q} = \begin{bmatrix} \mathbf{q}^{(F)} \\ \mathbf{q}^{(L)} \end{bmatrix}, \tag{9.2}$$

where $\mathbf{q}^{(F)} = \begin{bmatrix} x^{(F)} & y^{(F)} & z^{(F)} & \psi^{(F)} & \theta^{(F)} & \varphi^{(F)} \end{bmatrix}^T$,

$\mathbf{q}^{(L)} = \begin{bmatrix} x^{(L)} & y^{(L)} & z^{(L)} & \psi^{(L)} & \theta^{(L)} & \varphi^{(L)} \end{bmatrix}^T$.

It has been mentioned that the motion of the base (deck of the platform), that means the motion of the system $\{A\}$, in respect to the inertial coordinate system $\{\}$ has been assumed as known, described by pseudo-harmonic functions:

$$y_i = \sum_{j=1}^{n_i^{(A)}} A_{i,j}^{(A)} \sin\left(\omega_{i,j}^{(A)} t + \varphi_{i,j}^{(A)}\right) \qquad i = 1,\ldots,6, \tag{9.3}$$

where $A_{i,j}^{(A)}, \omega_{i,j}^{(A)}, \varphi_{i,j}^{(A)}$ – amplitude, angular frequency and phase angle of the input, respectively,

$n_i^{(A)}$ – number of harmonics of the series.

The application of homogenous transformations allows converting a position vector of the point defined in the system $\{A\}$ to system $\{\}$ according to relation:

$$\mathbf{r}_P^{\{\}} = {}_A^0\mathbf{T}\,\mathbf{r}_P^{\{A\}}, \tag{9.4}$$

where $\mathbf{r}_P^{\{\}} = \begin{bmatrix} x_P & y_P & z_P & 1 \end{bmatrix}^T$ – position vector of point P in the inertial system $\{\}$,

$\mathbf{r}_P^{\{A\}} = \begin{bmatrix} x_P^{\{A\}} & y_P^{\{A\}} & z_P^{\{A\}} & 1 \end{bmatrix}^T$ – position vector of point P in the system $\{A\}$,

${}_A^0\mathbf{T}$ – matrix of a homogenous transformation from the system {A} to the system {}.

The matrix ${}_A^0\mathbf{T}$ can be presented as product of six matrices, where each of them is a function of one variable dependent on time (9.4). Order of rotations included in the matrix ${}_A^0\mathbf{T}$ corresponds to Euler angles ZYX.

Kinetic and Potential Energy of the Frame and the Load

Kinetic and potential energy of the frame, as well as the load, can be determined using general algorithms presented in chapter 5. If one denotes the homogenous transformation matrix from the frame system $\{F\}$ to the deck system $\{A\}$ as $\tilde{\mathbf{T}}^{(F)}$ and from the load system $\{L\}$ as $\tilde{\mathbf{T}}^{(L)}$, the transformation matrices from the frame system and from the load system to the system { } can be calculated as:

$$\mathbf{T}^{(F)} = {}^{0}_{A}\mathbf{T}\,\tilde{\mathbf{T}}^{(F)}, \tag{9.5}$$

$$\mathbf{T}^{(L)} = {}^{0}_{A}\mathbf{T}\,\tilde{\mathbf{T}}^{(L)}. \tag{9.6}$$

Introducing notation of the Lagrange operator:

$$\varepsilon_k^{(b)} = \frac{d}{dt}\frac{\partial E^{(b)}}{\partial \dot{q}_k^{(b)}} - \frac{\partial E^{(b)}}{\partial q_k^{(b)}}, \tag{9.7}$$

where k is the number of the generalized coordinate, $b \in \{F, L\}$,

and using the transformation presented in chapter 5 one can obtain:

$$\varepsilon_k^{(b)} = tr\left\{\mathbf{T}_k^{(b)}\mathbf{H}^{(b)}\left[{}^{0}_{A}\ddot{\mathbf{T}}\tilde{\mathbf{T}}^{(b)} + 2\,{}^{0}_{A}\dot{\mathbf{T}}\dot{\tilde{\mathbf{T}}}^{(b)} + \sum_{i=1}^{6}\sum_{j=1}^{6}\mathbf{T}_{i,j}^{(b)}\dot{q}_i^{(b)}\dot{q}_j^{(b)} + \sum_{i=1}^{6}\mathbf{T}_i^{(b)}\ddot{q}_i^{(b)}\right]\right\}, \tag{9.8}$$

where $\mathbf{T}_k^{(b)} = \dfrac{\partial \mathbf{T}^{(b)}}{\partial q_k^{(b)}}$,

$$\mathbf{T}_{i,j}^{(b)} = \frac{\partial^2 \mathbf{T}^{(b)}}{\partial q_i^{(b)}\partial q_j^{(b)}}.$$

The above form requires repeated multiplication of matrices of 4×4 dimensions and then the calculation of the trace of the result matrices. In order to decrease the number of required numerical operations, the authors decided to derive formulae describing Lagrange operators in the explicit form.

The relation (9.8) can be presented in the following form:

$$\varepsilon_k^{(b)} = \mathrm{tr}\left\{{}^{0}_{A}\mathbf{T}\tilde{\mathbf{T}}_k^{(b)}\mathbf{H}^{(b)}\left[{}^{0}_{A}\ddot{\mathbf{T}}\tilde{\mathbf{T}}^{(b)} + 2\,{}^{0}_{A}\dot{\mathbf{T}}\dot{\tilde{\mathbf{T}}}^{(b)} + {}^{0}_{A}\mathbf{T}\ddot{\tilde{\mathbf{T}}}^{(b)}\right]^T\right\} =$$

$$= \underbrace{\mathrm{tr}\left\{{}^{0}_{A}\mathbf{T}^{T}\,{}^{0}_{A}\mathbf{T}\tilde{\mathbf{T}}_k^{(b)}\mathbf{H}^{(b)}\left[\ddot{\tilde{\mathbf{T}}}^{(b)}\right]^T\right\}}_{\varepsilon_{k,2}^{(b)}} + \underbrace{\mathrm{tr}\left\{{}^{0}_{A}\ddot{\mathbf{T}}^{T}\,{}^{0}_{A}\mathbf{T}\tilde{\mathbf{T}}_k^{(b)}\mathbf{H}^{(b)}\left[\tilde{\mathbf{T}}^{(b)}\right]^T\right\}}_{\varepsilon_{k,0}^{(b)}} \tag{9.9}$$

$$+ 2\,\underbrace{\mathrm{tr}\left\{{}^{0}_{A}\dot{\mathbf{T}}^{T}\,{}^{0}_{A}\mathbf{T}\tilde{\mathbf{T}}_k^{(b)}\mathbf{H}^{(b)}\left[\dot{\tilde{\mathbf{T}}}^{(b)}\right]^T\right\}}_{\varepsilon_{k,1}^{(b)}}.$$

Assuming that rotation angles of the frame and the load are small, the matrix $\tilde{\mathbf{T}}^{(b)}$ can be written as:

$$\tilde{\mathbf{T}}^{(b)} = \begin{bmatrix} 1 & -\psi^{(b)} & \theta^{(b)} & x^{(b)} \\ \psi^{(b)} & 1 & -\varphi^{(b)} & y^{(b)} \\ -\theta^{(b)} & \varphi^{(b)} & 1 & z^{(b)} \\ 0 & 0 & 0 & 1 \end{bmatrix} \tag{9.10}$$

or:

$$\tilde{\mathbf{T}}^{(b)} = \mathbf{I} + \sum_{j=1}^{6} \mathbf{D}_j q_j^{(b)}, \tag{9.11}$$

where $q_j^{(b)}$ – suitable elements of vectors $\mathbf{q}^{(F)}$ or $\mathbf{q}^{(L)}$,

and matrices $\mathbf{D}_j$ can be defined as:

for j=1,2,3:

$$\mathbf{D}_j = \begin{bmatrix} \mathbf{0} & \mathbf{a}_j \\ \mathbf{0} & 0 \end{bmatrix}, \tag{9.12}$$

where $\mathbf{a}_1 = \begin{bmatrix} 1 \\ 0 \\ 0 \end{bmatrix}$; $\mathbf{a}_2 = \begin{bmatrix} 0 \\ 1 \\ 0 \end{bmatrix}$; $\mathbf{a}_3 = \begin{bmatrix} 0 \\ 1 \\ 0 \end{bmatrix}$,

for j=4,5,6:

$$\mathbf{D}_j = \begin{bmatrix} \mathbf{R}_j & \mathbf{0} \\ \mathbf{0} & 0 \end{bmatrix}, \tag{9.13}$$

where $\mathbf{R}_4 = \begin{bmatrix} 0 & 0 & 0 \\ 0 & 0 & -1 \\ 0 & 1 & 0 \end{bmatrix}$, $\mathbf{R}_5 = \begin{bmatrix} 0 & 0 & 1 \\ 0 & 0 & 0 \\ -1 & 0 & 0 \end{bmatrix}$, $\mathbf{R}_6 = \begin{bmatrix} 0 & -1 & 0 \\ 1 & 0 & 0 \\ 0 & 0 & 0 \end{bmatrix}$.

In the paper [Urabś A., et al., 2010] it has been shown that:

for k=1,2,3:

$$\varepsilon_{k,2}^{(b)} = m^{(b)} \ddot{q}_k^{(b)}, \tag{9.14}$$

$$\varepsilon_{k,1}^{(b)} = \sum_{j=1}^{3} \dot{q}_j^{(b)} m^{(b)} \left(\mathbf{\Phi}_1^T \mathbf{\Phi}_0\right)_{j,k}, \tag{9.15}$$

$$\varepsilon_{k,0}^{(b)} = m^{(b)}, \tag{9.16}$$

for k=4,5,6:

$$\varepsilon_{k,1}^{(b)} = \sum_{j=4}^{6} \dot{q}_j^{(b)} \operatorname{tr}\left(\mathbf{\Phi}_1^T \mathbf{\Phi}_0 \mathbf{R}_k^{(b)} \mathbf{J}^{(b)} \left[\mathbf{R}_j^{(b)}\right]^T \right)_{j,k}, \tag{9.17}$$

$$\varepsilon_{4,2}^{(b)} = \left(J_{11}^{(b)} + J_{22}^{(b)}\right)\ddot{q}_4^{(b)} - J_{32}^{(b)}\ddot{q}_5^{(b)} - J_{31}^{(b)}\ddot{q}_6^{(b)}, \tag{9.18.1}$$

$$\varepsilon_{5,2}^{(b)} = -J_{23}^{(b)}\ddot{q}_4^{(b)} + \left(J_{11}^{(b)} + J_{33}^{(b)}\right)\ddot{q}_5^{(b)} - J_{21}^{(b)}\ddot{q}_6^{(b)}, \tag{9.18.2}$$

$$\varepsilon_{6,2}^{(b)} = -J_{13}^{(b)}\ddot{q}_4^{(b)} - J_{12}^{(b)}\ddot{q}_5^{(b)} + \left(J_{22}^{(b)} + J_{33}^{(b)}\right)\ddot{q}_6^{(b)}, \tag{9.18.3}$$

$$\varepsilon_{k,0}^{(b)} = \operatorname{tr}\left\{\mathbf{\Phi}_2^T \mathbf{\Phi}_0 \mathbf{R}_k^{(b)} \mathbf{J}^{(b)}\right\} + \sum_{j=4}^{6} q_j^{(b)} \operatorname{tr}\left\{ \mathbf{\Phi}_2^T \mathbf{\Phi}_0 \mathbf{R}_k^{(b)} \mathbf{J}^{(b)} \left[\mathbf{R}_j^{(b)}\right]^T \right\}, \tag{9.19}$$

where $m^{(b)}$ – mass of the body $b \in \{F, L\}$,

$\mathbf{J}^{(b)} = \begin{bmatrix} J_{11}^{(b)} & J_{12}^{(b)} & J_{13}^{(b)} \\ J_{21}^{(b)} & J_{22}^{(b)} & J_{23}^{(b)} \\ J_{31}^{(b)} & J_{32}^{(b)} & J_{33}^{(b)} \end{bmatrix}$ – elements of the matrix are defined in (5.11),

$\mathbf{\Phi}_0, \mathbf{\Phi}_1, \mathbf{\Phi}_2, \mathbf{S}_0, \mathbf{S}_1, \mathbf{S}_2$ – submatrices of ${}_A^0\mathbf{T}$, ${}_A^0\dot{\mathbf{T}}$, ${}_A^0\ddot{\mathbf{T}}$ respectively,

$${}_A^0\mathbf{T} = \begin{bmatrix} \mathbf{\Phi}_0 & \mathbf{S}_0 \\ \mathbf{0} & 1 \end{bmatrix}, \quad {}_A^0\dot{\mathbf{T}} = \begin{bmatrix} \mathbf{\Phi}_1 & \mathbf{S}_1 \\ \mathbf{0} & 0 \end{bmatrix}, \quad {}_A^0\ddot{\mathbf{T}} = \begin{bmatrix} \mathbf{\Phi}_2 & \mathbf{S}_2 \\ \mathbf{0} & 0 \end{bmatrix}.$$

Derivatives of potential energy of gravity forces of element of mass $m^{(b)}$ can be presented in the form of the vector:

$$\frac{\partial V_g^{(b)}}{\partial \mathbf{q}^{(b)}} = \left[m^{(b)} g\, t_{31} \quad m^{(b)} g\, t_{32} \quad m^{(b)} g\, t_{33} \quad 0 \quad 0 \quad 0\right]^T, \tag{9.20}$$

where $\mathbf{q}^{(b)}$ – vector of coordinates of the frame or the load (defined in (9.2)), respectively,

$m^{(b)}$ – mass of the frame or the load,

t_{31}, t_{33}, t_{33} – proper elements of the third row of the matrix ${}_A^0\mathbf{T}$.

Model of the Support

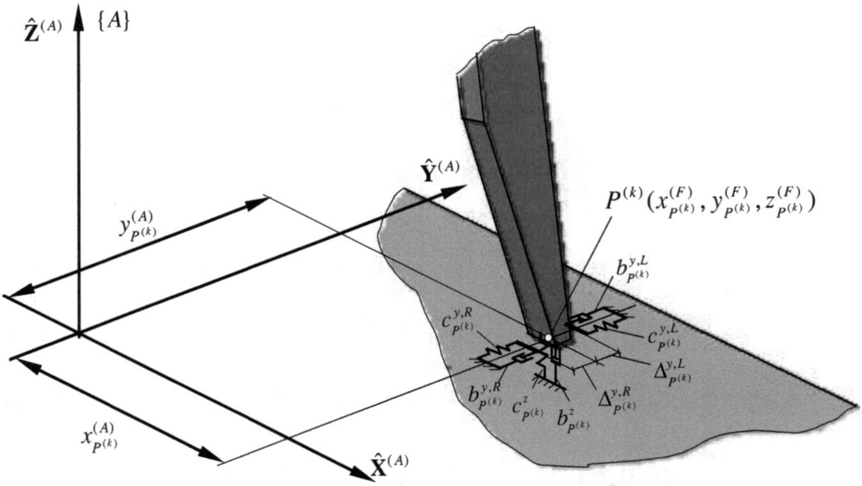

Fig. 9.6. Flexible connection of the frame to the deck

It has been assumed that the frame of the BOP crane is supported flexibly in four points denoted as $P^{(k)}$ $(k = 1,2,3,4)$. The crane is moving on a dedicated rail system in direction parallel to $\hat{\mathbf{X}}^{(A)}$ axis (Fig.9.6). Additionally, a constructional clearance can occur in $\hat{\mathbf{Y}}^{(A)}$ direction.
The reaction force, i. e. the reaction force of the base on the frame, is depicted by the vector:

$$\mathbf{F}^{(F)}_{P^{(k)}} = \left[F^{(F,x)}_{P^{(k)}} \quad F^{(F,y)}_{P^{(k)}} \quad F^{(F,z)}_{P^{(k)}} \right]^T . \tag{9.21}$$

The $F^{(F,z)}_{P^{(k)}}$ component can be calculated as:

$$F^{(F,z)}_{P^{(k)}} = F^{(F,z)}_{S,P^{(k)}} + F^{(F,z)}_{D,P^{(k)}} , \tag{9.22}$$

where $F^{(F,z)}_{S,P^{(k)}}$ – stiffness force,

$F^{(F,z)}_{D,P^{(k)}}$ – damping force.

The stiffness and damping forces are determined by relations:

$$F^{(F,z)}_{S,P^{(k)}} = -c^{z}_{P^{(k)}} \delta^{z}_{P^{(k)}} \Delta z_{P^{(k)}}, \tag{9.23.1}$$

$$F^{(F,z)}_{D,P^{(k)}} = -b^{z}_{P^{(k)}} \delta^{z}_{P^{(k)}} \Delta \dot{z}_{P^{(k)}}, \tag{9.23.2}$$

where $\delta^{z}_{P^{(k)}} = \begin{cases} 1 \text{ when } \Delta z_{P^{(k)}} < 0 \\ 0 \text{ when } \Delta z_{P^{(k)}} \geq 0 \end{cases}$

$\Delta z_{P^{(k)}} = z^{(A)}_{P^{(k)}} - z^{(A,0)}_{P^{(k)}}$, where $z^{(A,0)}_{P^{(k)}} = 0$, and $z^{(A)}_{P^{(k)}}$ is the z coordinate of the point $P^{(k)}$ in the system $\{A\}$, $\Delta \dot{z}_{P^{(k)}} = \dot{z}^{(A)}_{P^{(k)}}$,

$c^{z}_{P^{(k)}}, b^{z}_{P^{(k)}}$ – stiffness and damping coefficients of the connection in the $\hat{\mathbf{Z}}^{(A)}$ direction, respectively.

In the case of the component $F^{(F,y)}_{P^{(k)}}$, the possibility of occurrence of clearance in the anti-lift system is taken into account. Due to modelling clearance, two spring-damping elements acting in the $\hat{\mathbf{Y}}^{(A)}$ direction are introduced, as described in chapter 6.2.

The component $F^{(F,x)}_{P^{(k)}}$ from (9.21) can be expressed by:

$$F^{(F,x)}_{P^{(k)}} = -\operatorname{sgn}(\upsilon^{(A,x)}_{P^{(k)}}) S^{(F,x)}_{P^{(k)}} (F^{(F,y)}_{P^{(k)}}, F^{(F,z)}_{P^{(k)}}) \tag{9.24}$$

where $S^{(F,x)}_{P^{(k)}}$ – resisting force caused by rolling or sliding friction,

$\upsilon^{(A,x)}_{P^{(k)}}$ – component x of the velocity of the point $P^{(k)}$ in the coordinate system $\{A\}$.

After calculating suitable coordinates and velocity of points of support, generalized force of flexible connection of the frame and the deck can be written as:

$$\mathbf{Q}^{(F)}_{P^{(k)}} = \mathbf{U}^{(F)^T}_{P^{(k)}} \mathbf{F}^{(F)}_{P^{(k)}}, \tag{9.25}$$

where $\mathbf{U}^{(F)}_{P^{(k)}} = \begin{bmatrix} 1 & 0 & 0 & -y^{(F)}_{P^{(k)}} & z^{(F)}_{P^{(k)}} & 0 \\ 0 & 1 & 0 & x^{(F)}_{P^{(k)}} & 0 & -z^{(F)}_{P^{(k)}} \\ 0 & 0 & 1 & 0 & -x^{(F)}_{P^{(k)}} & y^{(F)}_{P^{(k)}} \end{bmatrix}$.

Generalizing the relation (9.25) to four supports one can obtain:

$$\mathbf{Q}_p^{(F)} = \sum_{k=1}^{4} \mathbf{Q}_{P^{(k)}}^{(F)} = \sum_{k=1}^{4} \mathbf{U}_{P^{(k)}}^{(F)^T} \mathbf{F}_{P^{(k)}}^{(F)} . \tag{9.26}$$

Modelling Clearance between the Load and Guides

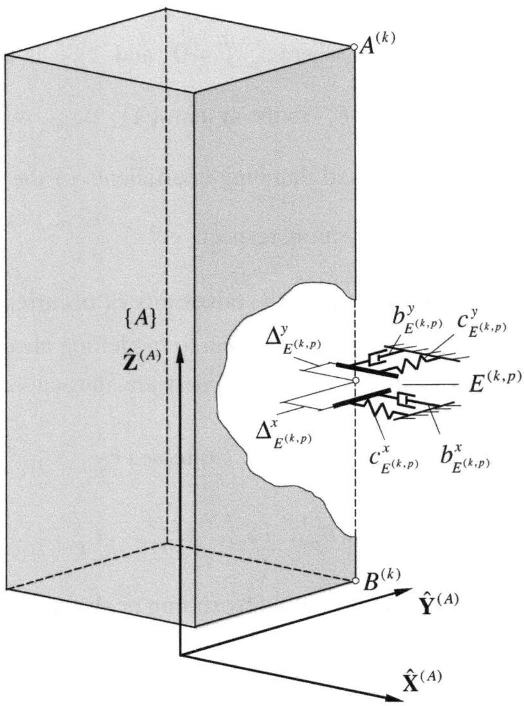

Fig. 9.7. Load and spring-damping elements with clearance

The guides have been replaced by spring-damping elements with clearance (SDE $E^{(kp)}$) that limited the movement of the load in $\hat{\mathbf{X}}^{(A)}$ and $\hat{\mathbf{Y}}^{(A)}$ directions (Fig. 9.7). It has been assumed that the load can contact with guides only along its edges and the number of spring-damping elements can be different for each edge. The manner of calculation of stiffness and damping forces coming from the each side is analogical to the one presented in chapter 5.3. Additionally, one has to determine equivalent coefficients of flexibility of elements modelling the guides. Suitable calculations have been executed by the means of the Finite Elements Method. They were presented in details in the doctoral thesis [Urbaś A., 2011].

Drive of Travel System

The input in the drive of the travel system has been modelled in two ways (Fig. 9.8): a kinematic input via a spring-damping element (flexible) and a force input (rigid). It has been assumed that the drive acts in points $P^{(1)}$ and $P^{(4)}$.

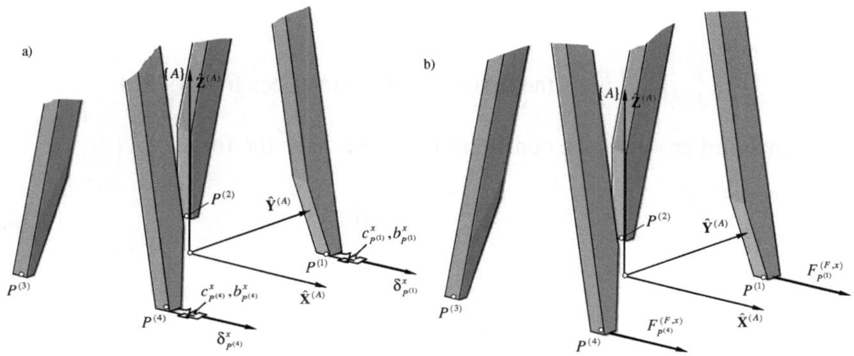

Fig. 9.8. The travel system of the crane: a) flexible, b) rigid

1. Kinematic input

In this case, the potential energy of elastic deformation and the dissipation function of the drive system can be calculated as:

$$\begin{aligned} V_t^{(i)} &= \frac{1}{2} c_{P(i)}^x \left[\delta_{P(i)}^x(t) - x_{P(i)}^{(A)} \right]^2 \\ D_t^{(i)} &= \frac{1}{2} b_{P(i)}^x \left[\dot{\delta}_{P(i)}^x(t) - \dot{x}_{P(i)}^{(A)} \right]^2 \end{aligned} \qquad i=1,4 \qquad (9.27)$$

where $\delta_{P(1)}^x(t)$, $\delta_{P(4)}^x(t)$ – assumed displacement (kinematic input),

$c_{P(i)}^x, b_{P(i)}^x$ – stiffness and damping coefficients of the drive of the travel system, respectively.

After determining coordinates $x_{P(i)}^{(A)}$ as function of elements of vector $\mathbf{q}^{(F)}$, one should place suitable derivatives in the equations of motion of the system.

2. Force input

In the case of force input, the unknown forces $F_{P(1)}^{(F)}$, $F_{P(4)}^{(F)}$ and suitable constrains equations have been introduced. Generally, the forces can be placed on the left side of the equations of motion of the system which can be written as:

$$\mathbf{A}\ddot{\mathbf{q}} - \mathbf{DF} = \mathbf{f}\,, \tag{9.28}$$

where $\mathbf{D} = \begin{bmatrix} \mathbf{0} & \mathbf{0} \\ \mathbf{U}_{p^{(1)},1}^{(F)^T} & \mathbf{U}_{p^{(4)},1}^{(F)^T} \end{bmatrix}$,

$\mathbf{F} = \begin{bmatrix} F_{p^{(1)}}^{(F)} \\ F_{p^{(4)}}^{(F)} \end{bmatrix}$,

$\mathbf{U}_{p^{(1)},1}^{(F)^T}, \mathbf{U}_{p^{(4)},1}^{(F)^T}$ – the first rows of the matrices from (9.25).

In the analysed problem, the constrains equations have the form:

$$x_{p^{(1)}}^{(D)} = \delta_{p^{(1)}}^{x}(t)\,, \tag{9.29.1}$$

$$x_{p^{(4)}}^{(D)} = \delta_{p^{(4)}}^{x}(t)\,. \tag{9.29.2}$$

Due to convenience of the computer implementation, they can be presented in the matrix and acceleration form:

$$\mathbf{D}^T\ddot{\mathbf{q}} = \ddot{\boldsymbol{\delta}} = \begin{bmatrix} \ddot{\delta}_{p^{(1)}}^{x}(t) \\ \ddot{\delta}_{p^{(4)}}^{x}(t) \end{bmatrix}. \tag{9.30}$$

Energy of Elastic Deformation and Energy Dissipation of the Ropes

The load is suspended on two ropes, so their energy of elastic deformation can be written as:

$$V_r^{(p)} = \sum_{p=1}^{2} \frac{1}{2} c_r^{(p)} \delta_r^{(p)} \left[\Delta l_{A_p B_p}^{(p)}\right]^2, \tag{9.31}$$

where $c_r^{(p)}$ – stiffness coefficient of the rope p,

$\Delta l_{A_p B_p}^{(p)}$ – deformation of the rope p,

$$\delta_r^{(p)} = \begin{cases} 0, & \text{when } \Delta l_{A_p B_p}^{(p)} \le 0 \\ 1, & \text{when } \Delta l_{A_p B_p}^{(p)} > 0 \end{cases}.$$

The derivatives of the potential energy of elastic deformations of the ropes have the form:

$$\frac{V_r^{(p)}}{\partial \mathbf{q}^{(F)}} = -c_r^{(p)} \delta_r^{(p)} \frac{\Delta l_{A_p B_p}^{(p)}}{l_{A_p B_p}^{(p)}} \mathbf{U}_{A_p}^{(F)T} \overline{\mathbf{r}}_{A_p B_p}^{(p)}, \tag{9.32.1}$$

$$\frac{V_r^{(p)}}{\partial \mathbf{q}^{(L)}} = -c_r^{(F)} \delta_r^{(p)} \frac{\Delta l_{A_p B_p}^{(p)}}{l_{A_p B_p}^{(p)}} \mathbf{U}_{B_p}^{(L)T} \overline{\mathbf{r}}_{A_p B_p}^{(p)}, \tag{9.32.2}$$

A similar reasoning may be conducted in the case of determining the dependency describing the energy dissipation function:

$$D_r^{(p)} = \sum_{p=1}^{2} \frac{1}{2} b_r^{(p)} \delta_r^{(p)} \left[\Delta \dot{l}_{A_p B_p}^{(p)}\right]^2, \tag{9.33}$$

where $d_r^{(p)}$ – damping coefficients of the rope p.

Hence the formulas:

$$\frac{\partial D_r^{(p)}}{\partial \mathbf{q}^{(F)}} = -b_r^{(p)} \delta_r^{(p)} \mathbf{U}_{A_p}^{(F)T} \dot{\overline{\mathbf{r}}}_{A_p B_p}^{(p)}, \tag{9.34.1}$$

$$\frac{\partial D_r^{(p)}}{\partial \mathbf{q}^{(L)}} = b_r^{(p)} \delta_r^{(p)} \mathbf{U}_{B_p}^{(L)T} \dot{\overline{\mathbf{r}}}_{A_p B_p}^{(p)}. \tag{9.34.2}$$

Taking into consideration all components of the Lagrange equations, we obtain the system of differential equations:

$$\mathbf{A}\ddot{\mathbf{q}} = \mathbf{f}(t, \mathbf{q}, \dot{\mathbf{q}}), \tag{9.35}$$

where $\mathbf{A} = \mathbf{A}(t, \mathbf{q})$ – a mass matrix.

In the case when the input in the drive of the travel system has been modelled as force input, equations (9.35) have to be completed by the constrains equations (9.30) and equations of motion have to be presented in the form (9.28). The fourth order Runge-Kutta method has been used to solve the system of equations.

9.1.2 Example of Numerical Calculations

The presented dynamic model of a BOP gantry allows for comprehensive analyses of the device's operation both under usual working conditions and intense waves. Much detailed discussion is contained in the thesis [Urbaś A., 2011].

In the current book, sample results of numerical simulations for phenomena occurring in a gantry's supporting structure are presented. Masses and geometrical parameters of the crane have been chosen based upon technical documentation

(2007). The main parameters are given below: mass of the frame 110 000 kg, mass of the load 550 000 kg, dimension of the load 4,8 m x 5,5 m x 20,3 m. Data concerning the motion of the deck that should be taken into calculation are also provided in the technical documentation (2007) (Table 9.1). In our simulations, the operational conditions have been assumed.

Table 9.1. Deck motion due to waves

Condition	Heading [deg]	Heave [m]	Pitch [rad]	Roll [rad]
Z1	0	0,1343	0,0023	0
Z2	45	0,1115	0,0008	0,0023
Z3	90	0,1140	0	0,0045

Table 9.2. Load cases analysed - gantry crane not moving

Symbol	Description	Clearance	Deck motion
Z1-M0-C0		0	Z1
Z2-M0-C0	No clearance in travel system	0	Z2
Z3-M0-C0		0	Z3
Z1-M0-C1		1cm	Z1
Z2-M0-C1	With clearance in travel system	1cm	Z2
Z3-M0-C1		1cm	Z3

Calculations for the BOP crane that does not move on the deck have been denoted according to the Table 9.2. The same denotations are used in the graphs.

In Fig. 9.9 there are presented time courses of general coordinates $\psi^{(L)}$ of the load of the BOP crane with and without clearance in the travel system.

The influence of clearance in the travel system for the reaction forces in the support system (the leg no. 1) is shown in Fig 9.10. The deck motions Z2 and Z3 are taken into consideration.

The biggest influence of clearance in the travel system on the dynamics of the BOP crane occurs for input Z3, so this input is taken into account for the next calculations. The influence of clearance in the travel system on the reaction forces will be analyzed. The travel velocity is defined by the relation:

$$\begin{aligned} v &= 3at^2 - 4bt^3 \quad \text{when } t \le T_r, \\ v &= v_n \quad \text{when } t > T, \end{aligned} \tag{9.36}$$

where $v_n = 3\dfrac{\text{m}}{\text{min}}$, $T_r = 6\text{sec}$, $a = \dfrac{v_n}{T_r^2}$, $b = \dfrac{v_n}{2T_r^3}$.

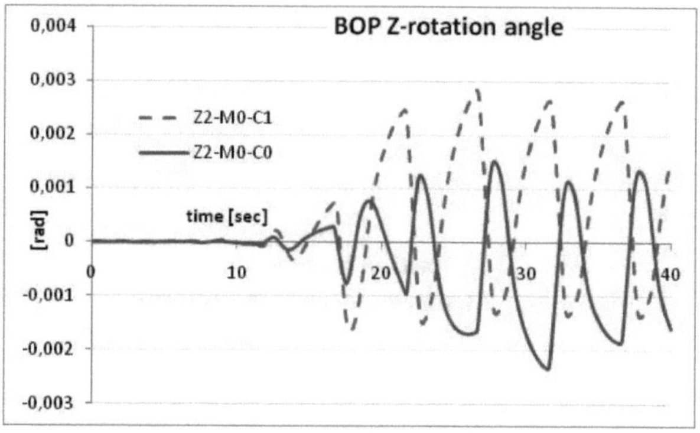

Fig. 9.9. Influence of clearance on roll angle of BOP

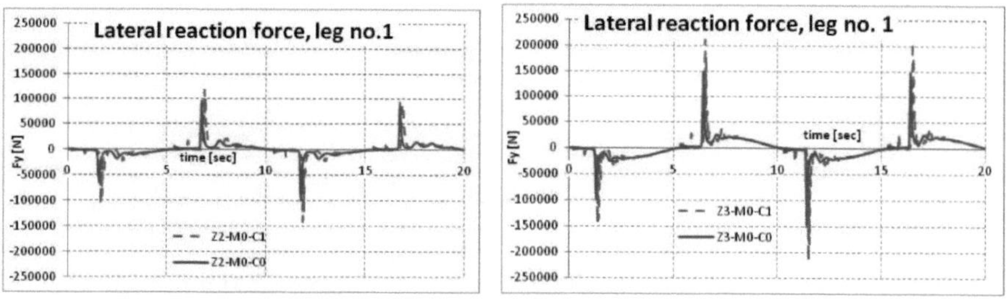

Fig. 9.10. Lateral reaction in leg no.1 (left - in sea conditions Z2, right - in sea conditions Z3)

On the up graph in Fig. 9.11, there is shown the drive force on the first gear (support no. 1) for kinematic and force input and for the case when no clearance occurs in the system. On the bottom graph, there is presented the influence of clearance on the drive force. One can notice that the clearance causes the occurrence of significant dynamic forces of short duration.

Required courses of drive forces acting on the legs 1 and 4 realizing the established travel of the crane are presented in Fig. 9.12. Kinematic and force inputs have been simulated.

The obtained results (values of forces) for assumed parameters are similar, but for kinematic input peak values they are bigger. These values depend on stiffness and damping coefficients taken into account during calculations.

Additional clearance in supporting system for the legs that aren't driven (i. e. 2 and 3), equal to 2 cm, has been taken into consideration, and the results are in Fig 9.13. The obtained values of dynamic forces prove significant influence of clearance on dynamic load of the drive system, the track-way and the whole construction of the crane.

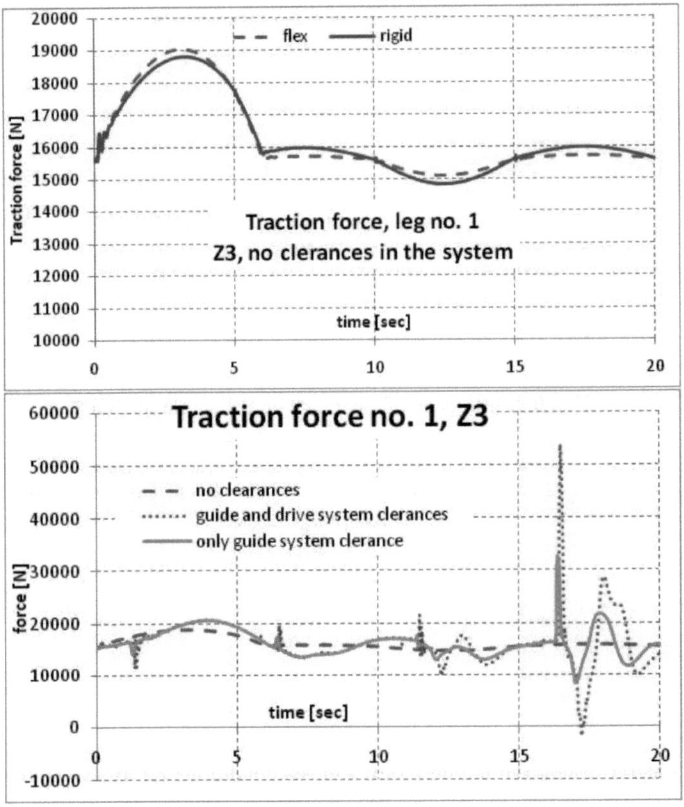

Fig. 9.11. Drive force on gear no. 1 when flexible and rigid model applied (up) influence of different clearances on drive force (bottom)

The mathematical models and the computer programs presented in this section make it possible to execute dynamical analysis of BOP cranes mounted on the floating base. They can be useful in calculating dynamic loads, dimensioning bearing elements of the crane and the track-way. They enable determination of static and dynamic loads by simulation for arbitrarily chosen sea waves conditions.

9.2 Offshore Column Crane with a Shock Absorber

This section discusses a mathematical model of an offshore column crane with a shock absorber [Krukowski J., Maczyński A., 2009], [Krukowski J., Maczyński A., 2010]. As already stated (chapter 2.4.2), in offshore crane design, two kinds of a shock absorber systems are used most commonly. The first one is mounted on

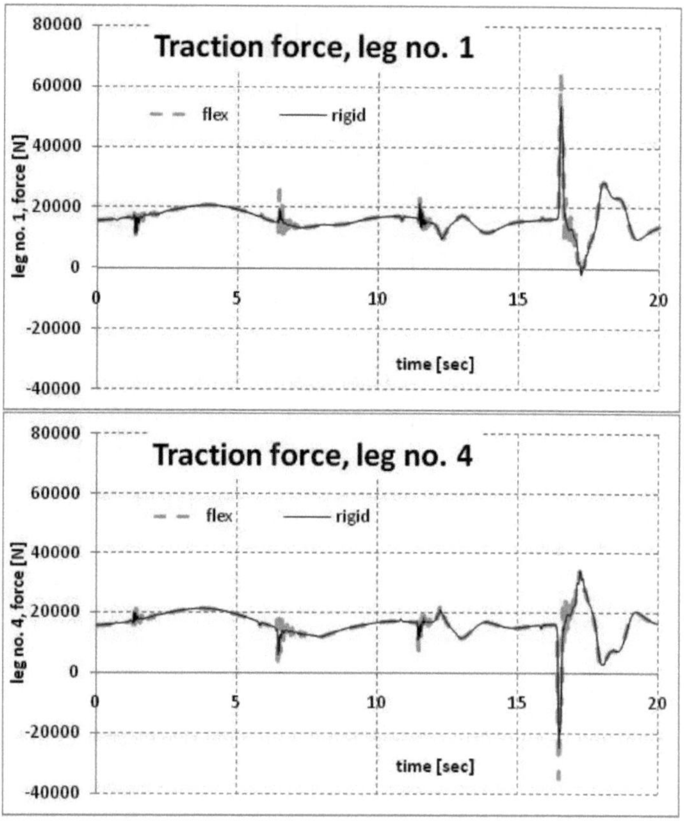

Fig. 9.12. Rigid and flexible drive, leg no. 1 (up) and no. 4 (bottom)

the crane's boom structure. The minimization of dynamic overload is obtained by leading the hoist rope through an additional sliding sheave connected with a hydraulic system. The other one, constituting the system of hydraulic accumulators, is mounted in the hook block.

The subject of the analyses presented in this book is the first type of the shock absorber because of its effectiveness, simple and compact construction. The hydraulic part of the shock absorber is shown in principle in Fig. 2.15. Let us remind that it is the system consisting of accumulator filled with gas and hydraulic cylinder. When force S applied to the piston is big enough, the piston is pulled out and oil is streaming from the cylinder to the accumulator. The end piston stroke length, Δ_2, must normally be shorter than the maximal possible piston stroke length, Δ_{max} minus a safety piston stroke length, Δ_{safe}, to make sure there is no risk for the piston to reach the bottom at normal operation. The safety piston stroke length, Δ_{safe}, shall normally not be less than about 50 mm.

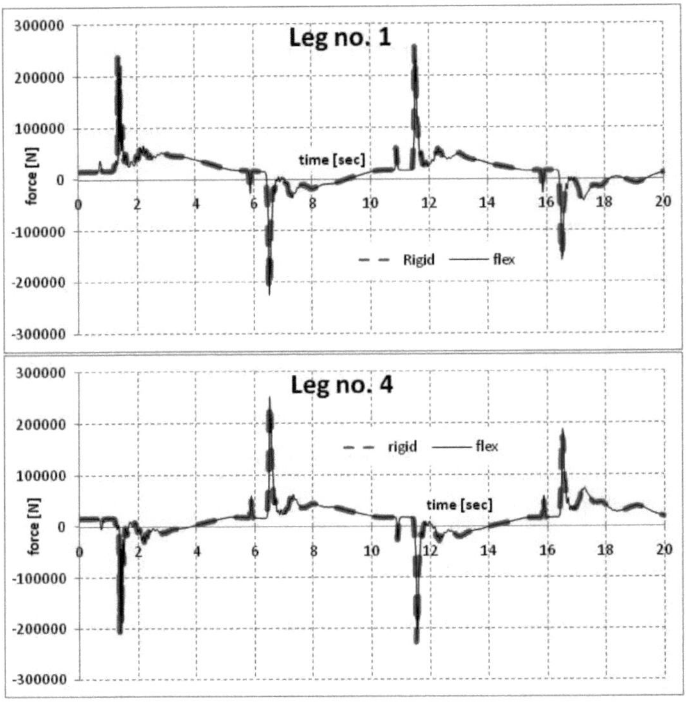

Fig. 9.13. Rigid and flexible drive on legs 1 and 4 - double size clearances in legs 2 and 3

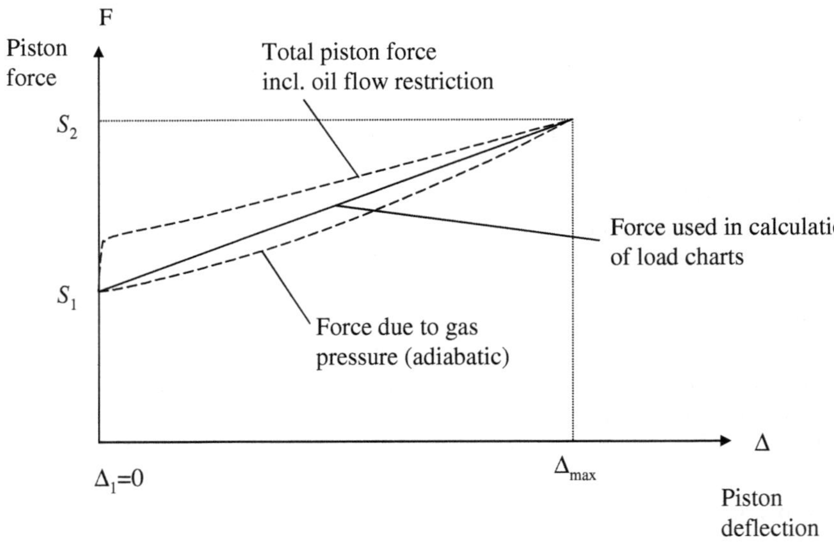

Fig. 9.14. Shock absorber characteristic

The main part of the force S on the piston, is balanced by the gas pressure in the accumulator. In addition, there is some oil pressure drop due to restriction when the oil passes the valves between the piston and the accumulator. This is illustrated in Fig. 9.14.

9.2.1 Model of the Offshore Crane

The subject of this section is the offshore pedestal crane equipped with the system reducing dynamic overload, situated on the boom (Fig. 9.15). The analysed crane type is, according to EN 13852-1 Annex L, the "Lattice boom type crane" or API Spec. 2C, type C. The main assumptions adopted at the design stage and the most important connections used during the derivation of equation of motion will be given below. Modelling the shock absorber was particularly emphasised. For the description of the system, joint coordinates and homogenous transformations were used based on Denavit-Hartenberg convention. The equations of motion were obtained using the Rigid Finite Element Method and the Lagrange equations of the second order.

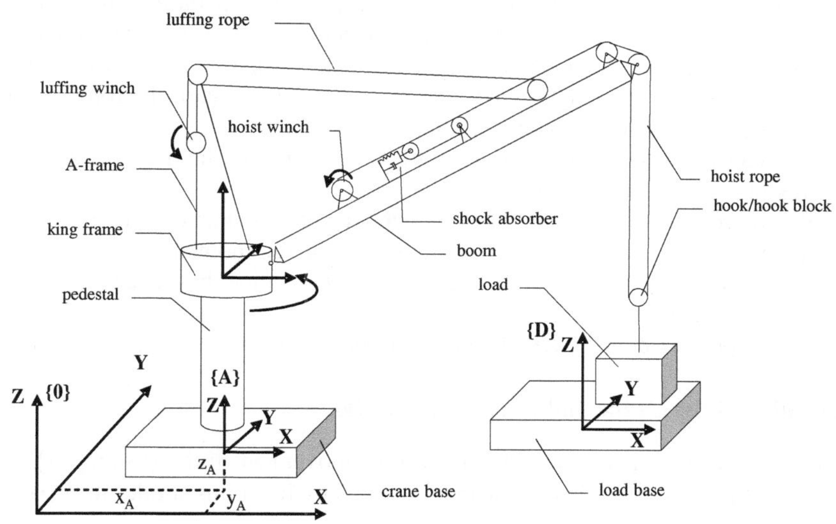

Fig. 9.15. Scheme of an offshore pedestal crane

While preparing the model, the following assumptions and subsystems were taken into consideration:

- the base of the crane (the platform of vessel) is a rigid body with 6 degrees of freedom; the movement is caused by the sea waves defined by pseudo-harmonic functions,

- the pedestal is modelled by means of the Rigid Finite Element Method using a modified approach (MRFEM) – chapter 8.2; hence, the flexibility of the pedestal could be included but also the treatment of the pedestal as a rigid structure is possible,
- the king frame, including the slewing part, is treated as a rigid structure with one degree of freedom with respect to the pedestal – the slew angle,
- the A-frame is modelled by means of MRFEM as a simplified, one-beam system having bending flexibility in the perpendicular direction to the A-frame plane; similarly to pedestal model, the A-frame can be treated as a rigid subsystem,
- the boom is modelled as a continuous system by means of the MFREM,
- the basic element of the shock absorber is the hydraulic cylinder, which is modelled as point mass (additionally including the mass of the moving sheave) connected to the boom by means of a spring damping system; the mass may slide only along the longitudinal axis of the boom; it is assumed that the characteristics of the spring is nonlinear,
- the hoist rope is modelled as a massless element with equivalent longitudinal flexibility; the damping is taken into account, with the assumption it is viscous, and that the damping coefficient has a constant value; with regards to significant changes of the hoist rope during crane operations, the value of rope stiffness coefficient has been made depended on the current rope length,
- the luffing rope is modelled similarly to the hoist rope; as a matter of fact that change of the rope length during crane operations is small, the rope stiffness coefficient is assumed to be constant,
- the load is treated as a material point; its contact with the deck of the supply vessel is taken into account,
- the drive function of the hoist winch can be assumed in two ways: as a kinematic excitation or force excitation by a given moment,
- the luffing winch drive and the slew of the crane has been adopted as a kinematic drive,
- the supply vessel is modelled identical to the crane base.

Modelling of the Crane and Cargo Base Motion

It is assumed that crane base motion and thus movement of the system $\{A\}$ with respect to the system $\{0\}$ is known and described by functions similar to (9.3):

$$y_i^{(A)} = \sum_{j=1}^{n_i^{(A)}} A_{i,j}^{(A)} \sin(\omega_{i,j}^{(A)} t + \varphi_{i,j}^{(A)}); \quad i = 1,...,6, \tag{9.37}$$

where $A_{i,j}^{(A)}, \omega_{i,j}^{(A)}, \varphi_{i,j}^{(A)}$ – respectively: amplitude, phase, frequency and forcing phase angle,

$n_i^{(A)}$ – number of harmonic series.

Movement of the cargo base, i.e. that of system $\{A\}$, will be described in the same way.

In further considerations, the coordinate system $\{0\}$ will be identified with the inertial coordinate system $\{\}$ and the following notation will be used for the homogeneous transformation matrix from coordinate system $\{p\}$ to the coordinate system $\{0\}$:

$$ {}_{p}^{0}\mathbf{T} = \mathbf{T}^{(p)}, \tag{9.38} $$

where p is the number of the member in the kinematic chain.

Homogeneous transformation matrix ${}_{A}^{0}\mathbf{T}$, taking into account the motion of the system $\{A\}$ in $\{\}$, can be presented as a product of six matrices, each being the function of only one variable dependent on time as described in chapter 5. It is to be noticed that if $\tilde{\mathbf{r}} = [\tilde{x} \quad \tilde{y} \quad \tilde{z} \quad 1]^T$ is a vector describing coordinates of the dm mass (point) in the local system $\{\}$', connected to any part of the system, the coordinates of such mass in the system $\{\}$ may be described with the equitation:

$$ \mathbf{r} = {}_{A}^{0}\mathbf{T}(t)\,\overline{\mathbf{T}}(\mathbf{q})\,\mathbf{r}' = \mathbf{T}\,\mathbf{r}', \tag{9.39} $$

where $\overline{\mathbf{T}}(\mathbf{q}) = {}_{\{\}'}^{A}\mathbf{T}(q_1,\ldots,q_n)$ – transformation matrix of coordinates from local coordinate system $\{\}$' into the $\{A\}$ coordinate system, dependent on the generalized coordinates $(q_1,\ldots,q_n)$ of the body,

$\mathbf{T} = {}_{A}^{0}\mathbf{T}(t)\overline{\mathbf{T}}(\mathbf{q})$.

Crane Pedestal

As mentioned before, the crane pedestal was discretized by means of MRFEM. The number of rigid finite elements, on which the pedestal was divided, equals n_1+1, where the first rigid finite element of the pedestal, RFE (1,0), is added to the vessel body. The generalized coordinates, describing the location of the second and other rigid elements modelling the pedestal with respect to its predecessors (coordinates describing mutual location of the rigid finite elements some times called flexible or elastic coordinates), may be presented as vectors:

$$ \tilde{\mathbf{q}}^{(1,i)} = \left[\varphi_x^{(1,i)} \quad \varphi_y^{(1,i)} \quad \varphi_z^{(1,i)}\right]^T = \left[\tilde{q}_x^{(1,i)} \quad \tilde{q}_y^{(1,i)} \quad \tilde{q}_z^{(1,i)}\right]^T, \tag{9.40} $$

where $\varphi_x^{(1,i)}$, $\varphi_y^{(1,i)}$, $\varphi_z^{(1,i)}$ are the rotation angles presented in Fig. 9.16.

The vector of generalized coordinates of the RFE yields:

$$ \mathbf{q}^{(1,1)} = \tilde{\mathbf{q}}^{(1,1)} = \left[q_1^{(1,1)} \quad q_2^{(1,1)} \quad q_3^{(1,1)}\right]^T, \tag{9.41.1} $$

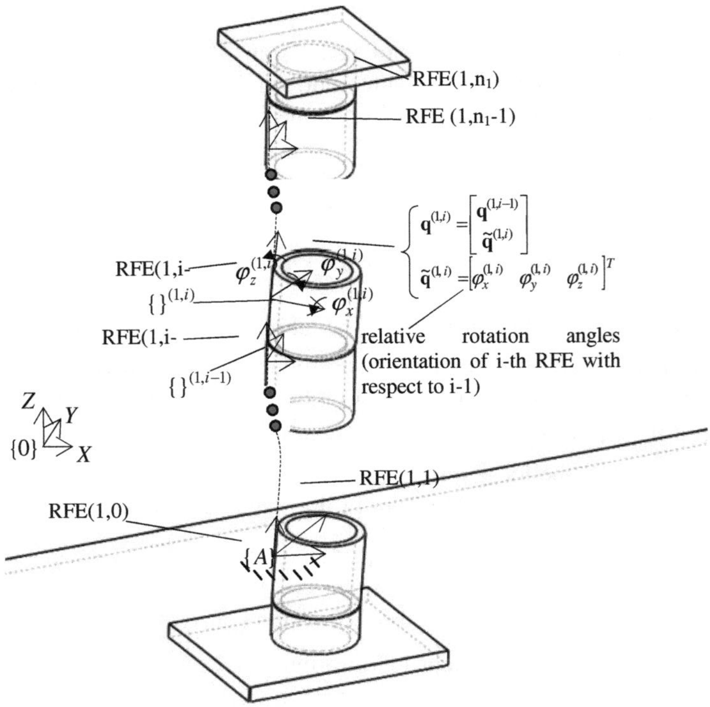

Fig. 9.16. Pedestal discretized by mean of MRFEM

$$\mathbf{q}^{(1,i)} = \begin{bmatrix} \mathbf{q}^{(1,i-1)} \\ \tilde{\mathbf{q}}^{(1,i)} \end{bmatrix} = \begin{bmatrix} q_1^{(1,i)} & q_2^{(1,i)} & \ldots & q_{3i}^{(1,i)} \end{bmatrix}^T \quad \text{for } i = 2,\ldots, n_1 . \tag{9.41.2}$$

In accordance to above consideration, during the derivation of the equations of motions, kinetic and potential energy of the RFE (1,0) have been omitted. On the basis of equations presented in previous sections, the kinetic energy of the body discretised by the MRFEM can be calculated as:

$$E_1 = \sum_{i=1}^{n_1} E_{(1,i)} , \tag{9.42}$$

where $E_{(1,i)} = tr\left\{\dot{\mathbf{T}}^{(1,i)} \mathbf{H}^{(1,i)} \dot{\mathbf{T}}^{(1,i)T}\right\}$,

$\mathbf{H}^{(1,i)}$ – inertia matrix of the RFE $(1,i)$ defined in its own coordinate system,

$\mathbf{T}^{(1,i)}$ – transformation matrix from coordinate system $\{1,i\}$ of RFE $(1,i)$ into the coordinate system $\{\}$,

$$\mathbf{T}^{(1,i)} = \mathbf{T}^{(1,i-1)}\,\tilde{\mathbf{T}}^{(1,i)} = {}^{0}_{A}\mathbf{T}\,\tilde{\mathbf{T}}^{(1,0)}\tilde{\mathbf{T}}^{(1,1)}\cdot\ldots\cdot\tilde{\mathbf{T}}^{(1,i-1)}\,\tilde{\mathbf{T}}^{(1,i)} \quad \text{for } i = 1,\ldots, n_1,$$

$\tilde{\mathbf{T}}^{(1,i)}$ – transformation matrix from coordinate system of RFE (1,i) into coordinate system of RFE (1,i-1).

For the Lagrange equations of the second order, the concept of Lagrange operators is introduced:

$$\varepsilon_i(E) = \frac{d}{dt}\frac{\partial E}{\partial \dot{q}_i} - \frac{\partial E}{\partial q_i}. \tag{9.43}$$

Such operators for other RFE (1,*i*) (*i*=1,....,) of the pedestal, can be written in the vector form as:

$$\boldsymbol{\varepsilon}_{\mathbf{q}^{(1,i)}}\left(E_{(1,i)}\right) = \mathbf{A}^{(1,i)}\ddot{\mathbf{q}}^{(1,i)} + \mathbf{e}^{(1,i)}, \tag{9.44}$$

where $\mathbf{A}^{(1,i)} = \left(a_{k,j}^{(1,i)}\right)_{k,j=1,\ldots,3n_1} = \operatorname{tr}\left\{\mathbf{T}_k^{(1,i)}\,\mathbf{H}^{(1,i)}\,\mathbf{T}_j^{(1,i)T}\right\}$,

$$\mathbf{e}^{(1,i)} = \left(e_k^{(1,i)}\right)_{k=1,\ldots,3n_1} = \sum_{j=1}^{3n_1}\sum_{l=1}^{3n_1}\operatorname{tr}\left\{\mathbf{T}_k^{(1,i)}\,\mathbf{H}^{(1,i)}\mathbf{T}_{j,l}^{(1,i)}\right\}\dot{q}_j^{(1,i)}\,\dot{q}_l^{(1,i)} +$$

$$+\operatorname{tr}\left\{\mathbf{T}_k^{(1,i)}\mathbf{H}^{(1,i)}\left[{}^{0}_{A}\ddot{\mathbf{T}}\overline{\mathbf{T}}^{(1,i)} + 2\,{}^{0}_{A}\dot{\mathbf{T}}\dot{\overline{\mathbf{T}}}^{(1,i)}\right]^T\right\},$$

$$\overline{\mathbf{T}}^{(1,i)} = \prod_{j=0}^{i}\tilde{\mathbf{T}}^{(1,j)},$$

$$\mathbf{T}_k^{(1,i)} = \frac{\partial \mathbf{T}^{(1,i)}}{\partial q_k^{(1,i)}},$$

$$\mathbf{T}_{j,l}^{(1,i)} = \frac{\partial}{\partial q_j^{(1,i)}}\left(\frac{\partial \mathbf{T}^{(1,i)}}{\partial q_l^{(1,i)}}\right).$$

The potential energy due to gravity forces of the pedestal's rigid finite elements is described by the relation:

$$V_{(1,i)}^{g} = m^{(1,i)}\,g\,\boldsymbol{\theta}_3\mathbf{T}^{(1,i)}\tilde{\mathbf{r}}_C^{(1,i)} \qquad \text{for } i=1,2,\ldots, n_1, \tag{9.45}$$

where $m^{(1,i)}$ – mass of the RFE (1,*i*),

$\tilde{\mathbf{r}}_C^{(1,i)}$ – vector of the mass centre of RFE (1,*i*) expressed in its own coordinate system.

The corresponding derivatives, which are the elements of the Lagrange equations, are:

$$\frac{\partial V_{(1,i)}^{g}}{\partial \mathbf{q}^{(1,i)}} = \mathbf{G}^{(1,i)}, \tag{9.46}$$

where $\mathbf{G}^{(1,i)} = \left(g_k^{(1,i)}\right)_{k=1,\dots,3n_1}$,

$g_k^{(1,i)} = m^{(1,i)} g\, \mathbf{\theta}_3 \mathbf{T}_k^{(1,i)} \tilde{\mathbf{r}}_C^{(1,i)}$.

It is known that in MRFEM the successive RFE are connected with each other by means of massless, elasto-damping elements (SDE). Potential energy of the elastic deformation SDE (1,*i*) is defined as follows:

$$V_{(1,i)}^{s} = \frac{1}{2}\left(c_{i,x}^{(1)}\left[\varphi_x^{(1,i)}\right]^2 + c_{i,y}^{(1)}\left[\varphi_y^{(1,i)}\right]^2 + c_{i,z}^{(1)}\left[\varphi_z^{(1,i)}\right]^2\right) = \frac{1}{2}\sum_{j=1}^{3} c_{i,j}^{(1)}\left[\tilde{q}_j^{(1,i)}\right]^2, \tag{9.47}$$

where $c_{i,x}^{(1)} = c_{i,1}^{(1)}$, $c_{i,y}^{(1)} = c_{i,2}^{(1)}$, $c_{i,z}^{(1)} = c_{i,3}^{(1)}$ where the adequate coefficients of the rotational stiffness of SDE (1,*i*).

Equation (9.47) can be presented in the form:

$$V_{(1,i)}^{s} = \frac{1}{2}\tilde{\mathbf{q}}^{(1,i)T} \mathbf{C}^{(1,i)}\, \tilde{\mathbf{q}}^{(1,i)}, \tag{9.48}$$

where $\mathbf{C}^{(1,i)} = \begin{bmatrix} c_{i,1}^{(1)} & 0 & 0 \\ 0 & c_{i,2}^{(1)} & 0 \\ 0 & 0 & c_{i,3}^{(1)} \end{bmatrix}$.

The required derivatives of the potential energy of elastic deformation with respect to generalized coordinates $\tilde{\mathbf{q}}^{(1,i)}$, have a simple form:

$$\frac{\partial V_{(1,i)}^{s}}{\partial \tilde{\mathbf{q}}^{(1,i)}} = \mathbf{C}^{(1,i)} \tilde{\mathbf{q}}^{(1,i)}. \tag{9.49}$$

It may additionally be assumed that in SDE (1,*i*) dissipation of the energy appears, which is described by means of equations:

$$D_{(1,i)} = \frac{1}{2}\left(b_{i,x}^{(1)}\left[\dot{\varphi}_x^{(1,i)}\right]^2 + b_{i,y}^{(1)}\left[\dot{\varphi}_y^{(1,i)}\right]^2 + b_{1,z}^{(1)}\left[\dot{\varphi}_z^{(1,i)}\right]^2\right) = \frac{1}{2}\sum_{j=1}^{3} b_{1,j}^{(1)}\left[\dot{\tilde{q}}_j^{(1,i)}\right]^2, \tag{9.50}$$

where $b_{i,x}^{(1)} = b_{i,1}^{(1)}$, $b_{i,y}^{(1)} = b_{i,2}^{(1)}$, $b_{i,z}^{(1)} = b_{i,3}^{(1)}$ are respective damping coefficients of the SDE (1,*i*).

Equation (9.50) may be also written as:

$$D_{(1,i)} = \frac{1}{2} \dot{\tilde{\mathbf{q}}}^{(1,i)^T} \mathbf{B}^{(1,i)} \dot{\tilde{\mathbf{q}}}^{(1,i)}, \tag{9.51}$$

where $\mathbf{B}^{(1,i)} = \begin{bmatrix} b_{i,1}^{(1)} & 0 & 0 \\ 0 & b_{i,2}^{(1)} & 0 \\ 0 & 0 & b_{i,3}^{(1)} \end{bmatrix}$,

and the adequate derivatives can be obtained from:

$$\frac{\partial D_{(1,i)}}{\partial \dot{\tilde{\mathbf{q}}}^{(1,i)}} = \mathbf{B}^{(1,i)} \dot{\tilde{\mathbf{q}}}^{(1,i)}. \tag{9.52}$$

King Frame/Slewing Part

Let us define the following vector of generalized coordinates for the slewing part:

$$\mathbf{q}^{(2)} = \begin{bmatrix} \mathbf{q}^{(1,n_1)} \\ \varphi_z^{(2)} \end{bmatrix} = \begin{bmatrix} q_1^{(2)} & q_2^{(2)} & \ldots & q_{n_2}^{(2)} \end{bmatrix}^T, \tag{9.53}$$

where $\varphi_z^{(2)}$ symbolizes the angle of rotation of the slewing part with respect to the pedestal.

The kinetic energy of the slewing part can be described as:

$$E_2 = tr\left\{\dot{\mathbf{T}}^{(2)} \mathbf{H}^{(2)} \dot{\mathbf{T}}^{(2)^T}\right\}, \tag{9.54}$$

where $\mathbf{H}^{(2)}$ – the inertial matrix of the slewing part.

Lagrange operators for the slewing part are formulated in the form:

$$\boldsymbol{\varepsilon}_{\mathbf{q}^{(2)}}(E_2) = \mathbf{A}^{(2)} \ddot{\mathbf{q}}^{(2)} + \mathbf{e}^{(2)}, \tag{9.55}$$

where $\mathbf{A}^{(2)} = \left(a_{i,l}^{(2)}\right)_{i,l=1,\ldots,n_2} = \mathrm{tr}\left\{\mathbf{T}_i^{(2)} \mathbf{H}^{(2)} \mathbf{T}_l^{(2)^T}\right\}$,

$$\mathbf{e}^{(2)} = \left(e_i^{(2)}\right)_{i=1,\ldots,n_2} = \mathrm{tr}\left\{\mathbf{T}_i^{(2)} \mathbf{H}^{(2)} \left[{}_A^0\ddot{\mathbf{T}} \overline{\mathbf{T}}^{(2)} + 2\, {}_A^0\dot{\mathbf{T}} \sum_{j=1}^{n_2} \overline{\mathbf{T}}_j^{(2)} \dot{q}_j^{(2)} + \right.\right.$$

$$\left.\left. + \sum_{l=1}^{n_2} \sum_{j=l}^{n_2} \delta_{l,j} \mathbf{T}_{l,j}^{(2)} \dot{q}_l^{(2)} \dot{q}_j^{(2)} \right]^T \right\},$$

$$\delta_{l,j} = \begin{cases} 1 & \text{for} \quad l = j \\ 2 & \text{for} \quad l \neq j \end{cases},$$

$$\overline{\mathbf{T}}^{(2)} = \tilde{\mathbf{T}}^{(1,0)}\,\tilde{\mathbf{T}}^{(1,1)} \cdot \ldots \cdot \tilde{\mathbf{T}}^{(1,n_1)} \cdot {}^{(1,n_1)}_{\ \ 2}\mathbf{T},$$

${}^{(1,n_1)}_{\ \ 2}\mathbf{T}\left(\varphi_z^{(2)}\right)$ – the transformation matrix form coordinate system of slewing part {2} to the last rfe coordinate system of pedestal $\{1,n_1\}$.

Potential energy of the gravity forces of the slewing body equals:

$$V_2^g = m^{(2)} g\, \boldsymbol{\theta}_3 \mathbf{T}^{(2)} \tilde{\mathbf{r}}_C^{(2)}, \tag{9.56}$$

where $m^{(2)}$ – mass of the slewing part,

$\tilde{\mathbf{r}}_C^{(2)}$ – position vector of the center of slewing part mass, expressed in the system {2}.

The necessary derivatives are defined bellow:

$$\frac{\partial V_2^g}{\partial \mathbf{q}^{(2)}} = \mathbf{G}^{(2)}, \tag{9.57}$$

where $\mathbf{G}^{(2)} = \left(g_k^{(2)}\right)_{k=1,\ldots,n_2}$,

$$g_k^{(2)} = m^{(2)} g\, \boldsymbol{\theta}_3 \mathbf{T}_k^{(2)} \tilde{\mathbf{r}}_C^{(2)}.$$

A-Frame and Boom

As mentioned above, the A-frame is modelled by means of the MRFEM in compliance with only bending flexibility in the perpendicular direction to the plane of the A-frame. Additionally, as for the pedestal, RFE (3,0) is added to the slewing part, and as a result, it does not have its own generalized coordinates (Fig. 9.17). Consequently, the following vectors of generalized coordinates for each rfe of the A-frame are defined:

- one-element vectors of the flexible coordinates:

$$\tilde{\mathbf{q}}^{(3,1)} = \left[\varphi_y^{(3,1)}\right]; \ldots; \tilde{\mathbf{q}}^{(3,n_3)} = \left[\varphi_y^{(3,n_3)}\right], \tag{9.58}$$

- coordinate vectors describing position of the rigid element with respect to the base coordinate system:

$$\mathbf{q}^{(3,i)} = \begin{bmatrix} \mathbf{q}^{(2)} & \tilde{\mathbf{q}}^{(3,1)} & \ldots & \tilde{\mathbf{q}}^{(3,i)} \end{bmatrix}^T = \begin{bmatrix} q_1^{(3,i)} & \ldots & q_{n_2+i}^{(3,i)} \end{bmatrix}^T \quad \text{for } i=1,2,\ldots,\ n_3. \tag{9.59}$$

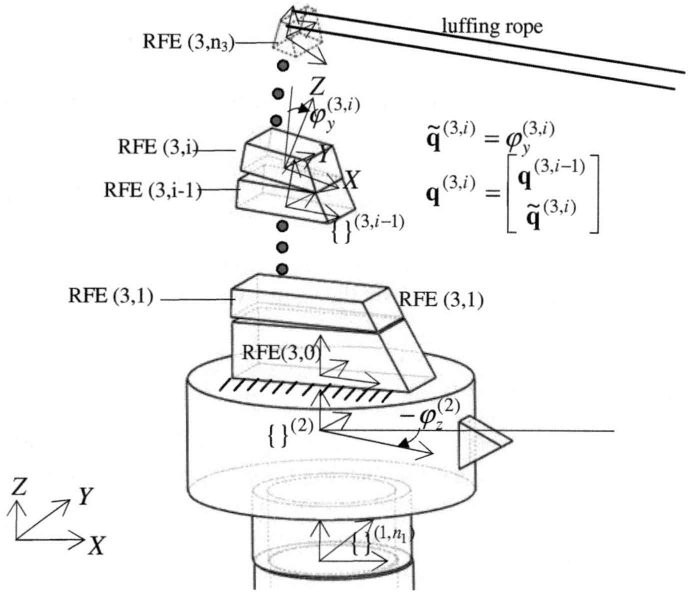

Fig. 9.17. Simplified model of flexible A-frame

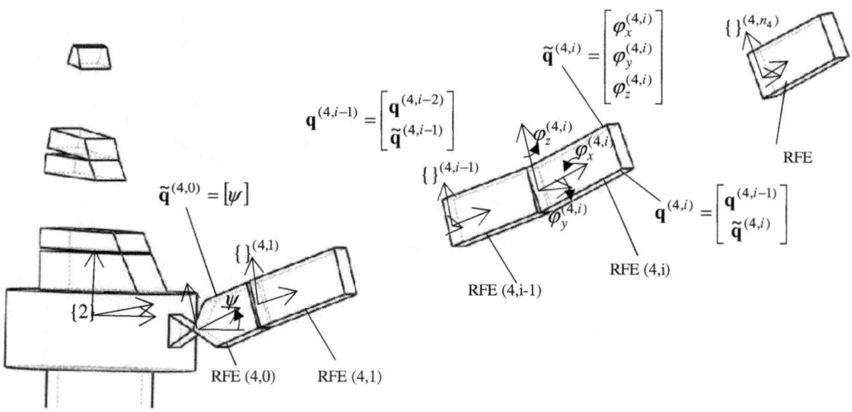

Fig. 9.18. Flexible boom discretized by mean of the MRFEM

In contrast to the pedestal and A-frame, in the case of boom it was assumed that there is a rotational connection defined by the boom angle ψ between the rotating part {2} and the RFE (4,0) (Fig 9.18). We can define the following vectors of generalized coordinates for the boom:

- vectors of the rigid and flexible coordinates:

$$\tilde{\mathbf{q}}^{(4,0)} = [\psi] = \left[\varphi_y^{(4,0)}\right];...;\ \tilde{\mathbf{q}}^{(4,i)} = \left[\varphi_x^{(4,i)} \quad \varphi_y^{(4,i)} \quad \varphi_z^{(4,i)}\right]^T \text{ for } i=1,2,...,\ n_4, \quad (9.60)$$

- coordinate vectors describing position of the rigid element with respect to the base coordinate system:

$$\mathbf{q}^{(4,i)} = \left[\mathbf{q}^{(2)T} \quad \tilde{\mathbf{q}}^{(4,0)T} \quad ... \quad \tilde{\mathbf{q}}^{(4,i)T}\right]^T = \left[q_1^{(4,i)} \quad ... \quad q_{3i+n_2+1}^{(4,i)}\right]^T \text{ for } i=0,1,...,\ n_4. \quad (9.61)$$

The necessary elements of the Lagrange equations related to the A-frame and boom subsystems were calculated in the same way as presented for the crane pedestal.

The Model of Shock Absorber

The model of shock absorber is presented in Fig. 9.19. Its basic element is sheave (3) possessing the mass m_{sA} mounted to the boom by means of a parallel spring-damping system. Relative motion of the sheave (3) is possible only along the longitudinal axis of the boom. The mass m_{sA} is enlarged due to movable parts of the hydraulic cylinder.

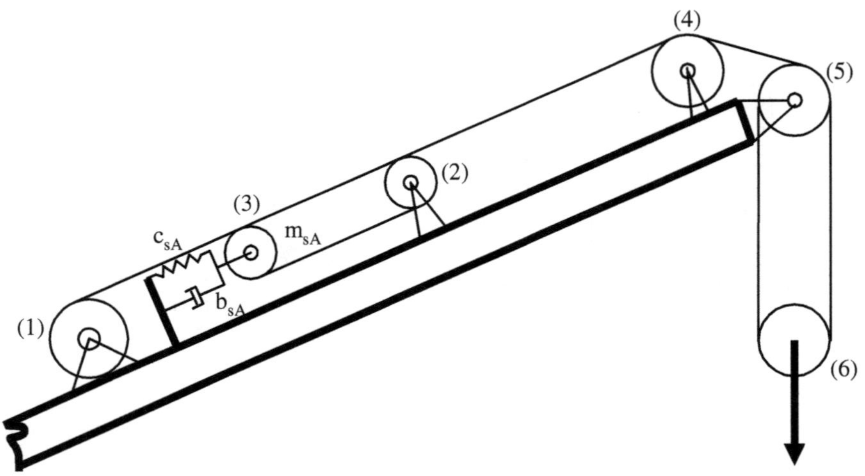

Fig. 9.19. Model of the shock absorber

The shock absorber is activated only if the hoist rope tension reaches specific value (and does not exceed design limits). Usually, in practical hand calculations, one assumes a multilinear characteristic (Fig. 9.14), but its first derivative is not determinable in points defining the working range. This is unfavourable in numerical simulations.

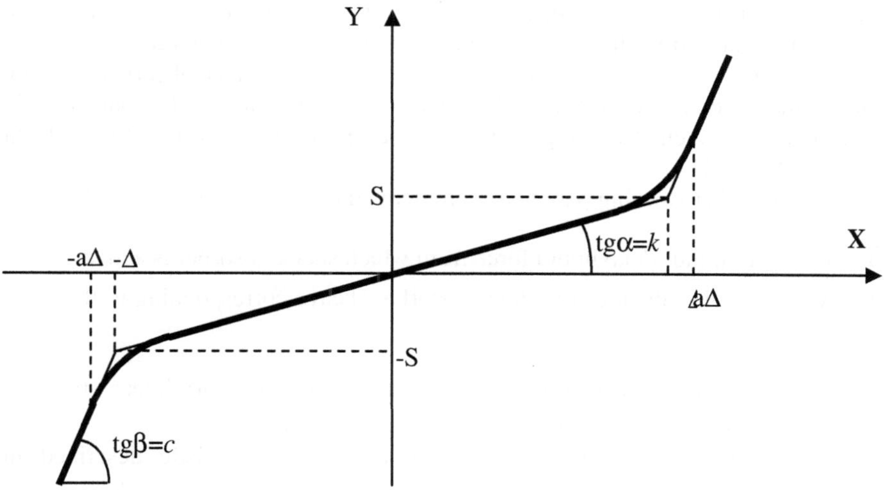

Fig. 9.20. Characteristic of an elastic element $c_{sA}=c_{sA}(x)$

In the presented model, characteristic shown in Fig. 9.20 was assumed [Krukowski J., Maczyński A., 2011]. It represents the characteristic of elastic element $c_{sA}=c_{sA}(x)$, and does not take into account the situation when the shock absorber sheave (3) is fixed to the boom structure. The curve given in Fig. 9.20 can be described as follow:

$$y=\begin{cases} S+c(x-\Delta) & \text{for} \quad x\geq a\Delta \\ kx+\alpha x^2 e^{\beta x} & \text{for} \quad 0\leq x< a\Delta \\ -S+c(x-\Delta) & \text{for} \quad x\leq -a\Delta \\ kx-\alpha x^2 e^{-\beta x} & \text{for} \quad -a\Delta < x < 0 \end{cases} . \tag{9.62}$$

By selecting appropriate values of α and β, one obtains a smooth transition curves at the point $x = a\Delta$ (and $x = -a\Delta$). Then, the following conditions must be fulfilled:

$$ka\Delta+\alpha(a\Delta)^2 e^{\beta a\Delta} = S + c(a\Delta-\Delta) = k\Delta + c(a-1)\Delta\,, \tag{9.63.1}$$

$$k+2\alpha a\Delta e^{\beta a\Delta}+\alpha a^2\Delta^2\beta e^{\beta a\Delta}=c\,. \tag{9.63.2}$$

After some transformation, parameters α and β are obtained:

$$\beta=\frac{2-a}{a\Delta(a-1)}, \tag{9.64.1}$$

$$\alpha=\sqrt{\frac{(a-1)(c-k)}{a^2\Delta e^{\beta a\Delta}}}\,. \tag{9.64.2}$$

The shock absorber is designed in such a way that it works only under the tensioning load. Up to the value of force S_1, the stiffness has a very high value. Within the limit of forces S_1 to S_2, the stiffness decreases (shock absorber working range), and beyond force S_2 the stiffness increases significantly. The characteristic of an elastic element from Fig. 9.20 must be appropriately scaled to the form shown in Fig. 9.21.

Shock absorber working parameters are defined by the following variables:

S_1, S_2 – minimum/maximum force from which shock absorber is active,

Δ_1, Δ_2 – displacement of the shock absorber sheave corresponding to the force S_1, S_2,

a – parameter specifying where the point of curvilinear part of characteristic is becoming rectilinear, $\alpha > 1$,

α, β – parameters defining the shape of the characteristic described in equation (9.64).

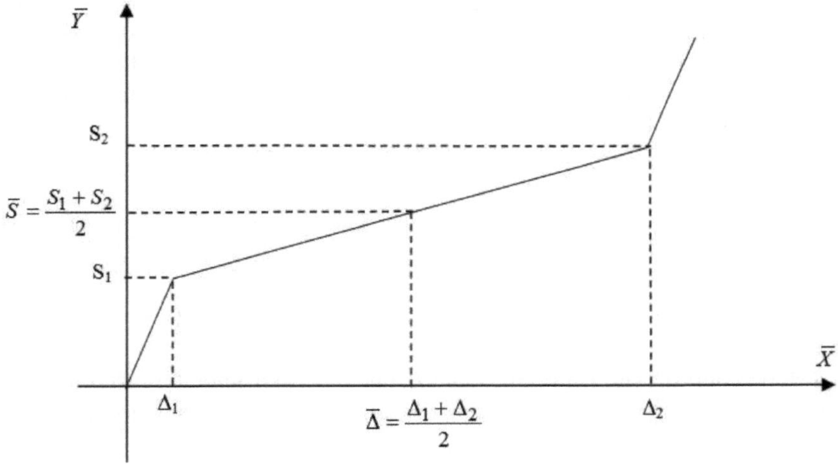

Fig. 9.21. Characteristic of shock absorber

From Fig. 9.21 it is easy to read, that the constans c and k are described by means of:

$$c = \frac{S_1}{\Delta_1}, \tag{9.65.1}$$

$$k = \frac{S_2 - S_1}{\Delta_2 - \Delta_1}, \tag{9.65.2}$$

and the values of x and y are determined as:

$$\begin{cases} y = \overline{y} - \overline{S} \\ x = \overline{x} - \overline{\Delta} \end{cases}. \tag{9.66}$$

where $\overline{x}$, $\overline{y}$, $\overline{\Delta}$ and $\overline{S}$ are shown in Fig. 9.21.

Hoisting and Luffing Ropes

The potential energy of elastic deformation and function of dissipation energy of the hoist rope and luffing rope can be described by the following equations:

$$V_l = \frac{1}{2}\delta c^{(l)} \Delta_l^2, \tag{9.67}$$

$$D_l = \frac{1}{2}\delta b^{(l)} \dot{\Delta}_l^2, \tag{9.68}$$

where $\delta = \begin{cases} 0 & \text{for } \Delta_l \le 0 \\ 1 & \text{for } \Delta_l > 0 \end{cases}$,

Δ_l – elongation of the hoist rope or luffing rope,

$c^{(l)}, b^{(l)}$ – stiffness and damping coefficients of rope, respectively.

Because of the possibility of the significant changes in the active length of the hoist rope during crane operation, the stiffness coefficient of the hoist rope is determining by means of:

$$c^{(l)} = \frac{E_6 F_6}{L_{6,0} - \alpha_{(6)} r_{(6)}}, \tag{9.69}$$

where $L_{6,0}$ – the initial length of hoist rope,

E_6 – Young's modulus of the wire rope core,

F_6 – cross section of the wire rope,

$\alpha_{(6)}$ – rotation angle of the hoist winch drum,

$r_{(6)}$ – radius of the hoist winch drum.

The stiffness coefficient $c^{(l)}$ of the luffing rope is considered as a constant value. A method for determining the necessary derivatives of equations (9.67) and (9.68) was described in chapter 9.1.

Load

The load was modelled as a material point. The weight of the hook block was added to the weight of the load. The vector of the generalized coordinates is defined as:

$$\mathbf{q}^{(L)} = \begin{bmatrix} x^{(L)} & y^{(L)} & z^{(L)} \end{bmatrix}^T = \begin{bmatrix} q_1^{(L)} & q_2^{(L)} & q_3^{(L)} \end{bmatrix}^T . \tag{9.70}$$

The kinetic and potential energy of the load are described by means of:

$$E_L = \frac{1}{2} m^{(L)} \left(\dot{x}^{(L)2} + \dot{y}^{(L)2} + \dot{z}^{(L)2} \right), \tag{9.71}$$

$$V_L^g = m^{(L)} \, g \, z^{(L)} \, , \tag{9.72}$$

where $m^{(L)}$ is the mass of the load.

On this basis, it is possible to write:

$$\boldsymbol{\varepsilon}_{\mathbf{q}^{(L)}} = \mathbf{A}^{(L)} \cdot \ddot{\mathbf{q}}^{(L)}, \tag{9.73}$$

$$\frac{\partial V_L^g}{\partial \mathbf{q}^{(L)}} = \begin{bmatrix} 0 & 0 & m^{(L)} g \end{bmatrix}^T, \tag{9.74}$$

where $\mathbf{A}^{(L)} = diag \left[m^{(L)}, m^{(L)}, m^{(L)} \right]$.

The developed computer software allows us to simulate the following cases:

1. load is in the air (water) – does not remain in contact with the deck of a supply vessel,
2. load remains stationary on board of the supply vessel; its coordinates are defined by the motion of the supply vessel,
3. load can by frozen to the deck, or other reason cause that the load is permanently connected to the supply vessel.

Drive Systems

Slewing, hoisting and luffing drive systems are modelled as the kinematic inputs. Therefore, the following function is known:

$$\phi_d = \phi_d(t), \tag{9.75}$$

where ϕ_d is respectively: slewing angle, hosting winch or luffing winch rotation angle.

From the perspective of planned applications of the presented model, the hosting machinery is one of the most significant drive system. Therefore, a second method of its modelling, using the force excitation, has been developed. Based on the analysis of literature (for example [Osiński M. et al., 2004]) as well as experience acquired from crane operators and designers, the hoist winch characteristic was assumed as shown in Fig. 9.22.

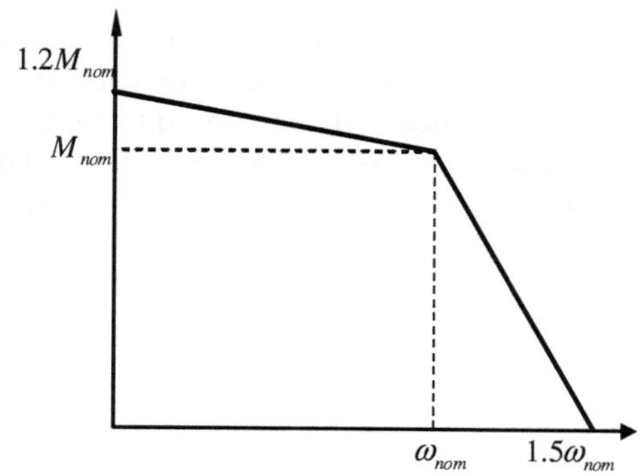

Fig. 9.22. Hoist winch characteristic

Agregation of the Equations of Motion

The equations of motion of the whole crane can be written as:

$$\mathbf{A}\,\ddot{\mathbf{q}} = \mathbf{F}\ , \tag{9.76}$$

where **A** – mass matrix,
q – vector of generalized coordinates,
F – the right side vector; its elements are designated as the partial derivatives of the kinetic energy, potential forces of gravity and flexibility, partial derivatives of function of dissipation energy and units derived from external forces.

The equations (9.76) were solved by a computer program using the fourth order of the Runge-Kutta method with fixed step integration. Before the integration of (9.76), initial conditions were calculated by solving the above equations assuming $\ddot{\mathbf{q}} = \dot{\mathbf{q}} = \mathbf{0}$. The resulting system of nonlinear algebraic equations was solved using the Newton's method.

9.2.2 Examples of Numerical Calculations

Example of simulation results obtained from the developed computer programme are presented in this section. Two load cases are considered:

LC-1: Hoisting of the load from a stationary deck.

LC-2: Hoisting from the deck which movement is described by the function:

$$z_P = 0.75\sin\left(\frac{2\pi}{6}t\right)\ [\mathrm{m}]. \tag{9.77}$$

A load of 18000 kg (including wire rope and hook block mass) is lifted from a supply boat deck. Assuming that the wire rope is loose at the beginning of the cycle (by a length of 1 m), some dynamic overload can be expected. The hoisting speed is assumed 0.4 m/s for quadruple operation, with the drum rotation characteristics consistent with Fig. 9.22. The shock absorber was defined by the following parameters: $S_1 = 97\,5\,00$ N, $S_2 = 1.4S_1$, $\Delta_1 = 0.02$ m, $\Delta_2 = 0.52$ m, $a = 1.1$.

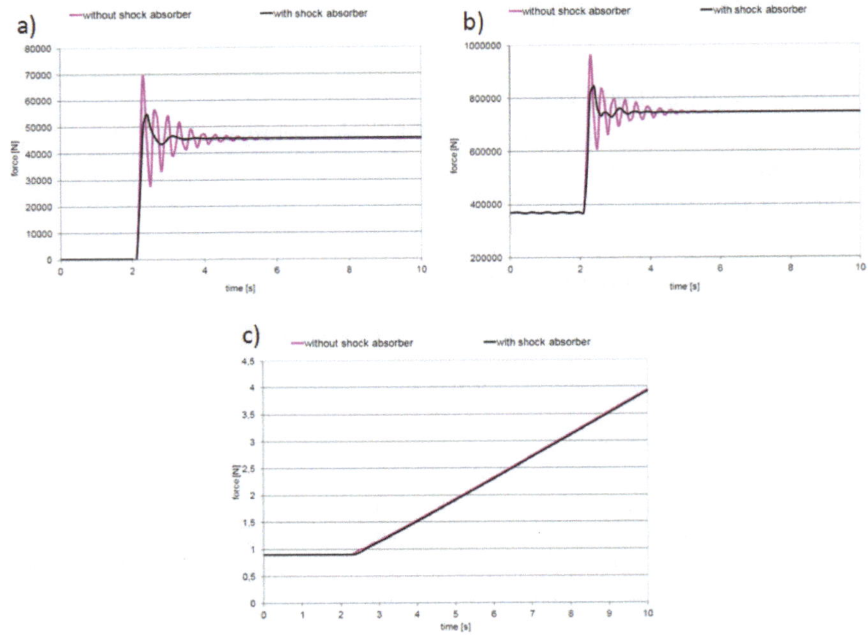

Fig. 9.23. LC-1 load case results: a) hoist rope force, b) luffing rope force,c) *z* coordinate of the load

In Fig. 9.23, there are presented time courses of the main hoist wire tension force, luffing wire force and *z*-coordinate of the load during lifting operation. Two crane models: working with and without shock absorber, are compared. The whole crane structure was assumed rigid.

The conditions assumed in the presented examples are rather theoretical – the winch acceleration during the first phase (when the rope is loose) produces a high dynamic peak load when the wire is suddenly pre-tensioned. This is one of the reason why a DAF (dynamic amplification factor) is applied when selecting various crane components. However, this scenario is simulated in order to show how effective the shock absorber could be. Even if the operator runs, be mistake, the winch without load (or there could be an imperfection in a drive system), thanks to the automatic overload protection system dynamic load in the hoist rope

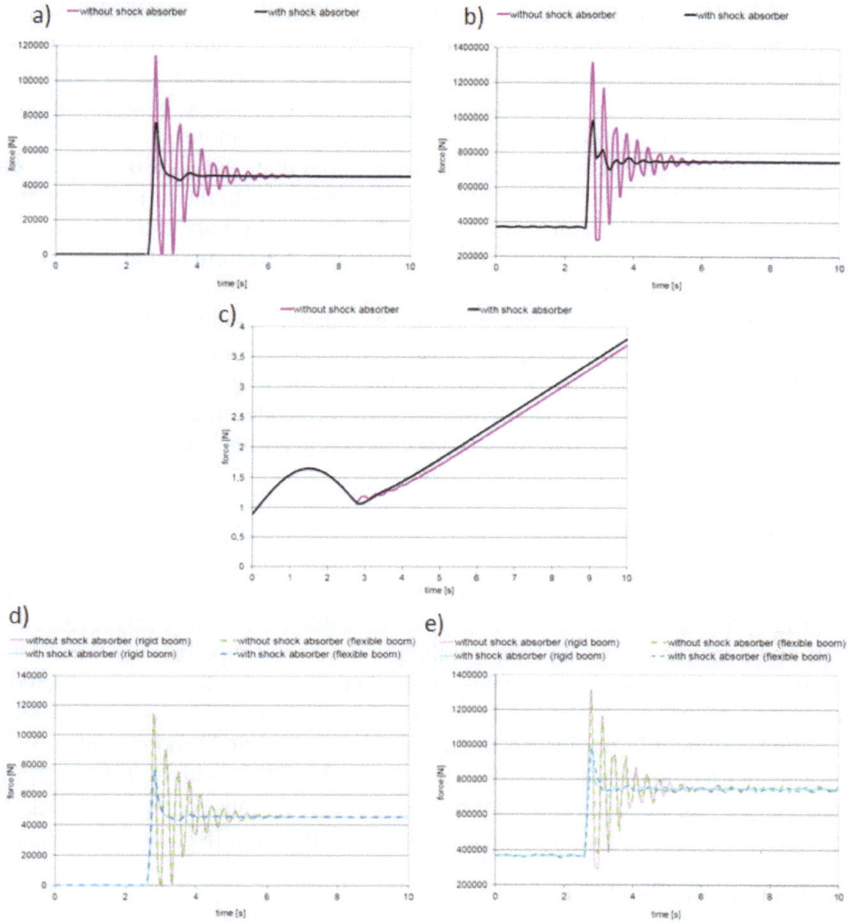

Fig. 9.24. LC-2 load case results: a) hoist rope force, b) luffing rope force, c) z coordinate of the load, d) hoist rope force (rigid and flexible crane jib), e) luffing rope force (rigid and flexible crane jib)

is reduced by approximately 100%. In some cases, without such a systems, the tension can be close to the breaking load of the wire, which if not breaks it at the accident time, makes its life time much shorter.

The plots shown in Fig. 9.24 were obtained for the load case LC-2. The results for the rigid crane gantry was compared with those obtained with flexible structure. Discretisation of the crane boom was performed using $n_4 = 7$ rigid finite elements.

The results of numerical simulations performed using the crane model having shock absorber installed confirm a significant decrease of dynamic overload experienced by the structural systems. Application of the shock absorber subsystem in real constructions would allow the crane to work in much more difficult conditions. Without such a systems, the same crane has to be de-rated, which makes it in a higher sea state less efficient handling tool, causing that the whole vessel or platform can not perform planned lifts, until the weather conditions improve. Consequently, the load chart of a crane equipped with an overload protection system will be much more different than the same construction without such a control device. Therefore, properly working shock absorber is now a relatively new technique in the offshore industry.

Taking into consideration flexibility of the boom does not significantly change the obtained results. Some slight differences are observed in the time history of the luffing rope force. It therefore appears that, for the preliminary calculations or for the bids purpose, the flexibility of the boom can be omitted. On the basis of a model with few degrees of freedom, an engineer obtains a quick software tool, supporting him during the design process. The calculation model presented enable us to determine the crane overload in various working conditions. That makes it possible to predict limiting weather conditions for a given crane design and specific operation scenarios. Implementation of the model in a standalone desktop application makes it attractive for various conceptual ad-hoc analyses.

9.3 Laying of Pipelines

The methods of analysing multibody systems, models of connections and materials presented in previous chapters were implemented in software suited for static and dynamic analyses of the installation process of pipelines for transporting oil and gas, of transfer lines (cables) and other types of infrastructure related to exploiting deposits of the seabeds. The current section discusses models operations commonly performed in reality. The constructed models and software are also indirectly verified. For this purpose, additional models in the ANSYS package are formulated and the results of calculations compared. Detailed derivations and a description of those models are offered in [Szczotka M., 2011b].

The Programme Pipelaysim

Based on the presented models, a computer programme supporting static and dynamic analysis of basic operations related to installing pipes. The programme is written in the C++ language (*Microsoft Visual Studio 2008* IDE), using elements of the *Delphi* package which are parts of *Borland Developer Studio 2006*. To produce graphics Silicon Graphics Inc's *OpenGL* library is used.

The main window of the programme *PipeLaySim* is shown in Fig. 9.25. It acts as a preprocessor. The user can, by means of standard interface components (GUI), define (or load from an external file) any parameters of the models and analysis options. In the main panel of the programme there are buttons assembled which enable running subsequent simulations for supported installation methods.

Results obtained from the calculations may be analysed in a built-in module for creating graphs or exported as text files and further processed in other programmes (e.g. in Excel). A functionality which may be found useful is passing the results of calculations in the form of scripts to the engineering computation system MATLAB. The software also supports concurrently displaying an

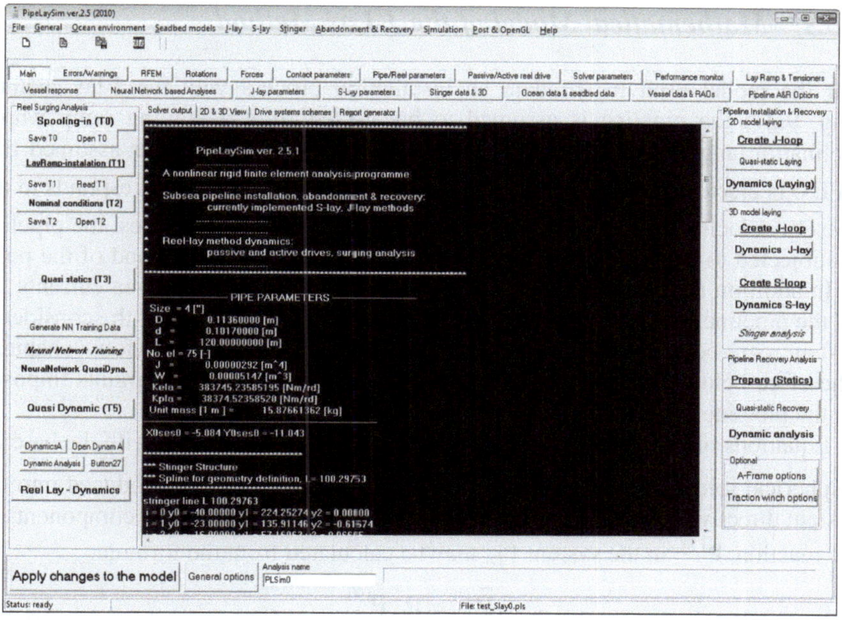

Fig. 9.25. Main window of the programme *PipeLaySim*

animation of the simulated system and saving it to multimedia files (for example *.avi). Sample postprocessor window with an animation produced using the *OpenGL* libraries is shown in Fig. 9.26.

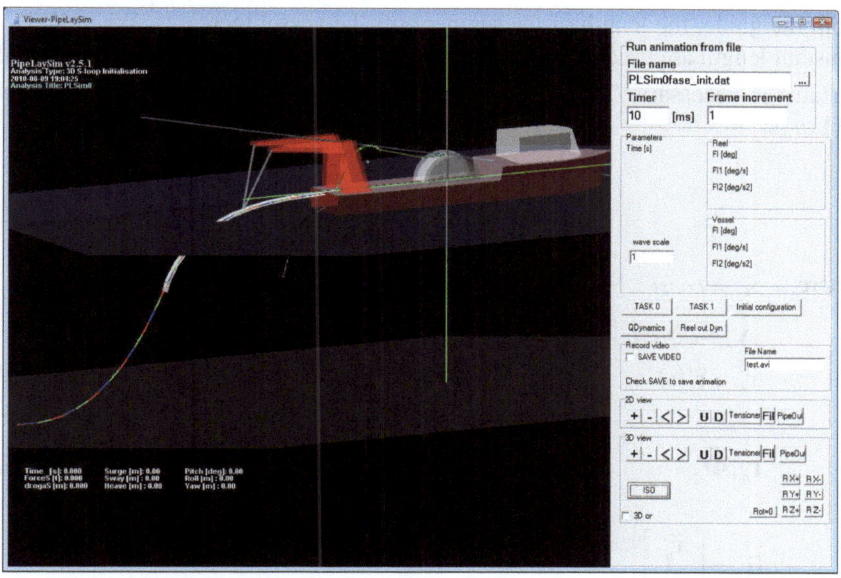

Fig. 9.26. Sample animation window in the programme *PipeLaySim*

9.3.1 Mathematical Model of the J-Lay Method

For the model used to simulate an installation process with the J-lay method (Fig. 9.27) the pipeline is assumed to be ejected from a guiding ramp whose inclination angle relative to the deck is $\alpha_T = const$. It is further assumed that RFE 0 is ejected with a known velocity $v_T(t)$ of laying which corresponds to the vessel's velocity. Because of the lifting movements caused by waves, the pipeline is subjected to forces due to the difference of velocities of laying and of the point S. In addition to that, hydrodynamic forces caused by waves and sea currents act on the pipeline. A detailed derivation of the equations of motion for the considered system can be obtained by using the dependencies from previous chapters. Therefore below only selected formulas related to modelling constraints imposed on the pipeline are given.

Equations of constraints related to the connection of the RFE 0 with the base by a spherical joint at the point H and the reactions $\mathbf{P}^{(0)}$ may be introduced into the system directly by using the dependencies for a spherical joint. The components of the reaction $\mathbf{P}^{(0)}$ in the system $\{\}_A$ may be calculated from the formula:

$$\tilde{\mathbf{P}}_A^{(0)} = \mathbf{R}_A^T \mathbf{P}^{(0)}, \tag{9.78}$$

where $\mathbf{R}_A$ – rotation matrix of the system $\{\}_A$ relative to $\{\}$,

$\tilde{\mathbf{P}}_A^{(0)}$ – reaction vector at the point H expressed in $\{\}_A$,

$\mathbf{P}^{(0)}$ – reaction vector at the point H expressed in the system $\{\}$.

The RFE with number n is placed in the ramp's guide. As the pipeline is ejected from the guide, the length of the segment off the vessel increases. When RFEs of constant length are used, incrementing the number n_{SES} (of RFEs and SDE in the system) is necessary. The general form of constraint equations imposed on the RFE n is:

$$\begin{aligned} \tilde{\mathbf{r}}_s &= \tilde{\mathbf{r}}_n + \tilde{\mathbf{R}}_n \tilde{\mathbf{r}}_s' = \tilde{\mathbf{a}}_s \\ \tilde{\mathbf{\Phi}}_n &= \tilde{\mathbf{\Lambda}}_s \end{aligned}, \tag{9.79}$$

where $\tilde{\mathbf{a}}_s = const$ – vector describing the position of the point S in $\{\}_A$,

$\tilde{\mathbf{\Lambda}}_s = const$ – vector describing the orientation of the guide in $\{\}_A$,

$\tilde{\mathbf{r}}_s' = \tilde{\mathbf{r}}_s'(t)$ – vector describing the coordinates of the point S in the system $\{\}_n$,

$\tilde{\mathbf{r}}_n, \tilde{\mathbf{\Phi}}_n$ – components of the vector $\tilde{\mathbf{q}}_n$ of generalized coordinates of the RFE n,

$$\tilde{\mathbf{q}}_n = \begin{bmatrix} \tilde{\mathbf{r}}_n \\ \tilde{\mathbf{\Phi}}_n \end{bmatrix}.$$

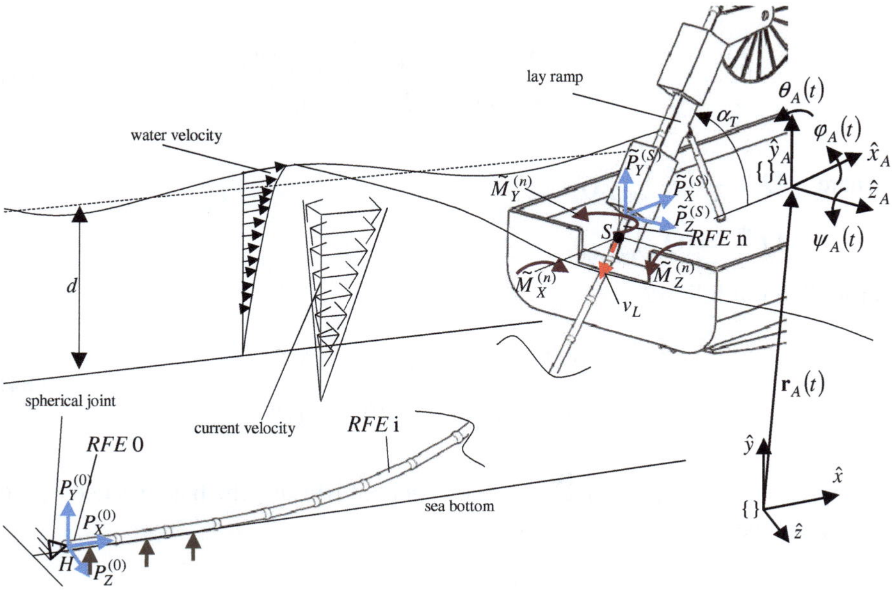

Fig. 9.27. Scheme of the system for installation of pipelines with the J-lay method

Differentiating (9.79) twice makes it possible to put the constraint equations in the accelerative form:

$$\begin{aligned} &\ddot{\tilde{\mathbf{r}}}_n + \ddot{\tilde{\mathbf{R}}}_n \tilde{\mathbf{r}}_s' + 2\dot{\tilde{\mathbf{R}}}_n \dot{\tilde{\mathbf{r}}}_s' + \tilde{\mathbf{R}}_n \ddot{\tilde{\mathbf{r}}}_s' = \mathbf{0}, \\ &\ddot{\tilde{\mathbf{\Phi}}}_n = \mathbf{0}. \end{aligned} \tag{9.80}$$

Since $\tilde{\mathbf{\Phi}}_n = \tilde{\mathbf{\Lambda}}_s = const$:

$$\begin{aligned} &\dot{\tilde{\mathbf{R}}}_n = \mathbf{0}, \\ &\ddot{\tilde{\mathbf{R}}}_n = \mathbf{0} \end{aligned} \tag{9.81}$$

so the equations (9.80) take the form:

$$\begin{aligned} &\ddot{\tilde{\mathbf{r}}}_n = -\tilde{\mathbf{R}}_n \ddot{\tilde{\mathbf{r}}}_s', \\ &\ddot{\tilde{\mathbf{\Phi}}}_n = \mathbf{0}. \end{aligned} \tag{9.82}$$

The equations of motion of the RFE n may be written as:

$$\begin{bmatrix} \tilde{\mathbf{m}}_n & \mathbf{0} \\ \mathbf{0} & \tilde{\mathbf{B}}_n \end{bmatrix} \begin{bmatrix} \ddot{\tilde{\mathbf{r}}}_n \\ \ddot{\tilde{\mathbf{\Phi}}}_n \end{bmatrix} - \begin{bmatrix} \tilde{\mathbf{P}}^{(n)} \\ \tilde{\mathbf{M}}^{(n)} \end{bmatrix} = \begin{bmatrix} \mathbf{Q}_n^{(\mathbf{r})} \\ \mathbf{Q}_n^{(\mathbf{\Phi})} \end{bmatrix}, \tag{9.83}$$

where $\mathbf{Q}_n^{(\mathbf{r})} = \left(\mathbf{Q}_n\right)_{i=1,2,3}$,

$\mathbf{Q}_n^{(\mathbf{\Phi})} = \left(\mathbf{Q}_n\right)_{i=4,5,6}$.

From (9.83) it follows:

$$\begin{aligned} \tilde{\mathbf{P}}^{(n)} &= \tilde{\mathbf{m}}_n \ddot{\tilde{\mathbf{r}}}_n - \mathbf{Q}_n^{(\mathbf{r})}, \\ \tilde{\mathbf{M}}^{(n)} &= \tilde{\mathbf{B}}_n \ddot{\tilde{\mathbf{\Phi}}}_n - \mathbf{Q}_n^{(\mathbf{\Phi})} = -\mathbf{Q}_n^{(\mathbf{\Phi})}. \end{aligned} \tag{9.84}$$

Components of the reaction $\tilde{\mathbf{P}}^{(n)}$ in the system of RFE n (which are needed e.g. to determine the tension) may be obtained from:

$$\tilde{\mathbf{P}}_n^{(n)} = \tilde{\mathbf{R}}_n^T \tilde{\mathbf{P}}^{(n)}. \tag{9.85}$$

The way of defining the vector $\tilde{\mathbf{r}}_s'$ giving the coordinates of the point S in the system $\{\}_n$ (Fig. 9.28) merits a further comment.

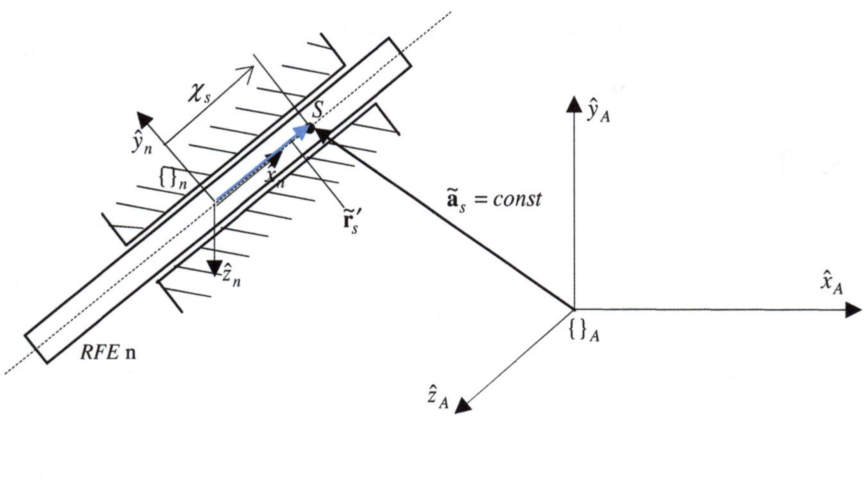

Fig. 9.28. Connection of the RFE n with the guide

In the situation of Fig. 9.28, the following holds:

$$\tilde{\mathbf{r}}'_s = \begin{bmatrix} \chi_s \\ 0 \\ 0 \end{bmatrix}, \tag{9.86}$$

where $\chi_s = \chi_s(t)$ is a function describing how the n-th *RFE* is ejected from the guide (velocity and possibly acceleration).

Static Analysis

Indirect verification was performed (by comparison with the calculations done employing the ANSYS package) on the J-lay system shown in Fig. 9.29. The total length of the analysed pipeline was 1000 m. Due to different positions of the initial point (*H*) of the pipeline attached with a joint to a rigid structure on the bottom the obtained curvatures of the pipeline and values of the forces and stresses differ. The models used are spatial, however, all the static forces act in the plane $\hat{x}\hat{y}$. Zero excitations were assumed ($\chi_s = const$, $\mathbf{q}_D = \mathbf{0}$, $H_S = 0$, no currents). The only forces acting on the pipeline were gravity and hydrostatic buoyancy. Data shown in Table 9.3 were assumed as input.

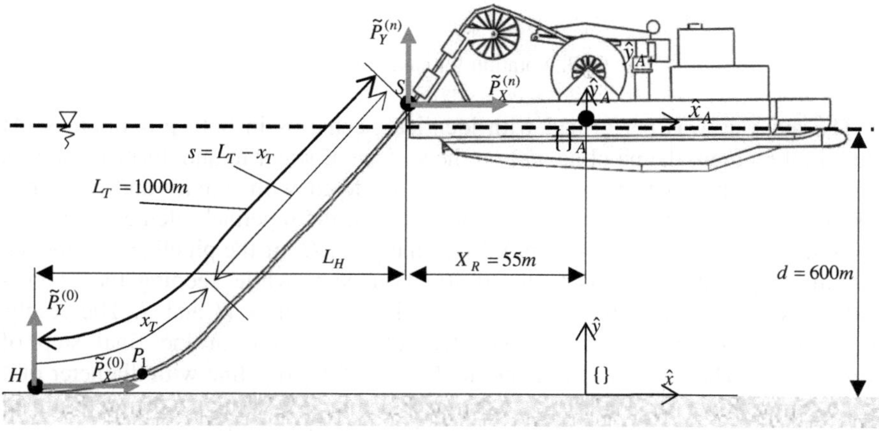

Fig. 9.29. Main parameters of a system in static analyses

Discretisation of the pipeline was performed for a few different numbers of finite elements (both for the programme *PipeLaySim* and the ANSYS package). Satisfactory correspondence was obtained already for the division into $n = 100$ elements (the results given below are for this number of elements). In ANSYS *PIPE288*, finite elements were used which are based on the *BEAM188* element

[Ansys Documentation, 2009] as well as linear shape functions. The element *PIPE288* supports input of hydrodynamic loads modelled using the commands: *OCDATA*, *OCTABLE*, *SOCEAN*. Results are shown on graphs and in tables using notation as in the following scheme (Fig. 9.30).

Table 9.3. Basic parameters assumed for static calculations

Parameter	Variant A	Variant B
Outer diameter of the pipeline [in]/side thickness [mm]	4/5.95	12/16.05
Content of the pipeline	Empty (air at atmospheric pressure) or filled (water under pressure equal to that outside)	
Distance L_H [m]	$L_H^{(1)}=700,\ L_H^{(2)}=725,\ L_H^{(3)}=750$	

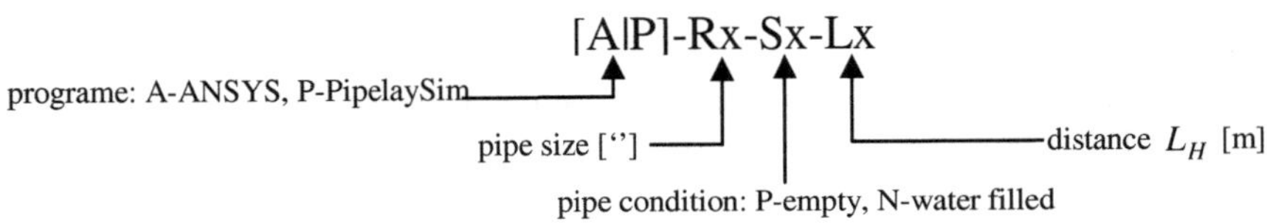

Fig. 9.30. Notation for cases of calculations

The analysis was performed with identical scenarios in both programmes. At the initial moment the pipeline was on the surface of water in undeformed state. In the first step, balance of the system was considered with gravity and buoyancy forces applied due to deflection and immersion of the pipeline's elements (with H being the loose end of the pipeline). Next, the point H of the pipeline was moved in multi-step static analysis to the destination point while keeping the point S motionless. Spherical joints were assumed in both points H and S. The results presented below correspond to the state of the system in the final step of computation. The reactions in the points H and S for a pipeline with diameter of 4 inches are shown in Table 9.4. Likewise, Table 9.5 contains the determined reactions of constraints for a pipeline with diameter of 12 inches.

Based on the performed comparative analyses a conclusion can be drawn that the results produced by the developed software are correct. The differences of forces calculated in the point S and of horizontal reactions in the point H are on average less than 0.5%. Also the values of the vertical reaction $\widetilde{P}_Y^{(0)}$ are in a satisfactory degree of accordance (the differences being 2–14% for pipes with diameters of 4 inches and 1–7% with diameters of 12 inches). The stated discrepancies are caused mainly by difficulties in modelling contact with the bottom which occur in the ANSYS environment.

Table 9.4. Reactions of constraints in points *H* and *S*, pipeline of 4 inches

Empty	[AIP]-R4-SP-L700	[AIP]-R4-SP-L725	[AIP]-R4-SP-L750
ANSYS	$\tilde{P}_X^{(n)}=19.64$ $\tilde{P}_Y^{(n)}=48.2$ $\tilde{P}_X^{(0)}=-19.79$ $\tilde{P}_Y^{(0)}=0.54$	$\tilde{P}_X^{(n)}=25.0$ $\tilde{P}_Y^{(n)}=51.66$ $\tilde{P}_X^{(0)}=-25.1$ $\tilde{P}_Y^{(0)}=0.53$	$\tilde{P}_X^{(n)}=32.3$ $\tilde{P}_Y^{(n)}=56.16$ $\tilde{P}_X^{(0)}=-32.4$ $\tilde{P}_Y^{(0)}=2.18$
PipeLaySim	$\tilde{P}_X^{(n)}=19.58$ $\tilde{P}_Y^{(n)}=48.38$ $\tilde{P}_X^{(0)}=-19.58$ $\tilde{P}_Y^{(0)}=0.49$	$\tilde{P}_X^{(n)}=24.86$ $\tilde{P}_Y^{(n)}=51.89$ $\tilde{P}_X^{(0)}=-24.86$ $\tilde{P}_Y^{(0)}=0.49$	$\tilde{P}_X^{(n)}=32.15$ $\tilde{P}_Y^{(n)}=56.24$ $\tilde{P}_X^{(0)}=-32.15$ $\tilde{P}_Y^{(0)}=2.23$
Filled	[AIP]-R4-SN-L700	[AIP]-R4-SN-L725	[AIP]-R4-SN-L750
ANSYS	$\tilde{P}_X^{(n)}=49.07$ $\tilde{P}_Y^{(n)}=121.6$ $\tilde{P}_X^{(0)}=-49$ $\tilde{P}_Y^{(0)}=0.31$	$\tilde{P}_X^{(n)}=62.55$ $\tilde{P}_Y^{(n)}=129.4$ $\tilde{P}_X^{(0)}=-62.5$ $\tilde{P}_Y^{(0)}=0.48$	$\tilde{P}_X^{(n)}=80.89$ $\tilde{P}_Y^{(n)}=140.6$ $\tilde{P}_X^{(0)}=-80.8$ $\tilde{P}_Y^{(0)}=-6.23$
PipeLaySim	$\tilde{P}_X^{(n)}=48.99$ $\tilde{P}_Y^{(n)}=122.06$ $\tilde{P}_X^{(0)}=-48.99$ $\tilde{P}_Y^{(0)}=0.27$	$\tilde{P}_X^{(n)}=62.49$ $\tilde{P}_Y^{(n)}=130.0$ $\tilde{P}_X^{(0)}=-62.49$ $\tilde{P}_Y^{(0)}=0.46$	$\tilde{P}_X^{(n)}=80.78$ $\tilde{P}_Y^{(n)}=141.2$ $\tilde{P}_X^{(0)}=-80.78$ $\tilde{P}_Y^{(0)}=-6.42$

The shape of the pipeline in the $\hat{x}\,\hat{y}$ plane is shown in Fig. 9.31. The presented results are for the empty pipeline with a diameter of 4 inches. Differences in the values are small and do not exceed 0.1%. Similar results were obtained for the pipeline with a diameter of 12 inches.

The influence of the pipeline's shape on the reduced stresses for the considered cases is presented in Fig. 9.32 (for a pipeline with diameter of 4 inches) and Fig. 9.33 (for a pipeline with diameter of 12 inches). In the analyses, the Huber-Mises-Hencky (HMH) hypothesis was assumed for the calculation of reduced stresses. The graphs show bending moments, axial forces and reduced stresses in sections along the relative length defined by the coordinate $\gamma = \frac{x_T}{L_T}$ ($\gamma = 0$ in

Table 9.5. Reactions of constraints in points *H* and *S*, pipeline of 12 inches

Empty	[A|P]-R12-SP-L700	[A|P]-R12-SP-L725	[A|P]-R12-SP-L750
ANSYS	$\tilde{P}_X^{(n)} = 130$	$\tilde{P}_X^{(n)} = 165.7$	$\tilde{P}_X^{(n)} = 214.1$
	$\tilde{P}_Y^{(n)} = 319.3$	$\tilde{P}_Y^{(n)} = 342.2$	$\tilde{P}_Y^{(n)} = 371.4$
	$\tilde{P}_X^{(0)} = -131$	$\tilde{P}_X^{(0)} = -166.1$	$\tilde{P}_X^{(0)} = -214.2$
	$\tilde{P}_Y^{(0)} = 3.72$	$\tilde{P}_Y^{(0)} = 2.65$	$\tilde{P}_Y^{(0)} = -14.28$
PipeLay Sim	$\tilde{P}_X^{(n)} = 128.7$	$\tilde{P}_X^{(n)} = 165.1$	$\tilde{P}_X^{(n)} = 213.8$
	$\tilde{P}_Y^{(n)} = 320.0$	$\tilde{P}_Y^{(n)} = 343.1$	$\tilde{P}_Y^{(n)} = 371.5$
	$\tilde{P}_X^{(0)} = -128.7$	$\tilde{P}_X^{(0)} = -165.1$	$\tilde{P}_X^{(0)} = -213.8$
	$\tilde{P}_Y^{(0)} = 3.58$	$\tilde{P}_Y^{(0)} = 2.48$	$\tilde{P}_Y^{(0)} = -15.1$
Filled	[A|P]-R12-SN-L700	[A|P]-R12-SN-L725	[A|P]-R12-SN-L750
ANSYS	$\tilde{P}_X^{(n)} = 394$	$\tilde{P}_X^{(n)} = 502.3$	$\tilde{P}_X^{(n)} = 649.15$
	$\tilde{P}_Y^{(n)} = 969.9$	$\tilde{P}_Y^{(n)} = 1038.1$	$\tilde{P}_Y^{(n)} = 1129.3$
	$\tilde{P}_X^{(0)} = -393.5$	$\tilde{P}_X^{(0)} = -501.8$	$\tilde{P}_X^{(0)} = -648.9$
	$\tilde{P}_Y^{(0)} = 5.55$	$\tilde{P}_Y^{(0)} = 5.9$	$\tilde{P}_Y^{(0)} = -41.6$
PipeLay Sim	$\tilde{P}_X^{(n)} = 393.4$	$\tilde{P}_X^{(n)} = 501.7$	$\tilde{P}_X^{(n)} = 648.78$
	$\tilde{P}_Y^{(n)} = 967.87$	$\tilde{P}_Y^{(n)} = 1040.2$	$\tilde{P}_Y^{(n)} = 1032.1$
	$\tilde{P}_X^{(0)} = -393.4$	$\tilde{P}_X^{(0)} = -501.7$	$\tilde{P}_X^{(0)} = -648.78$
	$\tilde{P}_Y^{(0)} = 5.88$	$\tilde{P}_Y^{(0)} = 6.01$	$\tilde{P}_Y^{(0)} = -41.9$

the point H and $\gamma = 1$ for the point *S*, Fig. 9.29). As it can be seen from the graphs, the RFE method gives close results also for reduced stresses, bending forces and moments in sections of the pipeline. Relative errors in all cases are below 1–1.5% (for the given number of elements) and definitely diminish with condensation of the division.

Analysing the graphs in Fig. 9.32 and Fig. 9.33 indicates that filling the pipeline with a liquid does not influence the forms of the bending moment (in the considered cases similar curvatures were obtained for an empty and filled pipeline). The axial force, which depends on the position of the vessel against the waves and density of the pipeline, has significant influence on the values of reduced stresses. Installation of pipelines when they are filled with air allows for reducing the axial forces and stresses.

a)

b)

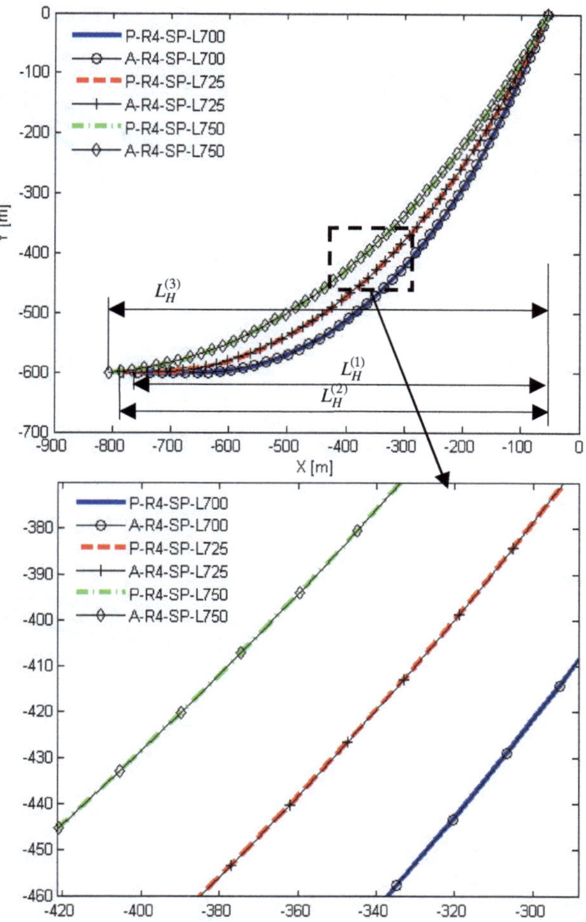

Fig. 9.31. Shape of a pipeline with diameter of 4 inches having reached static balance: a) shape of the pipeline obtained for different values of L_H , b) magnified fragment of the graph

The described method of static analysis and the obtained results may inform the determination of installation parameters of the pipeline concerned, taking into consideration the influence of depth, buoyancy and geometric traits of the system [Mohitpour M., et al., 2003], [Bai Y., Bai Q., 2005], [Palmer A. C., King R. A., 2008].

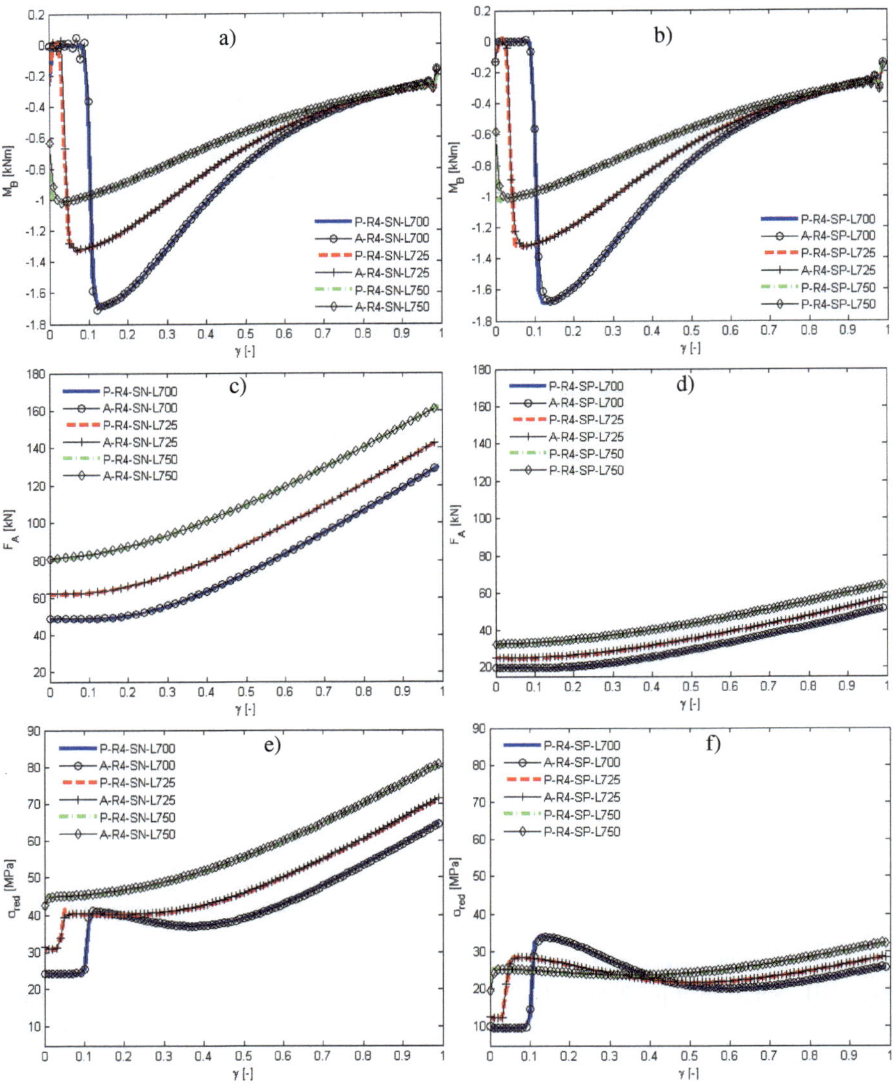

Fig. 9.32. A pipeline with diameter of 4 inches – values of moments, forces and stresses: bending moment for a filled (a) and empty (b) pipeline, axial force in a section of a filled (c) and empty (d) pipeline, reduced stress for a filled (e) and empty (f) pipeline

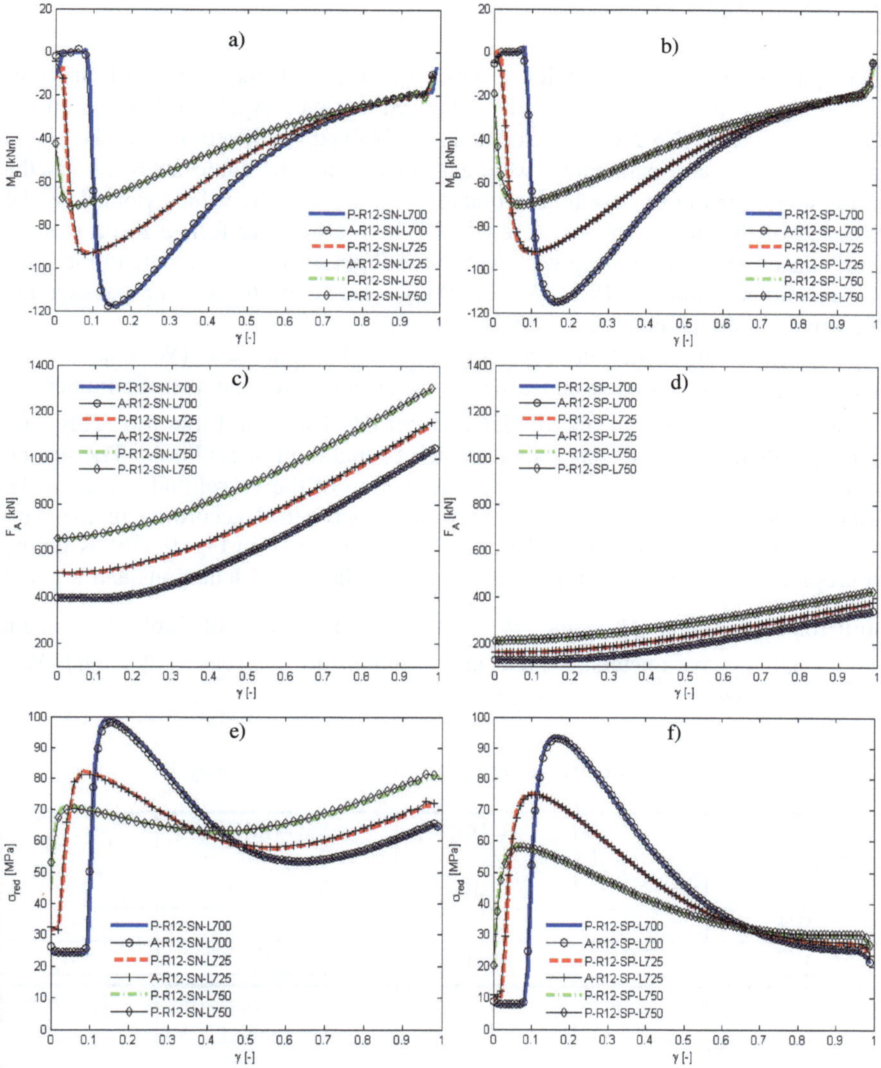

Fig. 9.33. A pipeline with diameter of 12 inches – values of moments, forces and stresses: bending moment for a filled (a) and empty (b) pipeline, axial force in a section of a filled (c) and empty (d) pipeline, reduced stress for a filled (e) and empty (f) pipeline

Dynamic Analysis

This part of the book contains a brief overview of the results of analyses pertaining to the dynamics of the analysed system. Again, the programmes *PipeLaySim* and ANSYS were used and the obtained results compared. The same geometry and mass parameters were assumed as for the static analyses. As the initial conditions in the dynamic problem (at time $t = 0$) the values yielded by the last step of the static analysis were taken. The fourth-order Runge-Kutta method with constant integration step was used in the *PipeLaySim* programme to integrate the equations of motion [Press W. H., 2002], whereas in the ANSYS package the Newmark method [Bathe K. J., 1996] was used.

Two types of excitation were applied. In the first case (W1), no waves $(H_S = 0\,\mathrm{m})$ was assumed and a harmonic excitation of the vessel's immersive motion (Fig. 9.34a) with amplitudes and periods listed in Table 9.6. This case corresponds to motion of the system immersed in a motionless liquid. The second type of excitation (W2) included both the motion of the vessel and waves of the water (calculations in both programmes were performed according to the Airy model of the wave). Graphs of the excitations are shown in Fig. 9.34b, assuming appropriate resizing of amplitudes in the initial phase of calculations and a phase shift for the variable x_D equal $\varphi_X^{(0)} = 90°$. The lower rows of Table 9.6 contain the remaining parameters, which are common to the cases W1 and W2. $\chi_s = const$ was assumed in both programmes.

Table 9.6. Parameters assumed in the dynamic analysis

Excitation	H_S [m]	A_X [m], $\varphi_X^{(0)}$ [deg]	A_Y [m], $\varphi_Y^{(0)}$ [deg]	Period T [s]
W1	0.0	0; 0	1; 0	8.0
W2	5.0	1; 90	2; 0	8.0
Coeff. C_D / C_A from (3.33)		1.0/1.0	Stiffness coeff. of the seabed	1.1e[5] N/m
Data set (geometry, diameter, content of the pipeline)		R4-SP-L700	Tangent resistance coefficients m and n (Table 3.3)	$m = 0.02$ $n = 0.04$

Fig. 9.35 presents time courses of coordinates of the point P_1 of the pipeline at the maximum of curvature (Fig. 9.29) determined by the coordinate $x_T = 250$ m. In both cases of the vessel's motion and waves, the graphs of displacements of the point are similar. Relative errors do not exceed 1%.

In Fig. 9.36, the velocities P_1 obtained from both programmes are presented. The produced graphs are virtually identical. The differences are due to integration methods and also to the accuracy with which the excitation is realized. In ANSYS, it was interpolated with a piecewise linear function (boundary conditions for the displacements given as tables).

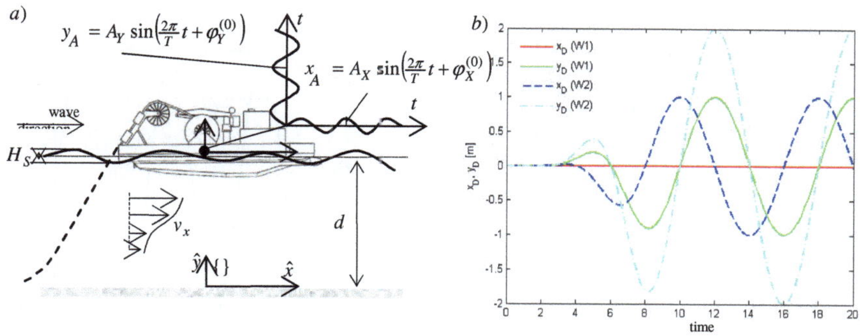

Fig. 9.34. Excitations of the vessel's motion in dynamic analysis: a) assumed conditions of waves and vessel's motion, b) graphs of longitudinal rolling x_A and heave y_A

Dynamic reactions at the point S (the point of connection with the guiding device onboard the vessel) are shown in Fig. 9.37. In the case of vertical reaction $R_Y^{(S)}$ the maximal relative error does not exceed 10% (W2 excitation).

The reduced stresses calculated along the pipeline's axis are shown in Fig. 9.38. The graphs were produced for the time $t = 10$ s taking dynamic forces into account.

On the comparative results of static and dynamic analyses presented a conclusion can be based that the proposed model and software are correct. Since actual objects (ships for laying pipelines) are hardly available and laboratory research is very costly and requires large pools and devices producing artificial waves, performing empirical tests is rather difficult. The authors are aware that results of measurements obtained from tests on actual objects may deviate from the values yielded by the process of numerical simulation, among other things due to the simplified description of interaction in the liquid – solid body system and the approximate model of waves. Yet, some verification is assured by comparing the results with that from another environment aimed at modelling and analyses (e.g. of the ANSYS type) which is commonly used and has been verified multiple times. This allows us to eliminate some possible errors in modelling and programming.

9.3.2 Installation of a Pipeline with the S-Lay Method

A mathematical model of a system for simulating the dynamics of the installation process with the S-lay method can be formulated by augmenting the model of the

J-lay method. The additional element is a specialized ramp guiding the pipeline (a *stinger*) (Fig. 9.39). In the model presented herein, the ramp is assumed to be a bent beam with variable section modelled with the classical finite element method connected by a joint to the vessel's deck at the point U and additionally with two supporting ropes.

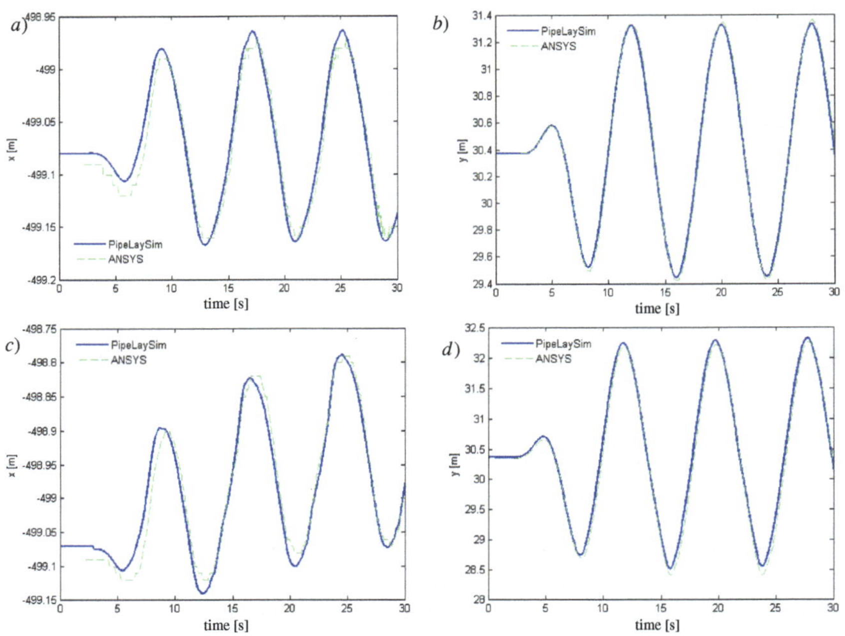

Fig. 9.35. Coordinates of the point xx of the pipeline: a) coordinate x (W1), b) coordinate y (W1), c) graph of the coordinate x (W2), d) graph of the coordinate y (W2)

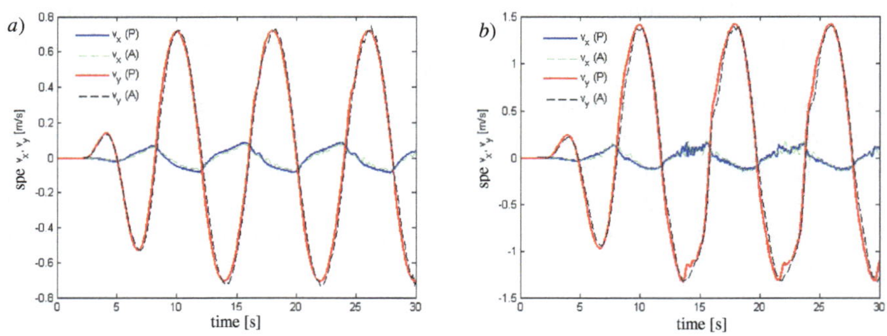

Fig. 9.36. Velocities of the point P_1 : a) W1 excitation, b) W2 excitation

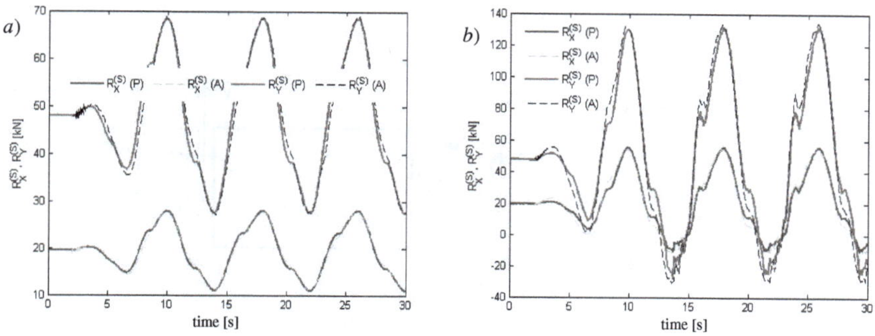

Fig. 9.37. Dynamic reactions at the point S: a) W1 excitation, b) W2 excitation

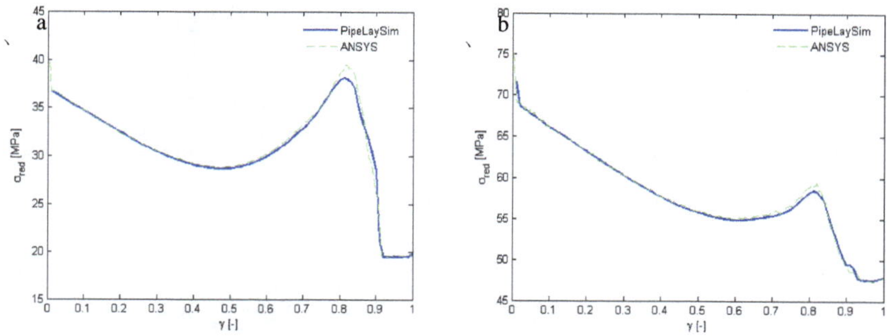

Fig. 9.38. Reduced stresses for the time t=10s: a) W1 excitation, b) W2 excitation

The model of the pipeline is similar to that used for the J-lay method. It is described in the previous chapter. A model of the ramp connected with the deck by a joint U and supporting ropes needs to be additionally formulated. Hence, the equations of motion of the system may be written in the form [Szczotka M., 2011b]:

$$\mathbf{A}^{(J)}\ddot{\tilde{\mathbf{q}}}^{(J)} = \mathbf{Q}^{(J)} + \mathbf{Q}^{(J,C)}, \tag{9.87}$$

$$\mathbf{A}^{(R)}\ddot{\tilde{\mathbf{q}}}^{(R)} = \mathbf{Q}^{(R)} + \mathbf{Q}^{(R,C)} + \mathbf{Q}^{(R,L)}, \tag{9.88}$$

where $\mathbf{A}^{(J)}$ – matrix of masses of the pipeline,

$\tilde{\mathbf{q}}^{(J)} = \begin{bmatrix} \tilde{\mathbf{q}}_0^{(J)} \\ \vdots \\ \tilde{\mathbf{q}}_n^{(J)} \end{bmatrix}$ – vector of generalized coordinates of the pipeline,

$\mathbf{Q}^{(J)}$ – vector of generalized forces acting on the pipeline,

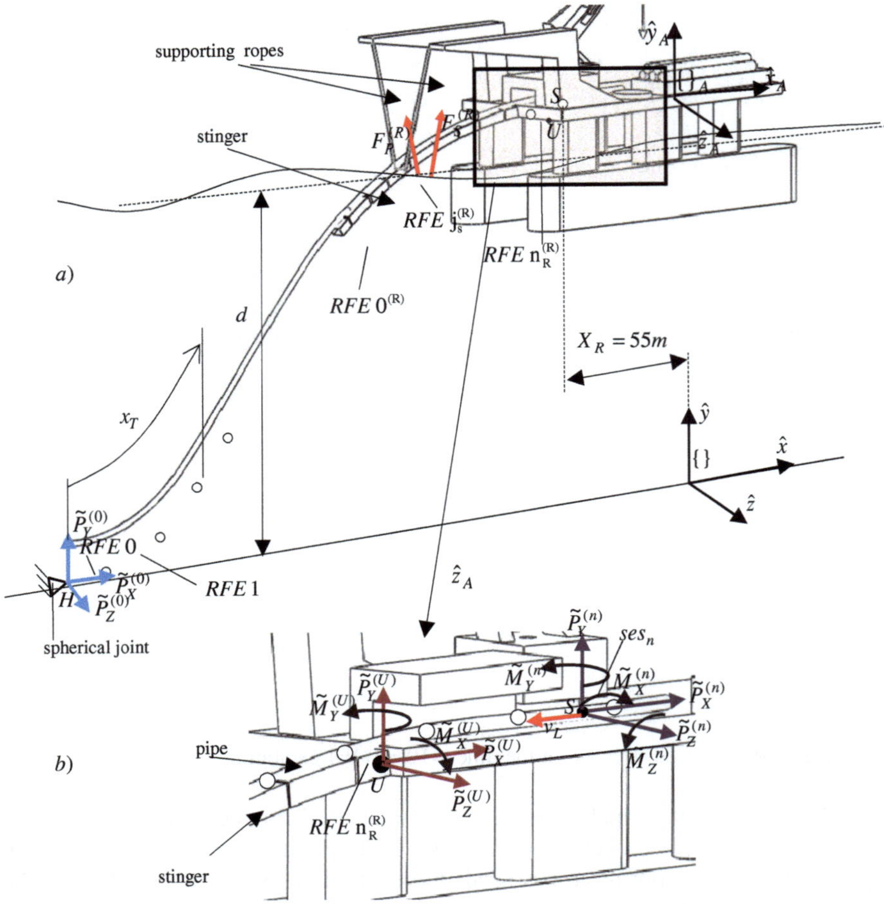

Fig. 9.39. Scheme of the model for the analysis of installing a pipeline with the S-lay method: a) positioning of the coordinate systems, b) reactions of constraints

$\mathbf{Q}^{(J,C)}$ – vector of generalized forces exerted by the ramp,

$\mathbf{A}^{(R)}$ – matrix of masses of the ramp,

$$\tilde{\mathbf{q}}^{(R)} = \begin{bmatrix} \tilde{\mathbf{q}}_0^{(R)} \\ \vdots \\ \tilde{\mathbf{q}}_{n_R}^{(R)} \end{bmatrix} \text{– vector of generalized coordinates of the ramp,}$$

$\mathbf{Q}^{(R)}$ – vector of generalized forces acting on the ramp,

$\mathbf{Q}^{(R,C)}$ – vector of generalized forces exerted by the pipeline,

$\mathbf{Q}^{(R,L)}$ – vector of generalized forces due to the actions of forces supporting the structure of the ramp.

The equations and reactions of constraints imposed on the pipeline are identical to those of the J-lay model. Whereas the joint at the connection of the ramp with the deck (a revolute connection) makes it necessary to include the reaction vector:

$$\tilde{\mathbf{P}}^{(U)} = \begin{bmatrix} \tilde{P}_X^{(U)} & \tilde{P}_Y^{(U)} & \tilde{P}_Z^{(U)} \end{bmatrix}^T, \tag{9.89}$$

and a vector of the pair of forces whose moment is:

$$\tilde{\mathbf{M}}^{(U)} = \begin{bmatrix} \tilde{M}_X^{(U)} & \tilde{M}_Y^{(U)} \end{bmatrix}^T. \tag{9.90}$$

By neglecting friction in the connection, $\tilde{M}_Z^{(U)} = 0$ is assumed.

The constraint equations take the form:

$$\tilde{\mathbf{r}}^{(U)} = \tilde{\mathbf{r}}_{n_R}^{(R)} + \tilde{\mathbf{R}}_{n_R}^{(R)} \tilde{\mathbf{r}}_{n_R}^{\prime(U)} = const, \tag{9.91}$$

$$\mathbf{\Lambda}_2 \tilde{\mathbf{\Phi}}_{n_R}^{(R)} = const, \tag{9.92}$$

where $\tilde{\mathbf{r}}_{n_R}^{\prime(U)}$ – vector of generalized coordinates of the point U in the system of the RFE n_R of the ramp,

$\tilde{\mathbf{r}}_{n_R}^{(R)}, \tilde{\mathbf{\Phi}}_{n_R}^{(R)}$ – vectors of components of the vector $\tilde{\mathbf{q}}_{n_R}^{(R)} = \begin{bmatrix} \tilde{\mathbf{r}}_{n_R}^{(R)} \\ \tilde{\mathbf{\Phi}}_{n_R}^{(R)} \end{bmatrix}$,

$\tilde{\mathbf{R}}_{n_R}^{(R)}$ – rotation matrix of the n_R of the ramp,

$\mathbf{\Lambda}_2 = \begin{bmatrix} 1 & 0 & 0 \\ 0 & 1 & 0 \end{bmatrix}$.

Following the procedure presented in the previous chapter the constraint equations may be put in an accelerative form allowing us to determine the vectors $\tilde{\mathbf{P}}^{(U)}$, $\tilde{\mathbf{M}}^{(U)}$ and $\ddot{\tilde{\mathbf{q}}}_{n_R}^{(R)}$. Actions of the following forces are also taken into account:

- in the ropes $F_S^{(R)}$, $F_P^{(R)}$ acting on REF $_{j_s}^{(R)}$ introduced by the vector $\mathbf{Q}^{(R,L)}$,
- contact forces, acting on elements of the pipeline and on the ramp, derived from $\mathbf{Q}^{(J,C)}$ and $\mathbf{Q}^{(R,C)}$.

Forces in the ropes may be determined using the model of a flexible rope with damping. The contact forces between the ramp and the pipeline are determined by assuming a series of spring-damping elements with clearance modelling the rollers guiding the pipeline. The forces of interaction of the ramp's structure with the water environment are approximated with the Morison equation keeping in mind the additional interactions occurring at the transition through the water surface.

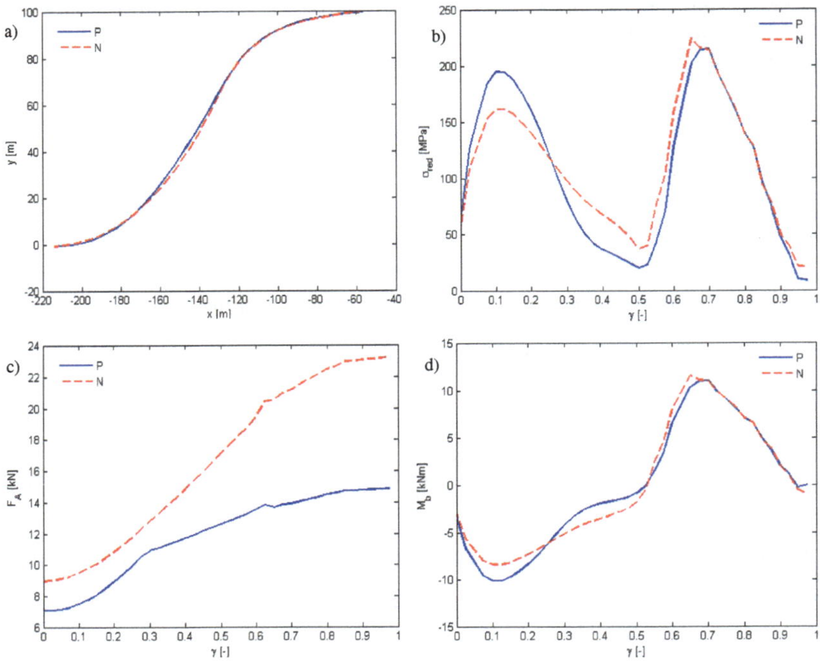

Fig. 9.40. Results obtained from static analysis of the S-lay system: a) shape of the pipeline, b) reduced stresses, c) graph of the axial force, d) bending moment

Sample calculations contained in this section were performed for an installation of a pipeline of 4 inches at the depth of $d = 100$ m. The shape of the pipeline after it has reached static balance is shown in Fig. 9.40a, where the coordinates on the graph are expressed in the inertial system {} depicted in Fig. 9.39a. Two options were considered: on the graphs P denotes a pipeline filled with air, N – a pipeline filled with water. Graphs of reduced stresses along the pipeline's axis (Fig. 9.40b) are different to those obtained in the J-lay method. Two places occur with considerable stresses due to bending. The first one is caused by the ramp's curvature (a section called *overbend*), the second results from the curvature of the pipeline above the bottom (*sagbend*).

Stresses in the pipeline may be controlled by changing: the immersion of the ramp, its shape and the value of the force stretching the pipeline. The bending moment being zero at some point (Fig. 9.40d) is also characteristic of this method.

Sample results of calculations of the dynamics are shown in Fig. 9.41. Harmonic motion of the vessel with period $T = 8\,\mathrm{s}$ and amplitudes $x_A = 0.5\,\mathrm{m}$, $y_A = 1.0\,\mathrm{m}$ was taken as the excitation. The coefficients of the Morison equation were assumed as in previous analyses.

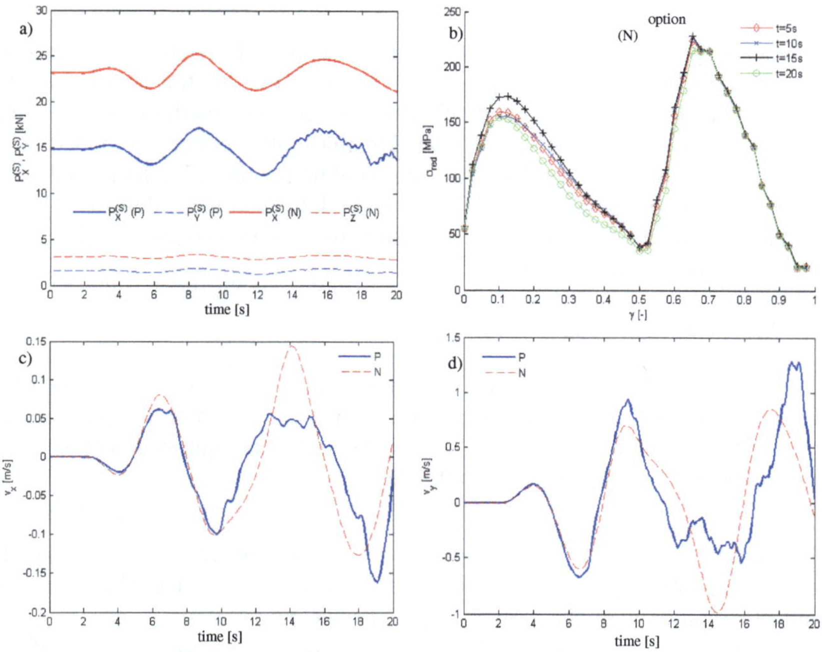

Fig. 9.41. Simulation of the dynamics of a system for S-lay installation: a) dynamic reactions at the point *S*, b) reduced stresses in selected points of time, c) graphs of the horizontal component of velocity in time, d) graphs of the vertical component of velocity in time

The dynamic reactions at the point *S* (Fig. 9.41a) (guiding the RFE *n* in the tensioner's mechanism) are not significantly different from the values obtained in the static problem due to the assumed length and size of the pipeline (the mass is fairly small). Additionally, as Fig. 9.41b implies, in which graphs of the reduced stresses at different points in time are presented, with a relatively rigid system of ramps the change of values of the stresses occurs in the lower segment of the pipeline only. The changes would be greater with a more flexible ramp (or suspension system), which may result, when the length of the system is significant, in considerable differences in the geometry of the lower segment of the pipeline (*sagbend*). A possibility also exists of controlling the lengths of the ropes in such a way that the growth of stresses caused by waves is eliminated. The velocities of the centre of mass of the RFE located approximately 4 m above the bottom of the sea are shown on graphs (Fig. 9.41 c and d). In the case of the S-lay method, the differences in the velocities of the pipeline in the lower part are greater for different contents of the pipeline (densities) than with the J-lay method.

9.3.3 Dynamics of a System for Installing Pipelines with the Reel Method

Dependencies presented in the preceding chapters allow us to formulate mathematical models and a computer programme suitable for simulating the operation of a device for laying pipelines with the reel method. The section presents a model for the analysis of the dynamics of a system equipped with a passive reel drive system. To discretize the pipeline the modified RFE method is used and the nonlinear dependency of stresses on deformations is described by an elasto-plastic characteristic. Models contained herein are investigated in [Szczotka M., 2010], [Szczotka M., 2011b].

9.3.3.1 Mathematical Model

In Fig. 9.42, a scheme is shown of a system consisting of a reel onto which the pipeline is wound and a specialized guiding ramp through which the pipe passes as it is unwound and lowered to the seabed. The ramp is equipped with devices controlling the tension and the speed of laying.

The following simplifying assumptions are made:

- the motion of the pipeline being unwound from the reel is kinematically forced by a device providing tension and guidance; influence of the immersed part of the pipeline on the motion may be neglected,
- swaying angle ψ_A of the vessel is the most important parameter of the lifting motion, therefore a simplification is proposed which reduces the problem to a planar system in which the vessel can move according to the known functions:

$$
\begin{aligned}
x_A &= \Im\left(t, \mathbf{RAO}, H_S, T_Z, \beta_f, S(\omega)\right),\\
y_A &= \Im\left(t, \mathbf{RAO}, H_S, T_Z, \beta_f, S(\omega)\right),\\
\psi_A &= \Im\left(t, \mathbf{RAO}, H_S, T_Z, \beta_f, S(\omega)\right),\\
z_A(t) &= 0, \quad \theta_A(t)=0, \quad \varphi_A(t)=0,
\end{aligned}
\tag{9.93}
$$

where $\Im$ – operator of transformation of the motion from the domain of frequency to the domain of time,

H_S, T_Z – height and period of waves,

β_f – wave's angle of attack,

$S(\omega)$ – defined by (3.27) or (3.28),

RAO – operator of the transition function,

- large deflections of the pipeline are taken into account by applying the modified RFE method including bending in the longitudinal plane of the vessel.

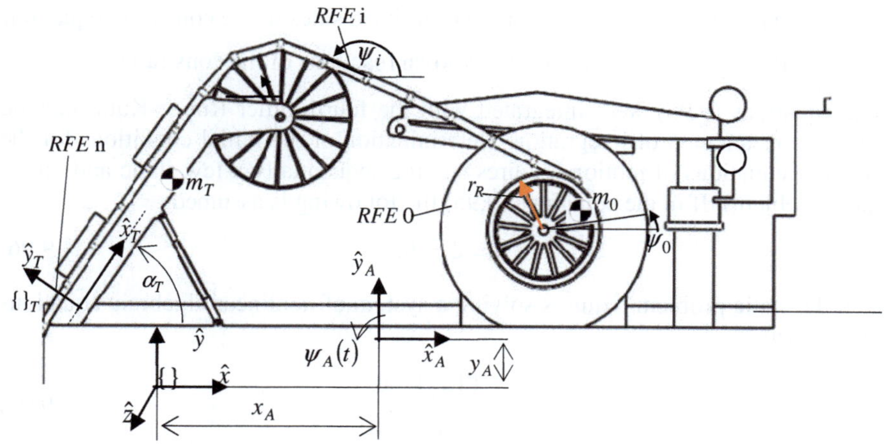

Fig. 9.42. Scheme of the reel system

The generalized coordinates of the considered system are therefore components of the following vector **q**, (an appropriate choice of mass, inertia and geometric parameters allows the reel to be treated as RFE 0):

$$\mathbf{q} = [\psi_0, \ldots, \psi_i, \ldots, \psi_n, \alpha_T]^T, \tag{9.94}$$

where ψ_i – inclination of the axes of the RFE i to the axes $\hat{x}_B$ of the inertial system,

α_T – inclination angle of the ramp's axis.

The equations of motion along with the constraint equations may be written as (detailed derivations are presented in the papers [Szczotka M., 2010], [Szczotka M., 2011b]):

$$\begin{aligned} \mathbf{A}\ddot{\mathbf{q}} + \mathbf{B}\dot{\mathbf{q}} - \mathbf{DP} &= \mathbf{Q} - \mathbf{G} - \mathbf{H}, \\ -\mathbf{S}\ddot{\mathbf{q}} &= \mathbf{W}, \end{aligned} \tag{9.95}$$

where $\mathbf{A} = \mathbf{A}(t, \mathbf{q})$ – matrix of inertia of the system,

$\mathbf{B} = \mathbf{B}(t, \mathbf{q}, \dot{\mathbf{q}})$,

$\mathbf{D}(\mathbf{q}), \mathbf{S}(\mathbf{q})$ – matrices of coefficients of reactions and constraint equations,

$\mathbf{Q}(t,\mathbf{q},\dot{\mathbf{q}})$ – vector of generalized forces,

$\mathbf{G}$ – vector taking the potential of gravity forces into account,

$\mathbf{H}=\mathbf{H}(t)$ – vector whose components depend on the base's lifting motion,

$\mathbf{W}(t,\mathbf{q},\dot{\mathbf{q}})$ – vector of the right-hand sides of the constraint equation,

$\mathbf{P}$ – vector of unknown reactions of the constraints.

The equations (9.95) were integrated with the fourth-order Runge-Kutta method with a constant step of integration. Determination of the initial conditions for the system's equations of motion requires solving sequentially a few static and quasi-static problems. If in the equations (9.95) the following is assumed:

$$\ddot{\mathbf{q}}=\dot{\mathbf{q}}=\mathbf{0}, \tag{9.96}$$

then the static problem requires solving a system of nonlinear algebraic equations of the form:

$$\begin{aligned}\mathbf{\Psi}(\mathbf{q}_s)&=\mathbf{0},\\ \mathbf{\Phi}(\mathbf{R}_s)&=\mathbf{0},\end{aligned} \tag{9.97}$$

where $\mathbf{\Psi}(\mathbf{q})=\mathbf{0}$ – equations of balance of the RFEs $1,\ldots,n$ and the guiding ramp,

$\mathbf{q}_s=[\psi_1,\ldots,\psi_n,\psi_T]^T$,

$\mathbf{R}_s=[U_E,N_E,M_n]^T$,

$\mathbf{\Theta}(\mathbf{R}_s)=\mathbf{0}$ – constraint equations.

Solving the system of equations (9.97) was done with the Newton method. The procedure preceding the calculations of the dynamics is depicted in Fig. 9.43. In the first stage, the pipeline is wound onto the reel. At this time plastic deformations may occur. The end of the pipe is transferred to the guiding ramp in the next stage. Having performed these calculations, we obtain the initial conditions assumed as the starting point of the dynamics.

In the computer programme, a possibility is also included to perform dynamic analysis with a simplified model in which oscillations (dynamics) are not included (the model is introduced in [Szczotka M., et al., 2007). In such case, an internal procedure solving the equations (9.97) with the Newton method determines the forces occurring during the unwinding of the pipeline when integrating the reel's equations of motion. The equation of the reel's dynamics may be obtained from the equations (9.95) assuming $n=0$ and taking into account the forces caused by deformations of the pipeline described by the system of equations (9.97). Note, however, that despite the minimal dimension of the model, the necessity of using the Newton method, which is sensitive to nonlinearity of the problem considered,

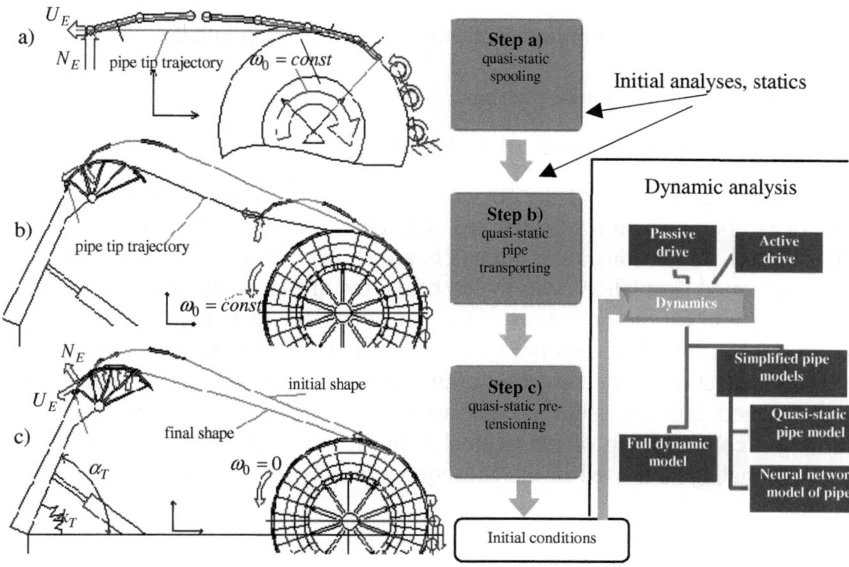

Fig. 9.43. Quasi-static analyses for determining the initial conditions: a) stage of winding the pipe onto the reel, b) transferring the pipe to the ramp, c) stretching the system to the nominal tension

lengthens the durations of computer simulations with respect to the full model of the dynamics. Those durations are approximately ten times longer, therefore using this model does not seem prudent. Hence, another approach to employing the simplified model is proposed. The discrete model of the pipeline is thereby replaced with an artificial neural network [Szczotka M., 2010]. Minimal computation time is then required to determine the forces due to the pipeline's work as it is being unwound, and use them in the equation of the reel's dynamics. Computational efficiency of a such model is particularly appealing. It also lends itself somewhat to real-time control. The results obtained from both variants of the model are presented later in this chapter.

9.3.3.2 Calculations for a Passive Drive of the Reel

The results of calculations performed for the system shown in Fig. 9.42 are presented below. The drive of the reel from which the pipeline is unwound is assumed to be passive and to exert a constant force applied at a dividing radius of the clockwork. To make the interpretation of the results more convenient, the simulations were performed assuming the following functions describing the motion of the vessel:

$$\begin{aligned} x_A &= A_X \sin(\omega t + \gamma_X), \\ y_A &= A_Y \sin(\omega t + \gamma_Y), \\ \psi_A &= A_\psi \sin(\omega t + \gamma_\psi), \end{aligned} \tag{9.98}$$

where A_X, A_Y, A_ψ – amplitudes of motion in the appropriate directions,

$\gamma_X, \gamma_Y, \gamma_\psi$ – initial phases,

ω – angular frequency of excitation.

Basic assumed parameters of the device are based on the documentation of one of the devices operating on the North Sea. The characteristic values are gathered in Table 9.7 (the dimensions and parameters are shown in Fig. 9.42). The objective of the performed simulations was to determine the influence of waves and braking force of the reel on the operation of the considered system. Typical settings and data used when installing pipelines were chosen. Table 9.8 shows the parameters of regular excitation of the vessel's motion according to (9.98), taking into account the mentioned characteristics of the vessel. All calculations were performed assuming a constant step of integration in the Runge-Kutta method Δh=0.001 s. The notational system used in the subsequent graphs is explained on the scheme presented in Fig. 9.44.

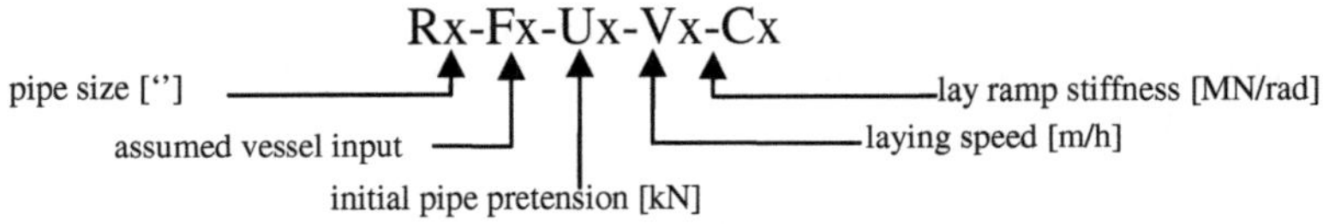

Fig. 9.44. Notations in the results presented on graphs

In Fig. 9.45, the results of calculations for a pipeline with diameter of 4 inches are shown. Graphs of tension indicate significant instability of operation caused by waves and lack of possibility to eliminate undesirably large overloads by changing the braking force of the reel (Fig. 9.45b and c). Fairly stable operation is guaranteed only for modest values of the amplitude of the swaying angle ψ_D when the braking force of the reel is increased (Fig. 9.45a and d). The decay of tension (values of the force U_E near zero in Fig. 9.45b and c) is due to excessive unwinding of the pipe from the reel caused by the increase in the reel's angular velocity (Fig. 9.46).

The graphs of the reel's velocity are shown in Fig. 9.46. The results obtained for F1 and F4 waves (little swaying, Fig. 9.46a and Fig. 9.46d) indicate fairly stable operation of the device in the range of swaying amplitudes from 0 to 1°. Lack of control of the reel's velocity under more intense waves leads to unwinding of great amounts of the pipeline and subsequently to abrupt arrest of the reel (jerk). Increasing the braking force makes it possible to reduce the maximum speed of the reel but it also causes high values of the axial forces which are dangerous to the personnel and the device.

Table 9.7. Main parameters of the device and sizes of pipelines used

Parameter	Value
Admissible mass of the reel together with the wound pipe	$2500 \cdot 10^3$ kg
Range of diameters of steel pipes installed	4"–18"
Designed value of pipes' tension	2500 kN
Capacity of the reel: (for max/min. diameter of the pipeline)	150 km / 7 km
Max/min. Winding diameter of pipes	$D_O = 25$ m / $D_I = 15$ m
Moment of inertia of the reel with the wound pipeline	$2.5 \div 3.0 \cdot 10^8$ kgm ^{2}
Length of the vessel	~100 m
Diameter of the gear wheel	$D_p = 25.7$ m
Link mass m_u of the pipeline for pipe diameter D	$D=4"$, $m_u = 16$ kg $D=8"$, $m_u = 42$ kg $D=12"$, $m_u = 128$ kg $D=16"$, $m_u = 240$ kg $D=18"$, $m_u = 340$ kg
Length of the guiding ramp	$L_R = 20$ m
Radius of the ramp's guiding wheel	$r_h = 8$ m
Mass of the ramp with devices	$120 \cdot 10^3$ kg
Distance between the reel and the joint attaching the ramp	$L_H = 55$ m
Ramp inclination angle	$\alpha_R = 60^0$

Table 9.8. Assumed parameters of the vessel's lifting motion

Description	A_X / γ_X, [m]/[°]	A_Y / γ_Y, [m]/[°]	A_ψ / γ_ψ, [m]/[°]	Period of the wave T [s]	Height of the wave H_s [m]	Direction of the wave β_f [°]
F1	0.12/90	0.4 / 75	0.15 / -45	6	3.0	30
F2	0/0	0.45 / 90	2.55 / 20	7	3.0	0
F3	1.0/-100	2.2 / 0	4.1 / 60	8	5.0	60
F4	0.27/-95	0.27 / 0	0.85 / 70	10	1.0	0

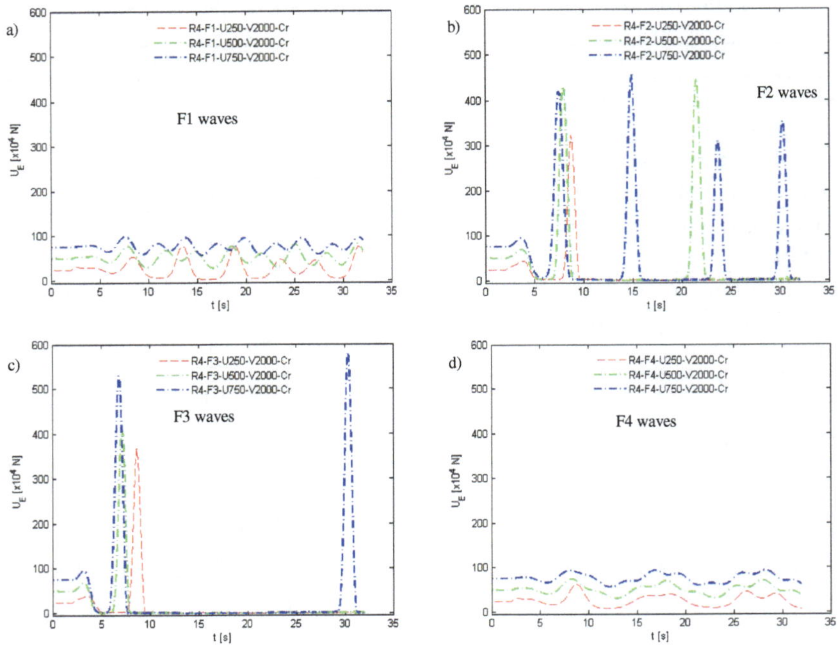

Fig. 9.45. Graphs of tension in the pipeline as it is unwound under waves and with different values of the barking force of the reel: a) ÷ d) correspond to F1 ÷ F4

Multiple factors influence the values of dynamic forces during the device's operation. Among those considered are: flexibility of the guiding ramp's suspension, speed of laying the pipeline, different diameters and load degrees of the reel, value of tension. The results are presented, among other things, in the papers [Szczotka M., 2010], [Szczotka M., 2011a]. In many cases it is impossible to eliminate or significantly reduce dangerously large dynamic forces under wave action with waves of height $H_s > 1$m without using an auxiliary control system. One is proposed in section 10.4.

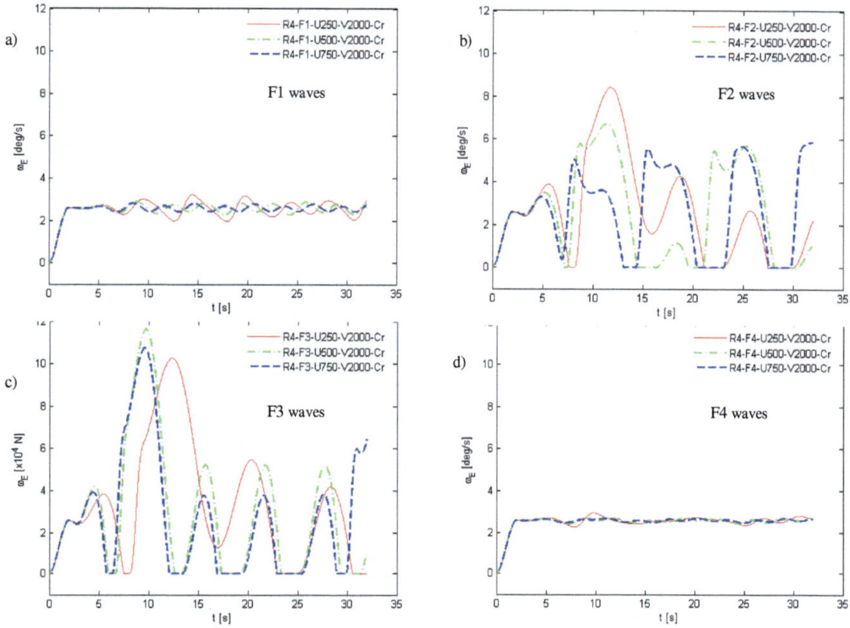

Fig. 9.46. Angular velocity of the reel for different waves conditions and values of braking forces of the reel: a) ÷ d) correspond to F1 ÷ F4

10 Selected Applications Related to Control of Offshore Structures

Dynamic analyses of mechanical systems are often considered together with problems related to their control. With traditional ways of operating machines it the operator who decides what the working motions are. In contemporary machines, it is becoming commonplace to support the process of control. Control systems based on microprocessor technology (programmable drivers, onboard computers) are supposed to facilitate human work or even replace it. They enable realization of various strategies unachievable with manual control. Automated control is used also in offshore structures, including cranes. The criteria of control strategies may be different, for example:

- minimal duration of motion,
- minimal consumption of energy,
- accuracy of load positioning, including minimization of oscillations after the motion has ended,
- minimization of dynamic loads,
- stabilization of the load's position,
- minimization of the influence of sea waves on the device's dynamics.

The present chapter describes the basics of the method of selecting the drive functions based on dynamic optimization. Control of the drum of a winch of an A-frame type crane allowing it to compensate for vertical movements of the base due to sea waves is presented. For an offshore jib crane, an auxiliary system is proposed enabling the load to be positioned in three directions. In the last part, a concept of active compensation of waves for a drum's drive of a device for laying pipelines is discussed.

10.1 Dynamic Optimization

As former considerations imply, the equation of dynamics of a multibody system may be presented in the following:

$$\mathbf{A}\ddot{\mathbf{q}} = \mathbf{f}\left(t, \mathbf{q}, \dot{\mathbf{q}}, M\right), \tag{10.1}$$

$$\mathbf{q}(0) = \mathbf{q}_0, \tag{10.2.1}$$

$$\dot{\mathbf{q}}(0) = \mathbf{q}_1, \tag{10.2.2}$$

E. Wittbrodt et al.: Rigid Finite Element Method, OEO 1, pp. 203–244.
springerlink.com

where $\mathbf{q} = [q_1 \quad \cdots \quad q_n]^T$ – vector of generalized coordinates of the system,

$\mathbf{f}$ – excitation vector due to these forces: elasticity, damping, centrifugal, Coriolis, gyroscopic, and to the drives,

M – function giving the drive force or moment thereof, henceforth assumed to be a specified function of time,

$\mathbf{q}_0, \mathbf{q}_1$ – vectors of initial values of generalized coordinates and velocities.

The equations (10.1) are typically nonlinear and require numerical integration. Duration of their integration is closely related to the number n of generalized coordinates. Assuming large n considerably lengthens the calculations, whereas n being too small disables the mathematical model from adequately reflecting the dynamic properties of the system. The number n of generalized coordinates should therefore hit the balance between computation time and accuracy. Its choice depends largely on the purpose of the model.

In some cases, which are described in chapter 6, it is possible to write the equations (10.1) as a combined system of ordinary differential and nonlinear algebraic equations:

$$\mathbf{A}\,\ddot{\mathbf{q}} - \mathbf{D}\,\mathbf{R} = \mathbf{f}\left(t, \mathbf{q}, \dot{\mathbf{q}}, M\right), \tag{10.3.1}$$

$$\mathbf{\Lambda}\left(\mathbf{q}, \dot{\mathbf{q}}\right) = \mathbf{0}\,, \tag{10.3.2}$$

gdzie $\mathbf{A}, \mathbf{q}, \mathbf{f}$ – defined as in (10.1),

$\mathbf{R} = [R_1 \quad \cdots \quad R_m]^T$ – vector of constraint reactions,

$\mathbf{D} = \mathbf{D}(\mathbf{q})$ – matrix of coefficients,

$\mathbf{\Lambda} = [\lambda_1(\mathbf{q}, \dot{\mathbf{q}}) \quad \cdots \quad \lambda_m(\mathbf{q}, \dot{\mathbf{q}})]^T$ – vector of constraint equations,

m – number of components of the vector of constraint reactions.

To solve the equations (10.3) completed with initial conditions, a procedure is often applied whereby the constraint equations (10.3.2) are put in accelerative form by differentiation:

$$\mathbf{D}^T\ddot{\mathbf{q}} = \mathbf{W}\left(t, \mathbf{q}, \dot{\mathbf{q}}\right). \tag{10.4}$$

The equations (10.3.1) and (10.4) replace (10.3.2) and may be written as:

$$\begin{bmatrix} \mathbf{A} & -\mathbf{D} \\ \mathbf{D}^T & \mathbf{0} \end{bmatrix} \begin{bmatrix} \ddot{\mathbf{q}} \\ \mathbf{R} \end{bmatrix} = \begin{bmatrix} \mathbf{F}(t, \mathbf{q}, \dot{\mathbf{q}}, M) \\ \mathbf{W}(t, \mathbf{q}, \dot{\mathbf{q}}) \end{bmatrix}. \tag{10.5}$$

Further reasoning assumes that the equations of dynamics of the system take the form (10.1). For such equations, as just shown, can be easily extended to a system with constraints.

In the case of controlling working motions of machines, it is important to choose the drive function in such a way dependent on time that the intended goal is achieved according to the set criteria. One method of finding appropriate drive functions is by optimization. The optimization task in this case is to choose the drive function *M*(*t*) satisfying:

minimization of the functional:

$$\Omega(\mathbf{q},\dot{\mathbf{q}},M), \tag{10.6}$$

maintaining the boundary conditions:

$$e_i(\mathbf{q},\dot{\mathbf{q}}) \le 0 \qquad \text{for} \quad i = 1,\ldots, n_e, \tag{10.7.1}$$

$$M_L(t) \le M(t) \le M_R(t), \tag{10.7.2}$$

where n_e – number of boundary conditions,

$M_L(t), M_R(t)$ – known conditions constraining the drive function $M(t)$.

Note that it is necessary to know the vectors $\mathbf{q},\dot{\mathbf{q}}$ corresponding to the function $M(t)$ in order to determine the functional Ω and the function e_i. This requires integration of the system's equations of motion (10.1) in each optimisation step. An optimization task formulated in this way is called a dynamic optimization task [Kręglewski T., et al., 1984], since its focus is the integration of equations of dynamics.

The problem of choosing the function *M*(*t*) can be reduced to a classical optimization problem by discretisation. Let $t \in \langle 0, T \rangle$ and:

$$M_i = M(t_i) \qquad \text{for} \quad i = 1,\ldots, p, \tag{10.8}$$

where *p* is defined in Fig. 10.1.

The value of *M*(*t*) for $t \in \langle t_{i-1}, t_i \rangle$ may then be determined with cubic splines (Fig. 10.1), using the formula:

$$M^{(i)}(t) = a_i(t - t_{i-1})^3 + b_i(t - t_{i-1})^2 + c_i(t - t_{i-1}) + M_{i-1}, \tag{10.9}$$

whereby the coefficients a_i, b_i, c_i are chosen such that:

$$M^{(i)}(t_i) = M_i \qquad \text{for} \quad i = 1,\ldots, p, \tag{10.10.1}$$

$$M^{(i)'}(t_i) = M^{(i+1)'}(t_i) \qquad \text{for} \quad i = 1,\ldots, p-1, \tag{10.10.2}$$

$$M^{(i)''}(t_i) = M^{(i+1)''}(t_i) \qquad \text{for} \quad i = 1,\ldots, p-1, \tag{10.10.3}$$

and:

$$M^{(0)'}(t_0) = 0 \qquad \text{or} \qquad M^{(0)''}(t_0) = 0, \tag{10.10.4}$$

$$M^{(p)'}(t_p) = 0 \qquad \text{or} \qquad M^{(p)''}(t_p) = 0. \tag{10.10.5}$$

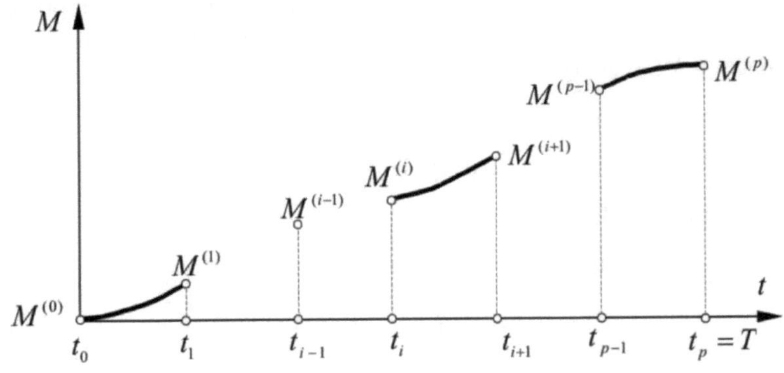

Fig. 10.1. Approximation of the function $M(t)$ with cubic splines

Equations (10.10) form a system of $3p$ linear algebraic equations with $3p$ unknowns $(a_1, b_1, c_1, ..., a_p, b_p, c_p)$. Its solution may be obtained easily by a recursive procedure.

If the drive function $M(t)$ is approximated by splines (10.9), then the optimization task consists in determining the p+1 values specified in (10.8). The decision variables in the considered task are therefore the components of the vector:

$$\mathbf{M} = [M_0 \quad \dots \quad M_p]^T. \tag{10.11}$$

Eventually, the problems of defining the optimization tasks of drive functions, considered in further examples, reduce to finding values $M_0, \dots, M_p$ which constitute the coordinates of the vector of decision variables (10.11) minimizing the functional:

$$\Omega(\mathbf{q}, \dot{\mathbf{q}}, M_0, ..., M_p), \tag{10.12}$$

and also satisfying the conditions:

$$e_i(\mathbf{q}, \dot{\mathbf{q}}) \leq 0 \qquad \text{for} \qquad i = 1, ..., n_e, \tag{10.13.1}$$

$$M_{iL} \leq M_i \leq M_{iR}, \; M_{iL} = M_L(t_i), \; M_{iR} = M_R(t_i) \quad \text{for} \quad i = 1, ..., p \tag{10.13.2}$$

The vectors $\mathbf{q}$ and $\dot{\mathbf{q}}$ are obtained by integrating the initial problem:

$$\mathbf{A}\ddot{\mathbf{q}} = \mathbf{f}(t, \mathbf{q}, \dot{\mathbf{q}}, M_0, ..., M_p), \tag{10.14.1}$$

$$\mathbf{q}(0) = \mathbf{q}_0, \quad \dot{\mathbf{q}}(0) = \mathbf{q}_1, \tag{10.14.2}$$

for $t \in \langle 0, T \rangle$.

Various methods may be used to solve this task. However, all of them are sensitive to the choice of the starting point, i.e. the initial values of $M_0, \dots, M_p$.

It can be easily noticed that the time cost of the optimization process directly depends on the time of integration of the equations of motion (10.14). Using for this case the mathematical models of the systems presented in preceding chapters would cause unacceptably long computations. It is the reason why the drive functions $M(t)$

are usually determined according to simplified models whose degrees of freedom are possibly few. This enables their possible application in real-time control. Verifying calculations which allow to assess the usefulness of the developed simplified models and control algorithms are carried out with respect to combined basic models, thus analogous to those described in earlier chapters.

10.2 Vertical Stabilization of Load of an A-Frame

In this chapter, two dynamic models of an A-frame are presented. In the first one, the flexibility of a frame is taken into account, while in the second this flexibility is omitted. In both cases the flexibility of rope is considered. The classical Rigid Finite Element Method has been used to discretise the frame – chapter 8.1. The algorithm of optimisation of the drive function for the drum of the hoisting winch is proposed. The goal of the optimisation is to ensure the stabilization of the load's position, i.e. to hold it at the required depth regardless of the ship's motion. In order to achieve appropriate numerical effectiveness, the optimisation problem has been solved using a simplified model of an A-frame.

10.2.1 A-Frame Model

The scheme of an A-frame and the most important points of it are presented in Fig. 10.2. The following denotations are used: F – supporting structure, P – pulley,

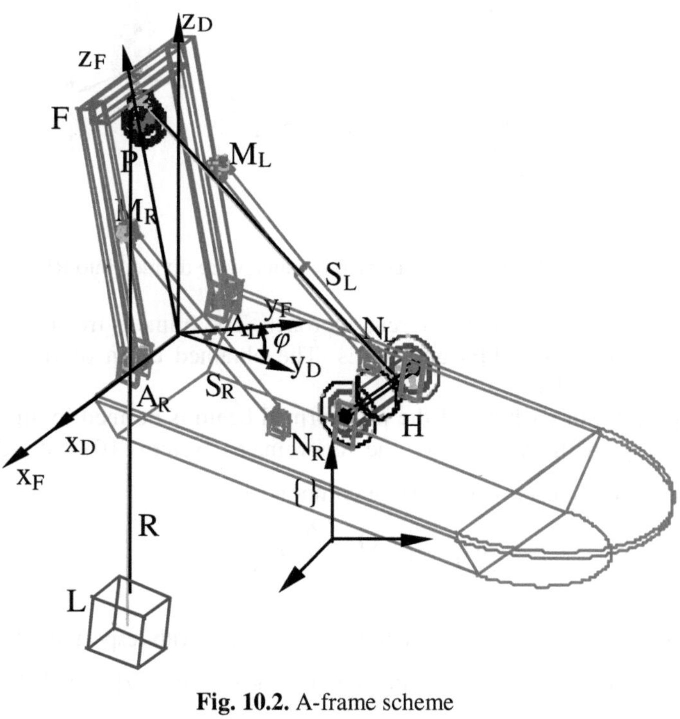

Fig. 10.2. A-frame scheme

R – rope, H – drum of the hoisting winch, L – load, S_R, S_L – right and left servomotor forces, N_R, N_L – connection points of servomotors to the A-frame, A_R, A_L – connection points of the A-frame to the deck, x_F, y_F, z_F and x_D, y_D, z_D – coordinate systems assigned to the supporting structure (frame) and to the deck, respectively.

The frame is the main element of the supporting structure in such cranes. In order to discretize the frame, the rigid finite element method can be applied.

In doctoral thesis [Fałat P., 2004], at first three beams were distinguished (right-1, top-2, left-3) in the frame. Then, each beam was divided into rigid finite elements and spring-damping elements (Fig. 10.3). This necessitates taking into account the reaction forces and moments at points B_L and B_B, and increases the number of constraint equations.

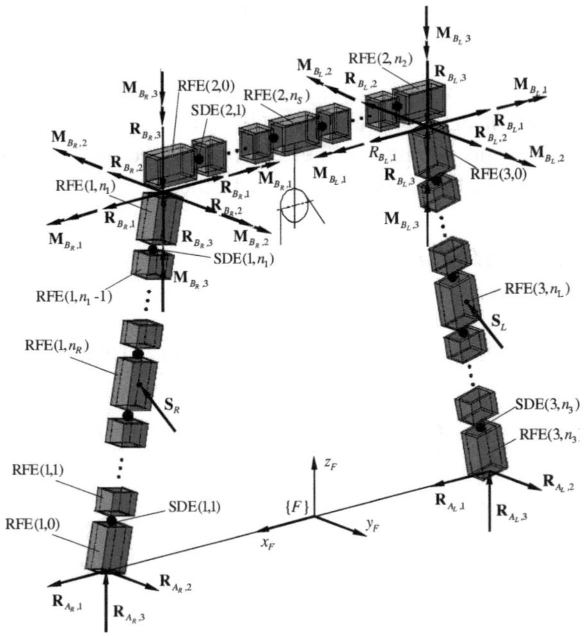

Fig. 10.3. A-frame divided into three beams which were divided into RFEs and SDEs

In this work we present a different approach. The frame is treated as one beam, which is divided into RFEs and SDEs. The obtained chain of rfes and sdes is presented in Fig. 10.4.

The position of each rfe of the undeformed beam is defined by the coordinate system ${}^{E}\{i\}$ with respect to the coordinate system $\{0\}$ of RFE 0, by a transformation matrix with constant components:

$$ {}_{E}^{0}\mathbf{T}_i = \begin{bmatrix} {}_{E}^{0}\mathbf{\Theta}_i & {}_{E}^{0}\mathbf{s}_i \\ \mathbf{0} & 1 \end{bmatrix}, \tag{10.15} $$

where ${}_{E}^{0}\mathbf{\Theta}_i$ is the matrix of cosines of the system ${}^{E}\{i\}$ with respect to $\{0\}$, and ${}_{E}^{0}\mathbf{s}_i$ is the vector of coordinates of the origin of the system ${}^{E}\{i\}$ in $\{0\}$ (Fig. 10.5).

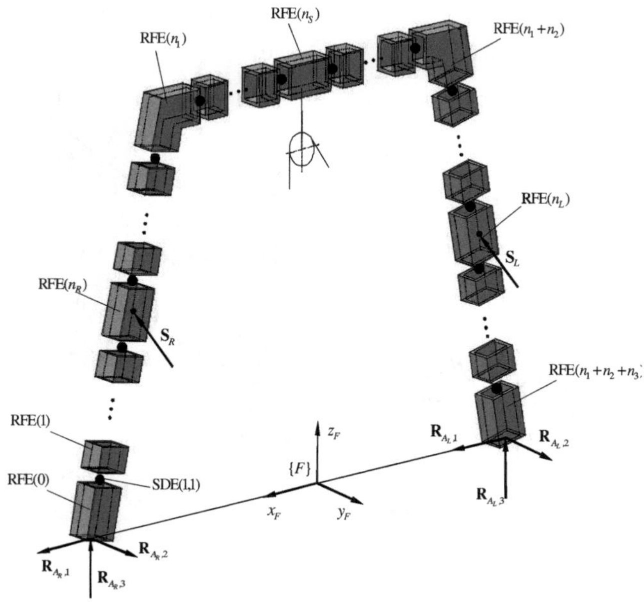

Fig. 10.4. A-frame as one beam, and its division into RFEs and SDEs

The coordinate system $\{i\}$ rigidly attached to RFE i moves together with the RFE when the beam is deformed. Its position in the coordinate system $^{E}\{i\}$ is defined by generalized coordinates of the ith element, which are the components of the vector:

$$\mathbf{q}_i = \begin{bmatrix} \mathbf{r}_i \\ \boldsymbol{\varphi}_i \end{bmatrix}, \tag{10.16}$$

where $\mathbf{r}_i = \begin{bmatrix} x_i & y_i & z_i \end{bmatrix}^T$ and $\boldsymbol{\varphi}_i = \begin{bmatrix} \varphi_i & \theta_i & \psi_i \end{bmatrix}^T$ are vectors of displacements and rotation angles presented in Fig. 10.5.

If we assume that angles $\varphi_i, \theta_i, \psi_i$ are small, then the transformation matrix from the local coordinate system $\{i\}$ to the system $^{E}\{i\}$ takes the following form:

$$\mathbf{T}_i = \begin{bmatrix} 1 & -\psi_i & \theta_i & x_i \\ \psi_i & 1 & -\varphi_i & y_i \\ -\theta_i & \varphi_i & 1 & z_i \\ 0 & 0 & 0 & 1 \end{bmatrix} = \mathbf{I} + \sum \mathbf{D}_j q_{i,j}, \tag{10.17}$$

where $\mathbf{q}_i = \begin{bmatrix} x_i & y_i & z_i & \varphi_i & \theta_i & \psi_i \end{bmatrix}^T$,

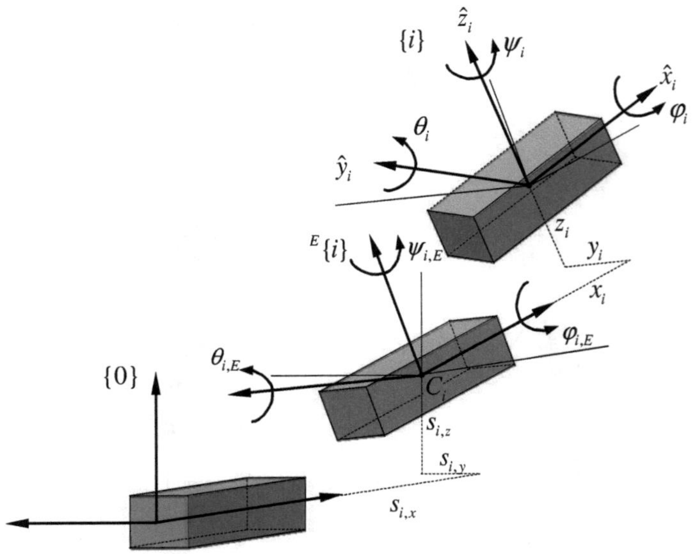

Fig. 10.5. The systems of ith RFE and generalized coordinates

$$\mathbf{D}_1 = \begin{bmatrix} 0 & 0 & 0 & 1 \\ 0 & 0 & 0 & 0 \\ 0 & 0 & 0 & 0 \\ 0 & 0 & 0 & 0 \end{bmatrix}, \quad \mathbf{D}_2 = \begin{bmatrix} 0 & 0 & 0 & 0 \\ 0 & 0 & 0 & 1 \\ 0 & 0 & 0 & 0 \\ 0 & 0 & 0 & 0 \end{bmatrix}, \quad \mathbf{D}_3 = \begin{bmatrix} 0 & 0 & 0 & 0 \\ 0 & 0 & 0 & 0 \\ 0 & 0 & 0 & 1 \\ 0 & 0 & 0 & 0 \end{bmatrix},$$

$$\mathbf{D}_4 = \begin{bmatrix} 0 & 0 & 0 & 0 \\ 0 & 0 & -1 & 0 \\ 0 & 1 & 0 & 0 \\ 0 & 0 & 0 & 0 \end{bmatrix}, \quad \mathbf{D}_5 = \begin{bmatrix} 0 & 0 & 1 & 0 \\ 0 & 0 & 0 & 0 \\ -1 & 0 & 0 & 0 \\ 0 & 0 & 0 & 0 \end{bmatrix}, \quad \mathbf{D}_6 = \begin{bmatrix} 0 & -1 & 0 & 0 \\ 1 & 0 & 0 & 0 \\ 0 & 0 & 0 & 0 \\ 0 & 0 & 0 & 0 \end{bmatrix}.$$

The transformation matrix $\mathbf{B}_i$ that allows us to transform coordinates from the local coordinate system $\{i\}$ to the inertial coordinate system $\{\}$ according to the relation:

$$\mathbf{r} = \mathbf{B}_i\, \mathbf{r}_i, \tag{10.18}$$

where $\mathbf{r}_i$ – vector of coordinates in local system $\{i\}$,

$\mathbf{r}$ – vector of coordinates in base system $\{\}$,

has the form:

$$\mathbf{B}_i = \mathbf{B}_i(t, \mathbf{q}_i) = \mathbf{T}_D\ \mathbf{T}_F\ {}^0_E\mathbf{T}_i\ \mathbf{T}_i = \mathbf{A}(t)\ \mathbf{P}_i(\mathbf{q}_i), \tag{10.19}$$

where $\mathbf{T}_D = \mathbf{T}_D(t)$ – defines the motion of the ship deck with respect to the base system $\{\}$,

$\mathbf{T}_F = \mathbf{T}_F(\varphi(t))$ – describes the rotation of the frame in the coordinate system of the deck $\{D\}$,

${}_E^0\mathbf{T}_i = \text{const}$ – defined in (10.15),

$\mathbf{T}_i = \mathbf{T}_i(\mathbf{q}_i)$ – presented in (10.17),

$\mathbf{A}(t) = \mathbf{T}_D\,\mathbf{T}_F$,

$\mathbf{P}_i = {}_E^0\mathbf{T}_i\,\mathbf{T}_i$.

In the case when the axes of the local coordinate system $\{i\}$ are chosen as principal central axes of the RFE, the mass and inertial features of the RFE i are defined by: its mass, m_i, and $J_{i,j}$ $(j = 1, 2, 3)$ which are mass moments of inertia with respect to the axis $\hat{x}_i$, $\hat{y}_i$, $\hat{z}_i$.

The equations of motion of the system considered can be obtained from Lagrange equations. This approach requires the kinetic and potential energy of the system to be defined. The kinetic energy of the RFE i can be calculated as:

$$E_i = \frac{1}{2}\operatorname{tr}\{\dot{\mathbf{B}}_i\,\mathbf{H}_i\,\dot{\mathbf{B}}_i^T\}, \tag{10.20}$$

where $\mathbf{H}_i$ – the pseudo-inertia matrix defined as in (5.11).

Following the considerations, we can obtain:

$$\frac{d}{dt}\frac{\partial E_i}{\partial \dot{\mathbf{q}}_i} - \frac{\partial E_i}{\partial \mathbf{q}_i} = \mathbf{M}_i\ddot{\mathbf{q}}_i + \mathbf{e}_i, \tag{10.21}$$

where $\mathbf{M}_i = diag\left[m_i, m_i, m_i, J_z, J_y, J_x\right]$,

$$e_{i,j} = e_{i,j}(t, \mathbf{q}_i, \dot{\mathbf{q}}_i) = \operatorname{tr}\left\{\mathbf{B}_{i,j}\mathbf{H}_i\left[\ddot{\mathbf{A}}\mathbf{P}_i + 2\dot{\mathbf{A}}\dot{\mathbf{P}}_i\right]^T\right\},$$

$$\mathbf{B}_{i,j} = \mathbf{A}\frac{\partial \mathbf{P}_i}{\partial q_{i,j}},$$

$$\frac{\partial \mathbf{P}_i}{\partial q_{i,j}} = {}_E^0\mathbf{T}_i\mathbf{D}_j = \text{const}.$$

The kinetic energy of the frame can be expressed by:

$$E = \sum_{i=0}^{n} E_i, \tag{10.22}$$

where $n = n_1 + n_2 + n_3 + 1$,

and it is possible to calculate:

$$\frac{d}{dt}\frac{\partial E}{\partial \dot{\mathbf{q}}_F} - \frac{\partial E}{\partial \mathbf{q}_F} = \mathbf{M}_F \ddot{\mathbf{q}}_F + \mathbf{e}_F , \tag{10.23}$$

where $\mathbf{M}_F = \text{diag}\left[\mathbf{M}_0, \ldots, \mathbf{M}_n\right]$,

$\mathbf{e}_F = \left[\mathbf{e}_0^T \ldots \mathbf{e}_n^T\right]^T$,

$\mathbf{q}_F = \left[\mathbf{q}_0^T \ldots \mathbf{q}_n^T\right]^T$.

The potential energy of deformation of SDEs can be expressed as follows:

$$V_F = \frac{1}{2}\mathbf{q}_F^T \mathbf{K}_F \mathbf{q}_F , \tag{10.24}$$

where $\mathbf{K}_F$ – the stiffness matrix with constant coefficients.

Similarly, one can calculate the dissipation of energy as:

$$D_F = \frac{1}{2}\dot{\mathbf{q}}_F^T \mathbf{L}_F \dot{\mathbf{q}}_F , \tag{10.25}$$

where $\mathbf{L}_F$ – the damping matrix with constant elements.

From what has been written above, one can calculate:

$$\frac{\partial V_F}{\partial \mathbf{q}_F} = \mathbf{K}_F \mathbf{q}_F , \tag{10.26.1}$$

$$\frac{\partial D_F}{\partial \dot{\mathbf{q}}_F} = \mathbf{L}_F \dot{\mathbf{q}}_F . \tag{10.26.2}$$

The potential energy of gravity forces of the frame can be calculated as:

$$V_g^F = \sum_{i=0}^{n} m_i g\, \boldsymbol{\theta}_3\, \mathbf{B}_i \mathbf{r}_{C,i} , \tag{10.27}$$

where $\mathbf{r}_{C,i} = [0 \quad 0 \quad 0 \quad 1]$.

So:

$$\frac{\partial V_g^F}{\partial \mathbf{q}_F} = \mathbf{G}_F , \tag{10.28}$$

where $\mathbf{G}_F = \left[\mathbf{G}_0^T, \ldots, \mathbf{G}_n^T\right]^T$,

$\mathbf{G}_i = \left[G_{i,1}, \ldots, \mathbf{G}_{i,6}\right]^T$,

$G_{i,j} = m_i\, g\, \boldsymbol{\theta}_3\, \mathbf{D}_j\, \mathbf{r}_{C,i}$,

$\mathbf{D}_j$ – defined in (10.17).

Energy of Load and Drum of the Hoisting Winch

The load is modelled as a particle. The vector of its generalized coordinates is expressed in the following form $\mathbf{q}_L = [x_L \quad y_L \quad z_L]^T$. The angle of rotation of the drum of the hoisting winch is denoted as φ_H. Kinetic energy of the load and the drum can then be calculated as:

$$T_R = \frac{1}{2} m_L \, \dot{r}_L^2 + \frac{1}{2} I_H \, \dot{\varphi}_H^2 \, , \tag{10.29}$$

where I_H – moment of inertia mass of the drum,
$\dot{r}_L^2 = \dot{x}_L^2 + \dot{y}_L^2 + \dot{z}_L^2$.

Potential energy of the load is determined as:

$$V_g^L = m_L \; g \; z_L \, . \tag{10.30}$$

Elastic Deformation of the Rope

The rope system of the A-frame is presented in the Fig. 10.6. It is assumed that the radii of pulleys are small compared to the dimensions of the whole mechanism, and also that the rope passes through points S and H – centres of the pulley and the drum, respectively. Because the radii of pulleys are small and the length of the rope may be hundreds of meters, this simplification can be seen as admissible.

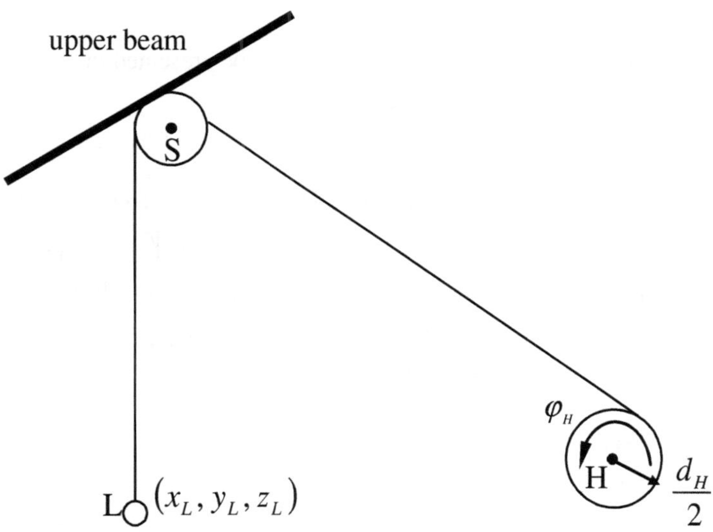

Fig. 10.6. Rope system of the A-frame

Potential energy of elastic deformation of the rope and its dissipation can be expressed in the following forms:

$$V_R = \frac{1}{2} c_R \delta_R \Delta_R^2 , \tag{10.31}$$

$$D_R = \frac{1}{2} d_R \delta_R \dot{\Delta}_R^2 , \tag{10.32}$$

where $\delta_R = \begin{cases} 0 & \text{if} \quad \Delta_R \le 0 \\ 1 & \text{if} \quad \Delta_R > 0 \end{cases}$,

$\Delta_R = |LS| + |SH| - l_0 - \varphi_H \dfrac{d_H}{2}$,

$|LS| = |\mathbf{r}_L - \mathbf{r}_S|$,

$|SH| = |\mathbf{r}_S - \mathbf{r}_H|$,

$c_R = \dfrac{E_R F_R}{l}$ – stiffness coefficient of the rope,

d_R – damping coefficient of the rope,

l_0, l – initial and current length of the rope, respectively,

E_R – Young's modulus of the rope material,

F_R – cross-section of the rope,

d_H – diameter of the drum.

Motion Equations

The vector of A-frame generalised coordinates can be presented in the form:

$$\mathbf{q} = \begin{bmatrix} \mathbf{q}_F \\ \mathbf{q}_R \end{bmatrix}, \tag{10.33}$$

where $\mathbf{q}_F$ – the vector of generalised coordinates of the discretised frame defined in (10.23) and vector $\mathbf{q}_R = [x_L \quad y_L \quad z_L \quad \varphi_H]^T$ contains generalised coordinates of the load and the angle of rotation of the drum.

Then, the equations of motion of the system can be written as:

$$\mathbf{M}\ddot{\mathbf{q}} + \mathbf{L}\dot{\mathbf{q}} + \mathbf{K}\mathbf{q} = \mathbf{Q} + \mathbf{DR} , \tag{10.34}$$

where $\mathbf{M} = \begin{bmatrix} \mathbf{M}_F & 0 \\ 0 & \mathbf{M}_L \end{bmatrix}$,

$\mathbf{M}_L = diag\,[m_L, m_L, m_L, I_H]$,

$\mathbf{L} = \begin{bmatrix} \mathbf{L}_F & 0 \\ 0 & 0 \end{bmatrix}$, $\mathbf{K} = \begin{bmatrix} \mathbf{K}_F & 0 \\ 0 & 0 \end{bmatrix}$,

$$\mathbf{Q} = \begin{bmatrix} -\frac{\partial V_g^F}{\partial \mathbf{q}_F} - \frac{\partial V_R}{\partial \mathbf{q}_F} - \frac{\partial D_R}{\partial \dot{\mathbf{q}}_F} - \mathbf{e}_F \\ -\frac{\partial V_g^L}{\partial \mathbf{q}_R} - \frac{\partial V_R}{\partial \mathbf{q}_R} - \frac{\partial D_R}{\partial \dot{\mathbf{q}}_R} - \mathbf{e}_L \end{bmatrix},$$

D, **R** – matrix and vector of reaction forces,

$\frac{\partial V_R}{\partial \mathbf{q}_F}, \frac{\partial V_R}{\partial \mathbf{q}_R}, \frac{\partial D_R}{\partial \dot{\mathbf{q}}_F}, \frac{\partial D_R}{\partial \dot{\mathbf{q}}_R}$ can be calculated as in chapter 9.1.1 and involves nonlinear terms.

Forces of reactions on the frame are presented in Fig. 10.4. Vector **R** of generalised forces then specifically includes:

- reaction $\mathbf{R}_{AL} = \left[R_{AL,x} \quad R_{AL,y} \quad R_{AL,z}\right]^T$,
- reaction $\mathbf{R}_{AR} = \left[R_{AR,x} \quad R_{AR,y} \quad R_{AR,z}\right]^T$,
- and forces in servomotors S_L and S_R.

These forces can be written in the vector form:

$$\mathbf{R} = \left[S_R \quad S_L \quad \mathbf{R}_{AR}{}^T \quad \mathbf{R}_{AL}{}^T\right]^T. \tag{10.35}$$

Finally, the mathematical model of an A-frame has been written in the form of a system of differential equations of the second order (10.34) and constraint equations in acceleration form:

$$\mathbf{D}^T \ddot{\mathbf{q}} = \mathbf{W}, \tag{10.36}$$

where $\mathbf{W} = \mathbf{W}(\mathbf{q}, \dot{\mathbf{q}})$.

In these equations, there are: $n_q = 6(1+n)+4$ (components of vector **q**) plus $n_R = 2+2\cdot 3 = 8$ (components of vector **R**) unknowns. So, the number of unknowns is equal to the sum of numbers of equations (10.34) and (10.36).

10.2.2 Optimisation Problem

One of the major problems connected with the design and control of cranes is the choice of the drive functions which ensure proper motion of the system. In the case of A-frame, a very important problem is the stabilisation of load position, regardless of motion of the ship caused by sea waves. Using the drive of the drum of the hoisting winch we can try to solve this problem. Time courses of drive functions can be defined in the optimisation process. In the book, the objective function is assumed to be in one of the following forms:

$$\Omega_1 = \int_0^{t_k} \left[z_L - h \right]^2 \to \min , \tag{10.37.1}$$

$$\Omega_2 = \max_{0 \le t \le t_k} \left| z_L - h \right| \to \min , \tag{10.37.2}$$

where z_L – load coordinate,
h – required depth.

This means that one expects that as the result of optimisation the course of the function $\varphi_H(t)$ will be obtained which minimizes the average or maximal value of deviation of load position from the required amount. During the optimisation process, the parameters of ship hull movement and coordinates of the winch position have been assumed to be known.

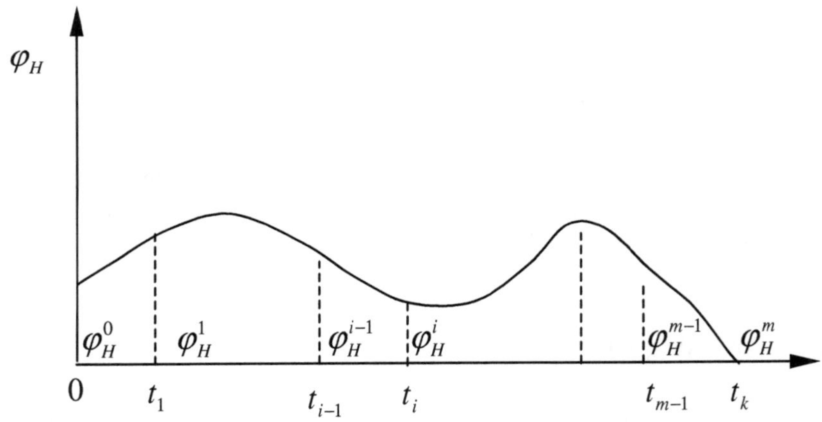

Fig. 10.7. The decisive variables

In this chapter, we assume that the function φ_H which describes the function $\varphi_H(t)$ defining the rotation angle of the winch drum has either the form:

$$\varphi_H(t) = a_i t^3 + b_i t^2 + c_i t + d_i , \text{ for } t \in < t_{i-1}, t_i > , \tag{10.38.1}$$

where $i = 1, \ldots, m$,
a_i, b_i, c_i, d_i – coefficients taken as shown in chapter 10.1 for spline functions of the third order,
t_i – point in interval $< 0, t_k >$ (Fig. 10.7),

or that introduced by [Maczyński A., 2005]:

$$\varphi_H(t) = A_0 + \sum_{i-=1}^{s} A_i \sin(\omega_i t + \alpha_{i,0}) \quad \text{for } t \in \langle t_{i-1}, t_i \rangle, \tag{10.38.2}$$

where A_i – amplitudes,

ω_i – frequencies,

$\alpha_{i,0}$ – phase angles.

As the decisive variables in the optimisation task we can choose:

$$\mathbf{X} = \left[\varphi_H^0, \varphi_H^1, \ldots, \varphi_H^m\right]^T \tag{10.39.1}$$

in the case (10.38.1), i.e. when spline functions are applied (Fig. 10.7), or:

$$X = \left[A_0, A_1, \omega_1, \alpha_{1,0}, \ldots A_s, \omega_s, \alpha_{s,0}\right]^T \tag{10.39.2}$$

in the case when a pseudo-harmonic response is assumed.

In either case, at every step of the optimisation, the equations of motion of the system have to be integrated for $t \in \langle 0, t_k \rangle$ in order to calculate the value of the functional $\Omega_{1,2}$ from (10.37). Such an approach requires high numerical efficiency in solving A-frame equations of motion. For that reason, the optimisation problem has been solved for the simplified model of an A-frame.

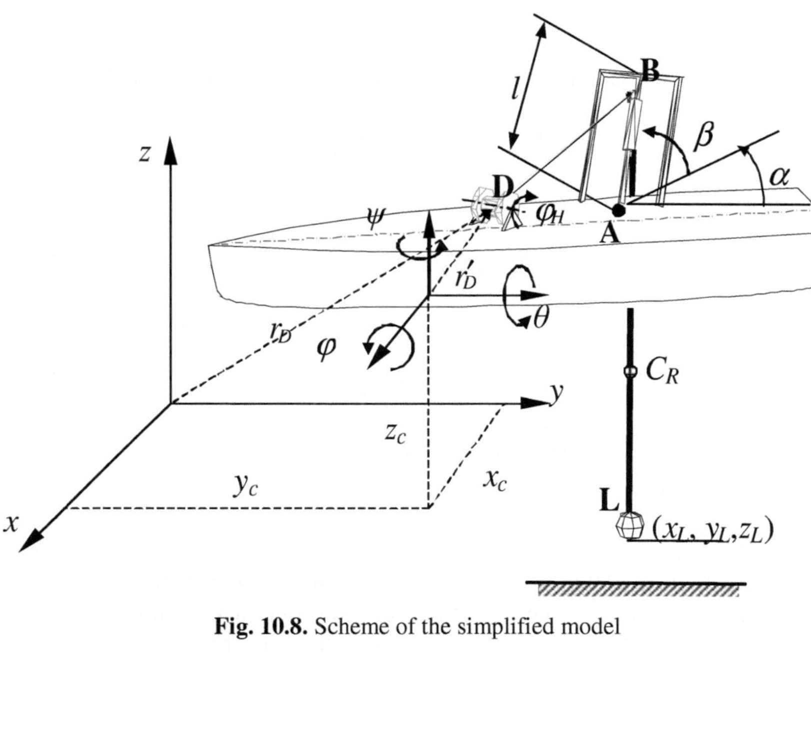

Fig. 10.8. Scheme of the simplified model

In the simplified model of an A-frame, ideal stiffness of the frame has been assumed (Fig. 10.8). However, flexibility of the rope has been taken into consideration. The water damping ratio has not been taken into account. Ship motion has been assumed to be determined, by known functions:

$$\begin{aligned} x_c &= x_c(t) \\ y_c &= y_c(t) \\ z_c &= z_c(t) \\ \varphi &= \varphi(t) \\ \theta &= \theta(t) \\ \psi &= \psi(t) \end{aligned} . \tag{10.40}$$

This means that matrix $\mathbf{T}_D$, from (10.19), has the form:

$$\mathbf{T}_D = \begin{bmatrix} c\psi c\theta & c\psi s\theta s\varphi - s\psi c\varphi & c\psi s\theta c\varphi + s\psi s\varphi & x_c \\ s\psi c\theta & s\psi s\theta s\varphi + c\psi c\varphi & s\psi s\theta c\varphi - c\psi s\varphi & y_c \\ -s\theta & c\theta s\varphi & c\theta c\varphi & z_c \\ 0 & 0 & 0 & 1 \end{bmatrix}, \tag{10.41}$$

where $c()=\cos()$ and $s()=\sin()$.

The frame angles are assumed to be constant.

Kinetic and potential energy of the system can be expressed in the form:

$$T = \frac{1}{2} m_L \left(\dot{x}_L^2 + \dot{y}_L^2 + \dot{z}_L^2 \right), \tag{10.42.1}$$

$$V = \frac{1}{2} \delta_R \, c_R \, \Delta_R^2 + m_L \, g \, z_L , \tag{10.42.2}$$

$$D = \frac{1}{2} \delta_R \, d_R \, \dot{\Delta}_R , \tag{10.42.3}$$

where m_L, δ_R, c_R, d_R – defined in (10.31),

$\Delta_R = |DB| + |BN| - l_0 + \varphi_H \, r_H$.

Lagrange's equations of the second order have been used to determine the equations of motion of the system. The details are presented in [Fałat P., et al., 2005]. These differential equations of the second order have been integrated using the Runge-Kutta method. The Nelder-Meads method has been applied in order to solve the optimisation task.

10.2.3 Numerical Simulations

It should be mentioned that the numerical model of the A-frame presented in chaper 10.2.2 has been used in the Norwegian company TTS-Aktro from Molde for a fast analysis of forces and stresses at the initial stage of choosing parameters of the system and for cost calculations. In order to verify the model, the results obtained using our program (RFEM) have been compared with those obtained using commercial FEM program (NASTRAN package) [Fałat P., et al., 2001]. There have been compared reactions in joints, stresses and deflections of beams obtained. Some examples are presented in Fig. 10.9.

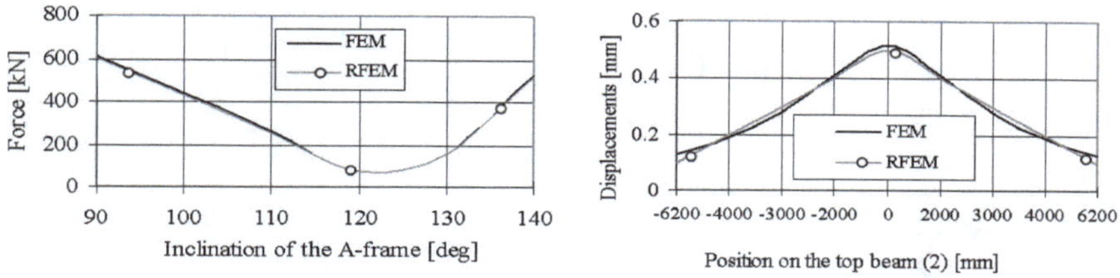

Fig. 10.9. Comparison of FEM and RFEM models

A comparison of the results obtained using RFEM model with those from Ansys-Adams systems in dynamical conditions can be find in [Fałat P., 2004] and same of them are presented Fig. 10.10.

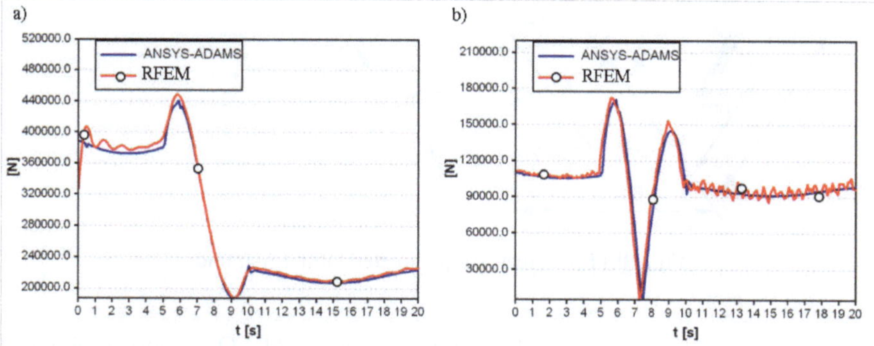

Fig. 10.10. Comparison of RFEM and Ansys-Adams models: a) vertical reaction in the A-frame leg, b) force in the servo-motor

Numerical simulations related to the load stabilisation problem have been carried out for the rectangular A-frame with lifting capacity up to 100 Mg. The main geometrical parameters of the crane are presented in Fig 10.11. The value of load coordinates z_L, for which the optimisation process has been carried out is h=-300 m, mass of load m_L=100 Mg, and the motion of the ship is defined as:

$$\begin{aligned} x_c(t) &= 1\sin\left(\frac{2\pi}{6}t\right) \\ y_c(t) &= 0 \\ z_c(t) &= 2\sin\left(\frac{2\pi}{12}t\right). \\ \varphi &= 0 \\ \theta &= 0 \\ \psi &= 0 \end{aligned} \tag{10.43}$$

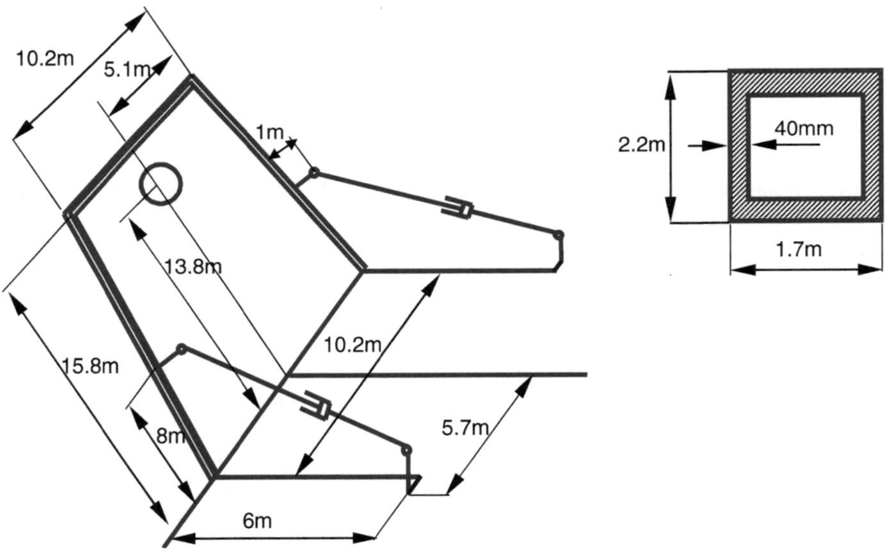

Fig. 10.11. Geometrical parameters of the A-frame

In the figures the following denotations are used: Ω_1, Ω_2 – curves obtained according to (10.37.1) and (10.37.2), respectively, S, H - curves obtained according to (10.39.1) and (10.39.2). Time courses of coordinate z_L obtained according to the full and the simplified model are shown in Fig. 10.12. In this case, the hoisting winch was motionless. The results of simulations are almost the same.

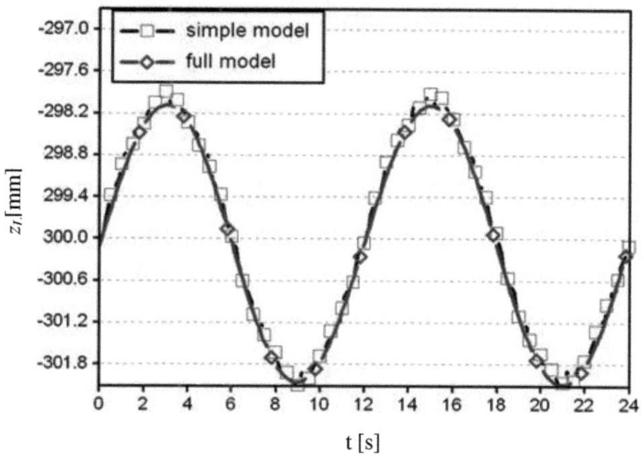

Fig. 10.12. Time courses of coordinates z_L

Because the simplified model is much more numerically efficient, the optimisation process has been solved for this model. Time courses of drive functions of the drum defined during the optimisation process are presented in Fig 10.13.

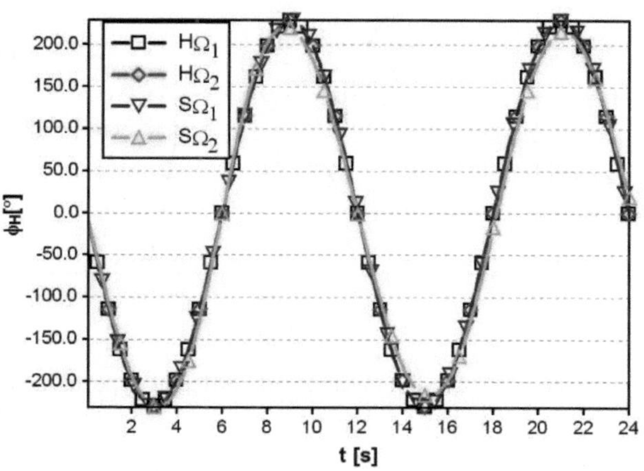

Fig. 10.13. Drive functions of drum after optimisation

As we can see, insignificant differences occurred between these drive functions. Drive functions obtained during optimisation have been taken as inputs of drum motion in the full model, so simulations presented below have been carried out according to the full model. Time courses of the coordinate z_L obtained when the drum of the hoisting winch was motionless and when its motion was determined by the function after optimisation (regardless of the type of the objective function and type of the drive function) are shown in Fig. 10.14. Amplitude of load oscillations has been decreased from 2 m to near zero.

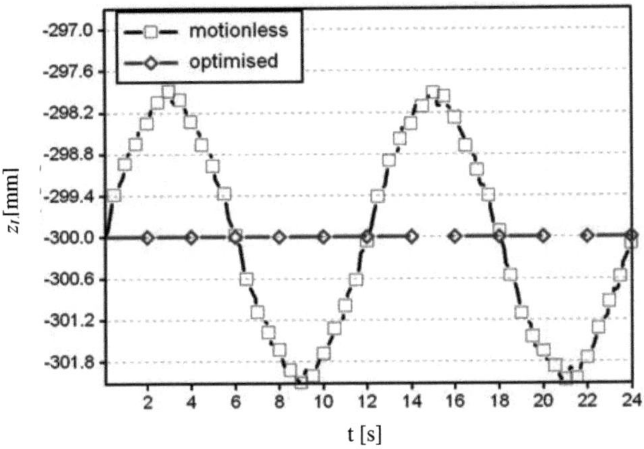

Fig. 10.14. Coordinate x_{L3} before and after optimisation

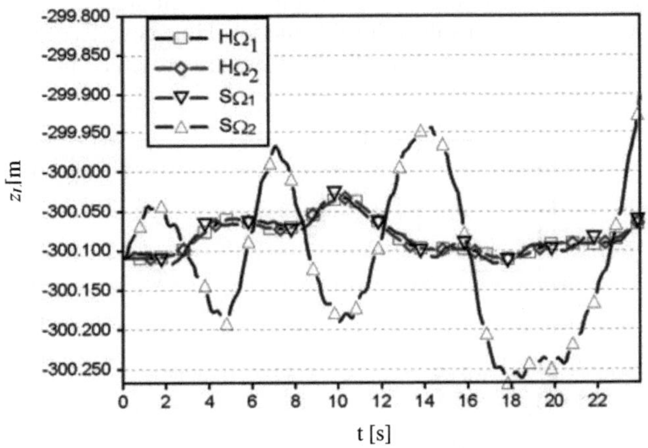

Fig. 10.15. Coordinate $x_{L,3}$ after optimisation for different type of objective and drive functions

Time courses of the coordinate z_L obtained for different types of objective functions and drive functions are presented in Fig. 10.15. The courses for the pseudo-harmonic drive function (10.38.2) and different types of objective functions are presented, in detail, in Fig. 10.16.

The model of an A-frame based on the finite element method has proved to be a useful instrument for carrying out dynamic analyses of this kind of cranes. This model is more numerically effective than the previous model presented in [Fałat P., 2004] (Fig. 10.3).

Numerical simulations presented in the chapter confirm the significant efficiency of the proposed method of optimisation drive function of the drum

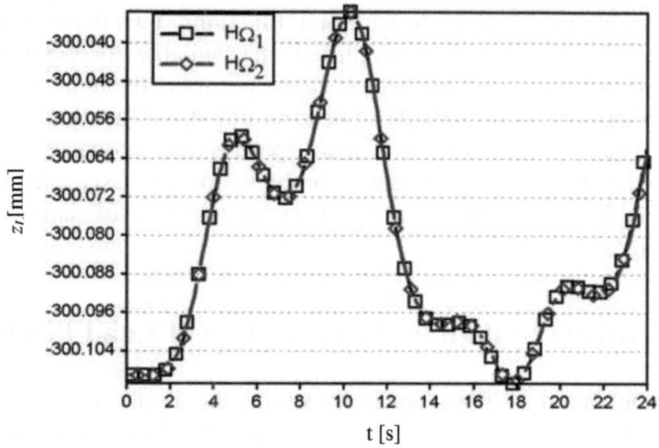

Fig. 10.16. Coordinate $x_{L,3}$ after optimisation for pseudo-harmonic drive function

where the main goal of the optimisation process is the stabilisation of the load position. Because the optimisation task has been carried out for the simplified model, the method is sufficiently effective.

For the motion of the ship discussed, the pseudo-harmonic drive functions are slightly better than spline functions. Amplitudes of load oscillations in the z_L direction are, for pseudo-harmonic functions, about 8 times smaller then for the spline function and the objective function Ω_1. When the objective function Ω_2 is taken, the results obtained are worst. However, when the system of waves is more complicated, the spline functions may be more useful.

Both objective functions, that is average and maximal value of deviation of load position from the demanded level, are acceptable in practice. There are no significant differences between results obtained for the two functions.

In real conditions, there are additional phenomena that can influence the quality of the stabilisation of the load position. There may be, for example, inaccurate definition of parameters of the crane. We should also remember that the rope interacts with the load and the environment mainly at low levels of depths, where water currents and waves are strong. Especially, in some conditions, a taut-slack phenomenon of a marine cable-body system can be significant [Huang S., 1999], [Jordan M. A., Bustamante J. L., 2007]. Vertical oscillations of the load induced by taut-slack phenomenon makes it more difficult to stabilise the load. An error-actuated control system for motion of the drum of the hoisting winch can minimise the impact of all those phenomena.

10.3 Stabilization of Crane Load with the Use of an Auxiliary System

In the present chapter, the authors take on the problem of load stabilization for a crane installed on a ship. The solution analyzed is inherently similar to the Knuckle Boom Telescopic Offshore Crane concept of TTS-Aktro. The base of the crane (i.e. the ship) is subject to general type of motion as a rigid body whose components are defined by pseudo-harmonic functions. Two dynamical models of a crane have been developed: a simplified one, suitable for determining drive functions, and a basic one, for which more accurate dynamical analyses are possible. Furthermore, two methods of determining drive functions of the crane's mechanisms ensuring stabilization of load's position are proposed. The first one involves solving the inverse kinematics problem for quasi-static conditions, the second one is based on dynamic optimization. Results of sample numerical simulations are included.

10.3.1 Auxiliary System for Stabilization of Load Position

Complete stabilization of load's position requires the ability to exert force on the load in three independent directions. In the vertical direction it is natural to use the hoisting winch drum's drive. For the stabilization in two remaining directions, the authors propose using an additional auxiliary system (Fig. 10.17). Its main component is a stiff element leading the rope along the GD segment. By inclining this element it is possible to move the point D of the hoist rope in two directions: tangential (τ) and radial (n). The directions τ and n have been defined relative to the trajectory of point G of the jib in its rotating motion. Dislocating point D is the means for influencing the load's motion and thus an attempt to stabilize it in the aforementioned directions becomes feasible. The proposed solution has the advantage of being applicable to stabilization of load's position or, when another strategy is selected, to other tasks, e.g. limiting swaying of the load during rotating movement of the crane's upper structure.

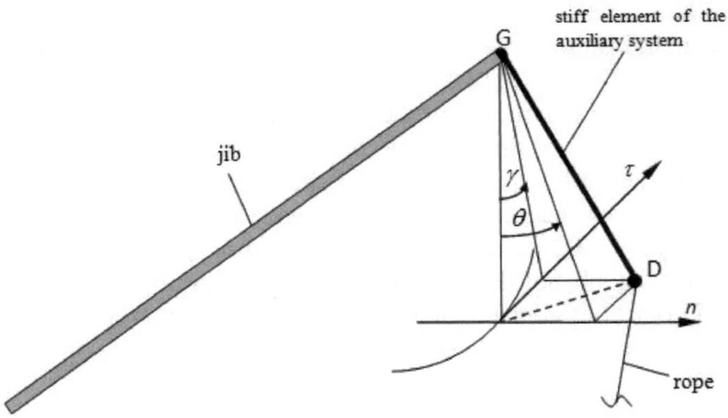

Fig. 10.17. Auxiliary system reducing load oscillations

10.3.2 Models of an Offshore Crane with the Auxiliary System

Depending on the purpose models of equipment are built which vary in complexity. Thus they represent the dynamic properties of the modelled object with different fidelity. In problems of control the speed of computation is very important. Real-time control is often required. This leads to possibly simple models being used which enable numerical efficiency. On the contrary, when the model is to be suitable for asserting validity of the equipment's operation, its basic dynamical behaviour must be reflected more accurately. Accuracy of the results obtained and their correspondence to reality is much more important than the time of computation in this case.

The authors have developed two models of an offshore crane with the auxiliary system: a simplified one, suitable for determining drive functions, and a basic one, for which more accurate dynamical analyses are possible. Simulations using the basic model enabled carrying out tests to confirm the suitability of the method proposed to the problem of load's position stabilization. In Table 10.1, there are compared the basic properties of both models. An outlined scheme of the simplified model is shown in Fig. 10.18.

The equations of motion in the basic model have been derived along the lines of [Wittbrodt E., et al., 2006] and [Adamiec-Wójcik I., et al., 2008]. Lagrange's equations of the second order have been used. The vector $\mathbf{q}$ of the generalized coordinates can be written as:

$$\mathbf{q}=\left[\mathbf{q}^{(A)^T} \quad \varphi \quad \psi \quad \mathbf{q}^{(J)^T} \quad \alpha \quad \gamma \quad \theta \quad \mathbf{q}^{(L)^T}\right]^T, \tag{10.44}$$

where $\mathbf{q}^{(A)}=\left[x_{org}^{(A)} \quad y_{org}^{(A)} \quad z_{org}^{(A)} \quad \varphi^{(A)} \quad \theta^{(A)} \quad \psi^{(A)}\right]^T$ – vector of generalized coordinates of the base (deck),

$\mathbf{q}^{(J)}=\left[q_1^{(J)} \quad \dots \quad q_k^{(J)} \quad \dots \quad q_m^{(J)}\right]^T$ – vector of generalized coordinates of the jib,

$\mathbf{q}^{(L)}=\left[x_L \quad y_L \quad z_L\right]^T$ – vector of generalized coordinates of the load,

φ – rotation angle of the crane's pedestal (upper structure) – slewing angle,

ψ – inclination angle of the undeformed jib,

α – rotation angle of the hoisting winch's drum.

γ,θ – angles of inclination of auxiliary system (Fig. 10.18).

Relationships which determine individual terms of the Lagrange's equations are obtained similarly as in the case of a mobile crane treated in [Maczyński A., Wojciech S., 2003].

To ensure that the crane's base moves according to the provisions of Table 10.1, the following condition must hold:

$$q_i^{(A)}=\alpha_i^{(A)}(t) \qquad \text{for } i=1,\dots,6. \tag{10.45}$$

Table 10.1. Comparison of models: simplified and basic

	Simplified model	Basic model
Form of description for system's points	absolute coordinates	coordinates relative to joints
Method of obtaining equations of motion	second Newton's law	Lagrange's equations of second order
Number of degrees of freedom	3	14+m where m is the total number of modes considered in the jib's model
Drives considered	1. hoisting winch drum's 2. auxiliary system's in directions τ and n	1. hoisting winch drum's 2. auxiliary system's in directions τ and n 3. of crane's upper structure's slewing 4. of reach changing (reach changing actuator)
Drive modeling method	kinematic driving	kinematic driving by a parallel system of a spring and a damper
Pedestal	modeled as a rigid body, fixed to the base (ship's deck)	
Load	modeled as an concentrated mass	
Rope	flexible	flexible with damping
Jib	rigid	capable of flexing – the jib has been discretized using modal method
Description of base's motion	pseudo-harmonic: $\alpha_i^{(A)} = \sum_{j=1}^{n_i^{(A)}} A_{i,j}^{(A)} \sin(\omega_{i,j}^{(A)} t + \varphi_{i,j}^{(A)})$ for $i = 1,\ldots,6$, where $\left[\alpha_1^{(A)} \;\ldots\; \alpha_6^{(A)}\right]^T = \left[x_{org}^{(A)} \; y_{org}^{(A)} \; z_{org}^{(A)} \; \varphi^{(A)} \; \theta^{(A)} \; \psi^{(A)}\right]^T = \boldsymbol{\alpha}$ $A_{i,j}^{(A)}$ - j^{th} amplitude in the ith direction, $\omega_{i,j}^{(A)}$ - j^{th} angular frequency in the ith direction, $\varphi_{i,j}^{(A)}$ - j^{th} phase angle in the ith direction.	
Equations of motion	$\begin{cases} m_L \ddot{x}_L = S \dfrac{x_D - x_L}{L_{DL}} + F_X + T_X \\ m_L \ddot{y}_L = S \dfrac{y_D - y_L}{L_{DL}} + F_Y + T_Y \\ m_L \ddot{z}_L = S \dfrac{z_D - z_L}{L_{DL}} + F_Z + T_Z - m_L g \end{cases}$ where $S = c_l (L_{DL} + L_D - L_0 + \alpha \cdot r_B) + b_l (\dot{L}_{DL} + \dot{\alpha} \cdot r_B)$, $L_{DL}^2 = (x_D - x_L)^2 + (y_D - y_L)^2 + (z_D -$, c_l, b_l – stiffness and damping coefficients of the rope, respectively, α – rotation angle of the hoisting winch's drum, r_B – radius of the hoisting winch's drum, L_0 – initial length of the rope.	Described later
Integration method for the equations of motion	Runge-Kutta method of fourth order	

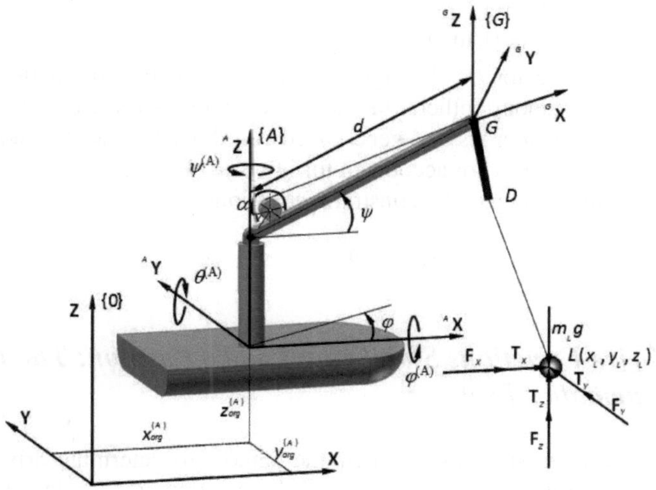

Fig. 10.18. Simplified model of an offshore crane

Forces and moments of force acting on the crane's base to make it move according to relationships (10.45) must be therefore introduced into the system. They are assumed to form the following vector:

$$\mathbf{R}^{(A)} = \left[F_x^{(A)}, F_y^{(A)}, F_z^{(A)}, M_x^{(A)}, M_y^{(A)}, M_z^{(A)}\right]^T . \tag{10.46}$$

Forces and moments of force are depicted in Fig. 10.19.

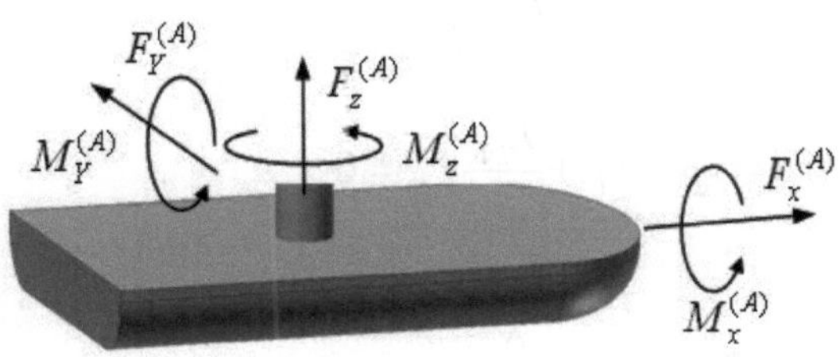

Fig. 10.19. External forces and their moments acting on the crane's base

The model of an offshore crane is ultimately described by the following equations of motion:

$$\mathbf{A}\ddot{\mathbf{q}} - \mathbf{D}\mathbf{R}^{(A)} = \mathbf{F} , \tag{10.47}$$

where $\mathbf{A}$ – mass matrix,

$\mathbf{R}^{(A)}$ – vector defined in (10.46),

$$\mathbf{D} = \begin{bmatrix} \mathbf{I}_{6\times 6} \\ \mathbf{0}_{\overline{m}\times 6} \end{bmatrix},$$

$\overline{m} = 14 + m - 6$,

$\mathbf{I}_{6\times6}$ – identity matrix of dimension 6x6,

$\mathbf{F}$ – vector on the right side of the equation of motion including among others the terms of Lagrange's equation related to velocities and generalized coordinates as well as non-potential forces not accounted for in vector $\mathbf{R}^{(A)}$,

which must be supplemented by constraint equations:

$$\mathbf{q}^{(A)} = \boldsymbol{\alpha}^{(A)}. \tag{10.48}$$

10.3.3 Drive Functions Stabilizing Load's Position: The Inverse Kinematics Problem

The first of the methods considered by the authors to determine drive functions which ensure stabilization of crane load's position is solving the inverse kinematics problem for quasi-static conditions. This task consists in choosing functions for the auxiliary system's and hoisting winch drum's motions so that the load stays in its initial position when the motion of the crane's base is taken into account. However, dynamic effects are not considered.

Before the desired drive functions can be determined, it is necessary to set the initial position of the mass m_L, i.e. the initial values of the coordinates x_L^0, y_L^0, z_L^0 that they have at the beginning of the stabilization. They provide the initial conditions for the problem of determining drive functions in quasi-static conditions. The equations of static equilibrium for the simplified model are:

$$\begin{aligned} f_1 &= x_D - x_L + \frac{L_{DL}}{S} F_X = 0 \\ f_2 &= y_D - y_L + \frac{L_{DL}}{S} F_Y = 0 \\ f_3 &= z_D - z_L + \frac{L_{DL}}{S} \left(F_Z - m_L g\right) = 0 \end{aligned} , \tag{10.49}$$

where $S = c_L \left[L_{DL} + L_D - L_0 + \alpha^0 r_B\right]$,

$L_{DL} = \sqrt{(x_D - x_L)^2 + (y_D - y_L)^2 + (z_D - z_L)^2}$,

$c_l = \dfrac{E_r A_R}{L_0 - \alpha r_B}$,

E_r – Young's modulus,

$r_B(t)$ – radius of the drum,

A_r – cross-sectional area of the rope.

Coordinates x_D, y_D, z_D can be obtained from:

$$\mathbf{r_D} = \begin{bmatrix} x_D \\ y_D \\ z_D \\ 1 \end{bmatrix} = \mathbf{T_P}\,\mathbf{T_G} \begin{bmatrix} x_D' \\ y_D' \\ z_D' \\ 1 \end{bmatrix} = \mathbf{T}\,\mathbf{r_D}', \tag{10.50}$$

where $z_D' = \dfrac{L_D \cos\theta \cos\gamma}{\sqrt{1-\sin^2\theta \sin^2\gamma}}$,

$x_D' = z_D'\, tg\,\theta$,

$y_D' = z_D'\, tg\,\gamma$,

$\mathbf{r_D}' = [x_D', y_D', z_D', 1]^T$,

$\mathbf{T_P}$ – transformation matrix from {P} coordinate system to the inertial coordinate system {0},

$\mathbf{T_G}$ – transformation matrix from {G} to {P} coordinate systems.

The matrices $\mathbf{T_P}$ and $\mathbf{T_G}$ appearing in (10.50) are defined as:

$$\mathbf{T_P} = \mathbf{T_P}(t) = \begin{bmatrix} \mathbf{R_P} & \mathbf{r_P} \\ \mathbf{0} & 1 \end{bmatrix}, \tag{10.51}$$

$$\mathbf{T_G} = \begin{bmatrix} c\varphi & s\varphi & 0 & d\,c\psi\,c\varphi \\ s\varphi & -c\varphi & 0 & d\,c\psi\,s\varphi \\ 0 & 0 & -1 & d\,s\psi \\ 0 & 0 & 0 & 1 \end{bmatrix} = const, \tag{10.52}$$

where d – length of the jib (Fig. 10.18),

$$\mathbf{R_P} = \begin{bmatrix} c\psi^{(A)} & -s\psi^{(A)} & 0 \\ s\psi^{(A)} & c\psi^{(A)} & 0 \\ 0 & 0 & 1 \end{bmatrix} \begin{bmatrix} c\theta^{(A)} & 0 & s\theta^{(A)} \\ 0 & 1 & 0 \\ -s\theta^{(A)} & 0 & c\theta^{(A)} \end{bmatrix} \begin{bmatrix} 1 & 0 & 0 \\ 0 & c\varphi^{(A)} & -s\varphi^{(A)} \\ 0 & s\varphi^{(A)} & c\varphi^{(A)} \end{bmatrix},$$

$$\mathbf{r_P} = \begin{bmatrix} x_P \\ y_P \\ z_P \end{bmatrix},$$

$c\varphi = \cos\varphi,\quad s\varphi = \sin\varphi$.

The unknown values of the coordinates x_L^0, y_L^0, z_L^0 are determined from (10.50) assuming $t = t_0$, $\alpha = \alpha^0, \theta = \theta^0$ and $\gamma = \gamma^0$, where $t_0, \alpha^0, \theta^0, \gamma^0$ are, respectively, the initial time (moment) of starting the stabilization and the values of angles α, γ, θ at the time $t = t_0$. Equations (10.50) form a system of 3 nonlinear equations with 3 unknowns $x_L = x_L^0$, $y_L = y_L^0$, $y_L = y_L^0$, which is solved with Newton's iteration method.

Having determined $x_L = x_L^0$, $y_L = y_L^0$ and $z_L = z_L^0$, the procedure continues determining the drive functions for quasi-static conditions. Let:

$$t_i = t_0 + i\Delta t \qquad i = 1,\ldots,p, \quad (10.53)$$

where $\Delta t = \dfrac{t_k - t_0}{p}$,

t_k – end time of load stabilization,

p – number of intervals into which the time interval $\langle t_0, t_k \rangle$ has been divided.

Assuming that the following conditions must hold at time $t = t_1,\ldots,t_p$:

$$\begin{aligned} x_L\big|_{t=t_i} &= x_L^0 \\ y_L\big|_{t=t_i} &= y_L^0 \\ z_L\big|_{t=t_i} &= z_L^0 \end{aligned} \qquad i = 1,\ldots,p, \quad (10.54)$$

equations (10.49) of static equilibrium for $t = t_i$ can be written as:

$$\begin{aligned} f_1\left(\alpha^{(i)},\theta^{(i)},\gamma^{(i)}\right) &= x_D - x_L^0 + \frac{L_{DL}}{S} F_X = 0 \\ f_2\left(\alpha^{(i)},\theta^{(i)},\gamma^{(i)}\right) &= y_D - y_L^0 + \frac{L_{DL}}{S} F_Y = 0 \\ f_3\left(\alpha^{(i)},\theta^{(i)},\gamma^{(i)}\right) &= z_D - z_L^0 + \frac{L_{DL}}{S} (F_Z - m_L\, g) = 0 \end{aligned} \qquad i = 1,\ldots,p, \quad (10.55)$$

where $\mathbf{r_D} = \begin{bmatrix} x_D \\ y_D \\ z_D \\ 1 \end{bmatrix} = \mathbf{T}^{(i)} \begin{bmatrix} x_D' \\ y_D' \\ z_D' \\ 1 \end{bmatrix} = \mathbf{T}^{(i)}\, \mathbf{r_D}'$,

$\mathbf{T}^{(i)} = \mathbf{T}(t_i)$,

S – defined in (10.49).

Equations (10.55) form a system of $3p$ nonlinear algebraic equations with $3p$ unknowns $\alpha^{(i)}$, $\gamma^{(i)}$, $\theta^{(i)}$. These equations have also been solved by applying Newton's iteration method taking the starting point for subsequent iterations to be:

$$\begin{aligned} \alpha^{(i),0} &= \alpha^{(i-1)} \\ \gamma^{(i),0} &= \gamma^{(i-1)} \quad , \\ \theta^{(i),0} &= \theta^{(i-1)} \end{aligned} \tag{10.56}$$

The continuous functions $\alpha = \alpha(t),\ \gamma = \gamma(t),\ \theta = \theta(t)$ have been obtained by connecting points with splines.

As it has been described above, the courses angles of the hoisting winch and the auxiliary system are calculated using the simplified model and omitting dynamic phenomena. However, these courses compensating sea waves can be successfully applied in dynamic simulations using both models. The further calculations are carried out using the basic model of the crane. The crane is assumed to have a 14 m long jib and an $L_D = 5$ m long auxiliary system. The angle ψ was 50°, the load's mass 10000 kg, its z coordinate -20 m. For this data, the distance from point G of the jib to the load is 31.75 m. Although, as previously mentioned, offshore cranes usually operate from watercrafts conveniently positioned against waves, a general motion of the base has been considered for sample excitations. The results follow [Balachandran B., et al., 1999]:

$$\begin{aligned} x_{org}^{(D)} &= 0.6\left(\sin 0.52t + \frac{1}{4}\sin 1.04t + \frac{1}{9}\sin 1.56t\right)[\text{m}], \\ y_{org}^{(D)} &= 0.6\left(\sin(0.52t + 3.14) + \frac{1}{4}\sin(1.04t + 3.14) + \frac{1}{9}\sin(1.56t + 3.14)\right)[\text{m}], \\ z_{org}^{(D)} &= 1.2\left(\sin(0.52t + 3.14) + \frac{1}{4}\sin(1.04t + 3.14) + \frac{1}{9}\sin(1.56t + 3.14)\right)[\text{m}], \\ \varphi^{(D)} &= 3\left(\sin(0.52t + 3.14) + \frac{1}{4}\sin(1.04t + 3.14) + \frac{1}{9}\sin(1.56t + 3.14)\right)[^\circ], \\ \theta^{(D)} &= 5\left(\sin 0.52t + \frac{1}{4}\sin 1.04t + \frac{1}{9}\sin 1.56t\right)[^\circ], \\ \psi^{(D)} &= 5\left(\sin 0.52t + \frac{1}{4}\sin 1.04t + \frac{1}{9}\sin 1.56t\right)[^\circ]. \end{aligned} \tag{10.57}$$

By the assumption of such motion of the crane's base, the analysis of the general case was possible to perform, which required using auxiliary system's drives in both directions: radial and tangential. The time of stabilization (observation of the load) was 60 s. The coordinates x, y, z of the load, respectively, for the auxiliary system and hoisting winch's drum both idle and controlled according to the drive functions determined are shown in Figs. 10.20, 10.21 and 10.22. Time courses of dynamic coefficient of the force in the rope for both cases are presented in Fig. 10.23. The dynamic coefficient is defined as:

$$\eta = \frac{S}{m_L\, g}, \tag{10.58}$$

where S is the force in the rope.

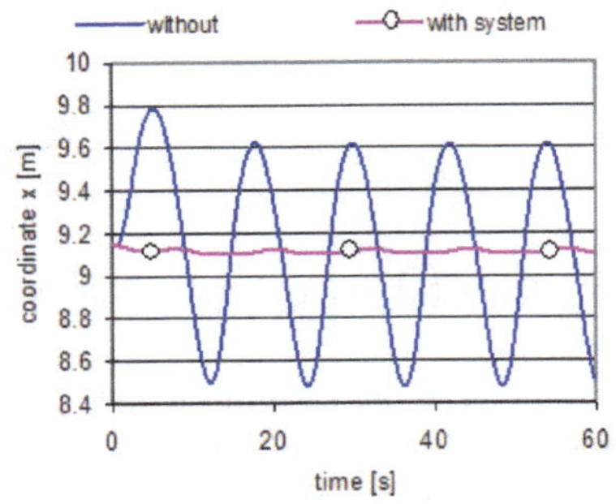

Fig. 10.20. Time course of x_L coordinate for general motion of the base

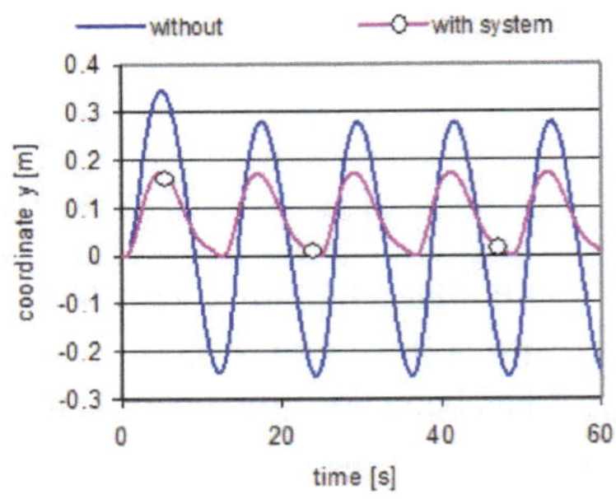

Fig. 10.21. Time course of y_L coordinate for general motion

The results obtained show that using the proposed method yields highly effective stabilization of the load. The outstanding quality of stabilization along the axes **X** and **Z** is worth emphasizing. Stabilizing the load's position has also decreased the dynamic coefficient of the force in the rope, thus alleviating dynamic strains in the crane's structure.

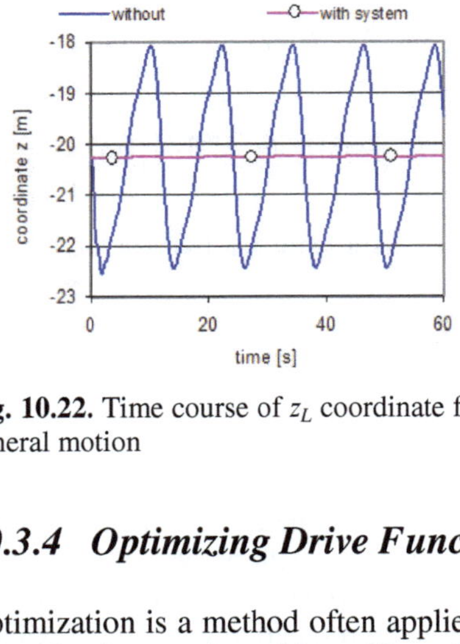

Fig. 10.22. Time course of z_L coordinate for general motion

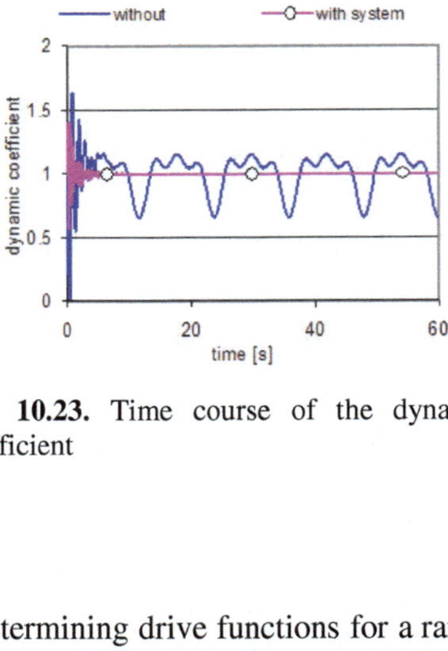

Fig. 10.23. Time course of the dynamic coefficient

10.3.4 Optimizing Drive Functions

Optimization is a method often applied in determining drive functions for a range of mechanisms and machines. Naturally, the optimization criteria may vary. Some examples are: minimal duration of motion, minimal energy consumption, minimal dynamic strains, approximating desired trajectories for selected points, etc.

In the considered case of stabilizing offshore crane load's position, the optimization task is to determine the functions $\alpha = \alpha(t)$, $\gamma = \gamma(t)$, $\theta = \theta(t)$, so that despite the crane base's motion (i.e. despite the variability of the matrix $\mathbf{T}(t)$

of (10.50)) the load remains possibly close to its initial position. Hence the definition of the goal function:

$$\Omega(\alpha,\gamma,\theta)=\int_{t_0}^{t_k}\left[\left(x_L-x_L^0\right)^2+\left(y_L-y_L^0\right)^2+\left(z_L-z_L^0\right)^2\right]dt\ , \tag{10.59}$$

where x_L^0, y_L^0, z_L^0 are initial coordinates of the load determined as in the previous section, from (10.49).

Since it has been assumed that the crane base's motion is defined by pseudo-harmonic functions, it seems natural to seek the drive functions $\alpha=\alpha(t)$, $\gamma=\gamma(t)$, $\theta=\theta(t)$ of the from:

$$\begin{aligned}\alpha&=\alpha^0+\sum_{i=1}^{n_\alpha}A_{\alpha,i}\sin\left(\omega_{\alpha,i}t+\varphi_{\alpha,i}\right)\\ \gamma&=\gamma^0+\sum_{i=1}^{n_\gamma}A_{\gamma,i}\sin\left(\omega_{\gamma,i}t+\varphi_{\gamma,i}\right)\ ,\\ \theta&=\theta^0+\sum_{i=1}^{n_\theta}A_{\theta,i}\sin\left(\omega_{\theta,i}t+\varphi_{\theta,i}\right)\end{aligned} \tag{10.60}$$

where $\alpha^0,\gamma^0,\theta^0$ – initial values of angles α,γ,θ,

$A_{j,i},\omega_{j,i},\varphi_{j,i}$ – i^{th} amplitude, angular frequency and phase angle of the jth drive, satisfying $j\in\{\alpha,\gamma,\theta\}$.

Decision variables of an optimization task thus stated can be written as vectors:

$$\mathbf{X}_\alpha=\begin{bmatrix}\mathbf{X}_{\alpha,1}\\ \vdots\\ \mathbf{X}_{\alpha,i}\\ \vdots\\ \mathbf{X}_{\alpha,n_\alpha}\end{bmatrix},\quad \mathbf{X}_{\alpha,i}=\begin{bmatrix}A_{\alpha,i}\\ \omega_{\alpha,i}\\ \varphi_{\alpha,i}\end{bmatrix}\qquad i=1,\ldots,n_\alpha, \tag{10.61}$$

$$\mathbf{X}_\gamma=\begin{bmatrix}\mathbf{X}_{\gamma,1}\\ \vdots\\ \mathbf{X}_{\gamma,i}\\ \vdots\\ \mathbf{X}_{\gamma,n_\gamma}\end{bmatrix},\quad \mathbf{X}_{\gamma,i}=\begin{bmatrix}A_{\gamma,i}\\ \omega_{\gamma,i}\\ \varphi_{\gamma,i}\end{bmatrix}\qquad i=1,\ldots,n_\gamma, \tag{10.62}$$

$$\mathbf{X}_\theta = \begin{bmatrix} \mathbf{X}_{\theta,1} \\ \vdots \\ \mathbf{X}_{\theta,i} \\ \vdots \\ \mathbf{X}_{\theta,n_\theta} \end{bmatrix} , \quad \mathbf{X}_{\theta,i} = \begin{bmatrix} A_{\theta,i} \\ \omega_{\theta,i} \\ \varphi_{\theta,i} \end{bmatrix} \qquad i = 1,...,n_\theta , \tag{10.63}$$

and finally:

$$\mathbf{X} = \begin{bmatrix} \mathbf{X}_\alpha \\ \mathbf{X}_\gamma \\ \mathbf{X}_\theta \end{bmatrix} . \tag{10.64}$$

Vector (10.64) has:

$$n = 3\left(n_\alpha + n_\gamma + n_\theta\right) \tag{10.65}$$

coordinates. The optimization consists in determining such combination of decision variables satisfying these constraints:

$$\begin{matrix} A_{\alpha,i,\min} \le A_{\alpha,i} \le A_{\alpha,i,\max} & A_{\gamma,i,\min} \le A_{\gamma,i} \le A_{\gamma,i,\max} & A_{\theta,i,\min} \le A_{\theta,i} \le A_{\theta,i,\max} \\ \omega_{\alpha,i,\min} \le \omega_{\alpha,i} \le \omega_{\alpha,i\max} & \omega_{\gamma,i,\min} \le \omega_{\gamma,i} \le \omega_{\gamma,i\max} & \omega_{\theta,i,\min} \le \omega_{\theta,i} \le \omega_{\theta,i\max} \\ \varphi_{\alpha,i,\min} \le \varphi_{\alpha,i} \le \varphi_{\alpha,i,\max} & \varphi_{\gamma,i,\min} \le \varphi_{\gamma,i} \le \varphi_{\gamma,i,\max} & \varphi_{\theta,i,\min} \le \varphi_{\theta,i} \le \varphi_{\theta,i,\max} \end{matrix} , \tag{10.66}$$

that the goal function (10.59) attains its minimal value.

To calculate the goal function, the coordinates x_L, y_L, z_L must be defined. This requires integration of the crane model's equations of motion for $t \in \langle t_0, t_k \rangle$. As mentioned before, to achieve reasonable numerical efficiency the optimization task was solved using the simplified model. Nelder-Mead simplex method was employed for the optimization [Wit R., 1986].

Judicious choice of initial approximations of functions to optimize is an important requirement. They should already be near the solution. The case of pseudo-harmonic functions involves an additional difficulty in selecting the number of harmonic components, i.e. the values n_α, n_γ, n_θ. This problem is briefly accounted for in the following. The initial approximation was chosen to be defined as a sum of n_j harmonic components ($j \in \{\alpha, \gamma, \theta\}$) obtained from Fourier analysis applied to drive functions which are the solutions of the inverse kinematics problem for quasi-static conditions. It should be clearly stated that the method proposed for determining optimized drive functions requires that they be preset with the procedure of chaper 10.3.3 beforehand.

With Fourier analysis [Kruszewski J., Wittbrodt E., 1992], an arbitrary periodic function can be represented as a series:

$$x(t) = x_0^0 + \sum_{\nu=1}^{l} x_\nu^o \sin\left(\nu\,\omega t + \varphi_{x\nu}\right), \tag{10.67}$$

where l – number of terms (harmonic components) in the series,
ν – index of harmonic component,
t – time,
x_0^0 – constant term of the series,
$x_\nu^0, \varphi_{x\nu}$ – amplitude and phase angle of the νth component, respectively,
$\omega = \dfrac{2\pi}{T}$ – lowest angular frequency,
T – function's period.

Individual coefficients of a Fourier series are given by the following formulae:

$$x_0^0 = \frac{1}{N}\sum_{i=0}^{N-1} x_i\,, \tag{10.68}$$

$$x_\nu^0 = \sqrt{a_\nu^2 + b_\nu^2}\,, \tag{10.69}$$

$$\varphi_\nu^0 = arctg\left(\frac{a_\nu}{b_\nu}\right), \tag{10.70}$$

where $a_\nu = \dfrac{2}{N}\sum_{i=0}^{N-1} x_i \cos\dfrac{2\pi\,\nu\,i}{N}$,

$b_\nu = \dfrac{2}{N}\sum_{i=0}^{N-1} x_i \sin\dfrac{2\pi\,\nu\,i}{N}$,

$\nu = 1,2,\ldots,l$,

N – number of sampling points for the function $x(t)$,

whereas l, the number of harmonic components, must satisfy this condition:

$$2l + 1 < N\,. \tag{10.71}$$

The hoisting winch drum's drive may obviously have different number of harmonic components in the initial approximation from either radial or tangential direction of the auxiliary system. It is, however, natural to assume that the number of components present in the initial approximation remains the same in the optimized function.

Drive functions optimized using different numbers of harmonic components ($n_j = 2$ and $n_j = 4$) with the method of chaper 10.3.3 (being the input of Fourier analysis) are compared below to their initial approximations. Parameters describing geometry and mass distribution of the crane were identical to those considered in chaper 10.3.3 and the excitation of the base was given by (10.57). The graphs in Figs. 10.24, 10.25 are for the drive function of the hoisting winch's drum, those in Figs. 10.26, 10.27 for the rotation of the auxiliary system's rigid component in the radial plane and those in Figs. 10.28 and 10.29 in the tangential plane.

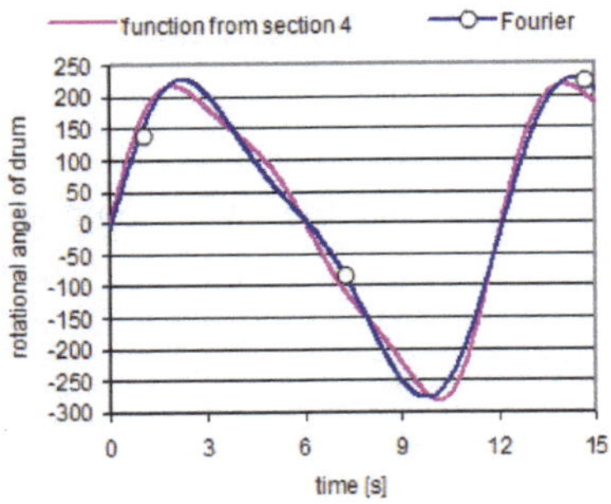

Fig. 10.24. Comparison of the hoisting winch drum's analytic drive function and its initial approximation (2 components)

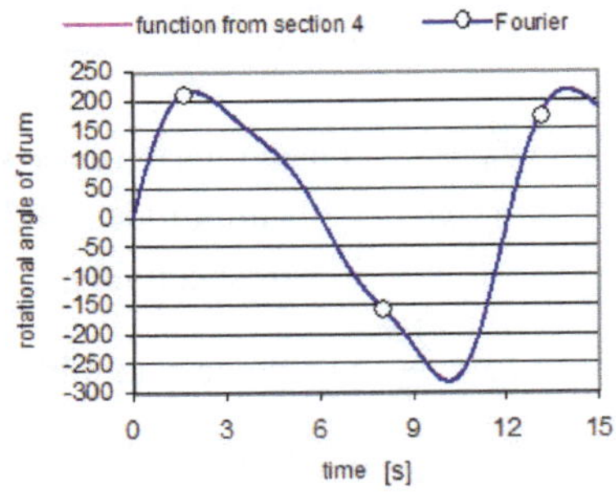

Fig. 10.25. Comparison of the hoisting winch drum's analytic drive function and its initial approximation (4 components)

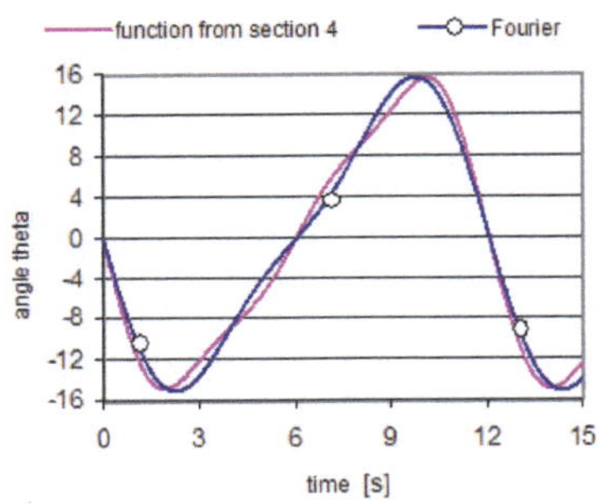

Fig. 10.26. Comparison of the θ angle's analytic drive function and its initial approximation (2 components)

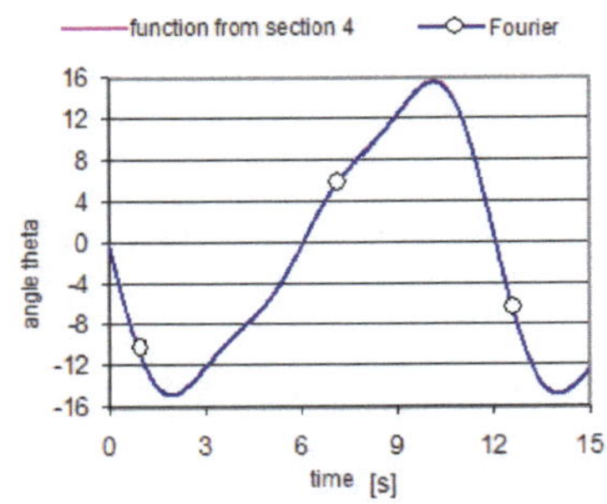

Fig. 10.27. Comparison of the θ angle's analytic drive function and its initial approximation (4 components)

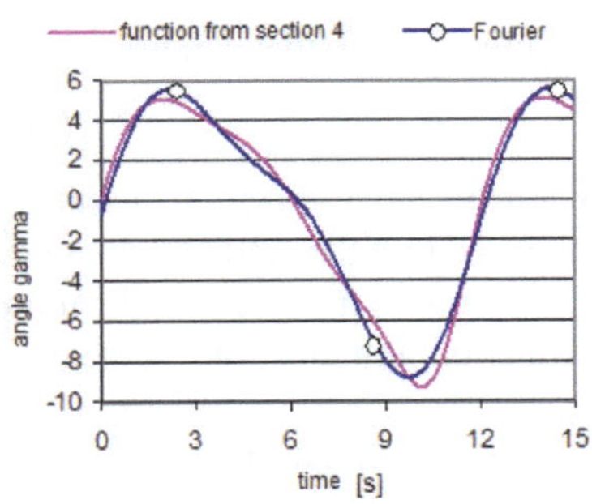

Fig. 10.28. Comparison of the γ angle's analytic drive function and its initial approximation (2 components)

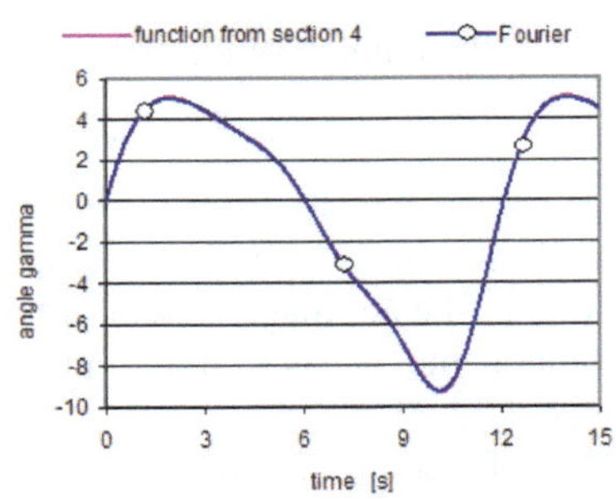

Fig. 10.29. Comparison of the γ angle's analytic drive function and its initial approximation (4 components)

As it can be seen, taking $n_j = 4$ leads to very good approximation of drive functions with a Fourier series. Thence follows reasonableness of seeking the initial approximation in the class of functions of the form (10.60).

Next, stabilization of load's position was considered for durations of 20 and 60 s. The value of the goal function (10.59) for idle auxiliary system and hoisting winch's drum was 47.07 for stabilization time 20 s and 142.10 for 60 s. Comparison of the goal function's values for initial approximations and their corresponding optimized drive functions for different numbers of harmonic components are presented in Table 10.2. Number of iterations and time of computation used by the optimization process are also compared.

Table 10.2. Comparison of the goal function's values

Duration of stabilization [s]	Number of components			Value of the goal function		Number of iterations	Approximated computation time taken by the optimization
	hoisting winch's drum	θ angle	γ angle	initial approxi-mation	optimized function		
20	2	2	2	0.579625	0.388667	3053	6.5 min
20	4	4	4	0.000571	0.000208	3835	9 min
20	6	6	6	0.000054	0.000011	10894	33 min
60	2	2	2	1.677738	1.226042	1699	13 min
60	4	4	4	0.001667	0.000555	3673	27 min
60	6	6	6	0.000121	0.000490	10369	85 min

The results presented (of calculations for the simplified model) made it possible to draw the following conclusions:

- in the Fourier analysis taking into consideration 4 harmonic components is sufficient for obtaining satisfying drive functions determined for quasi-static conditions,
- initial approximations stabilize the load's position well; indeed, taking as few as 2 harmonic components of each initial approximation gives a reduction of the goal function's value by a factor of about 100 compared to the case of idle hoisting winch's drum and auxiliary system,
- substantial influence of the number of components in the initial approximation on the quality of stabilization is clearly noticeable,
- only slightly does optimization improve stabilization of the load.

These conclusions led the authors to abandoning further work with the idea of determining drive functions for stabilization of load's position using optimization.

The reason is that the process is lengthy, requires determining drive functions for quasi-static conditions, and the improvement obtained in the quality of load stabilization is modest.

We would like to underlain that above conclusions are valid only for the assumed pseudo-harmonic base motion. In the case of more complicated functions defining base motion, the second approach, when optimization methods are applied, seems to be better, since it is more general. The decision variable could be the values of spline functions defining drives [Maczyński A., Wojciech S., 2003].

10.3.5 Control System

The proposed system of stabilization of load's position could be augmented with a closed-loop control system. Its task would be to minimize the influence of the jib and the hoisting system's flexibility neglected in the model, of inaccurate knowledge of parameters describing the geometry and mass distribution of the crane as well inaccurate knowledge of waves parameters, eventually of potential external disturbances.

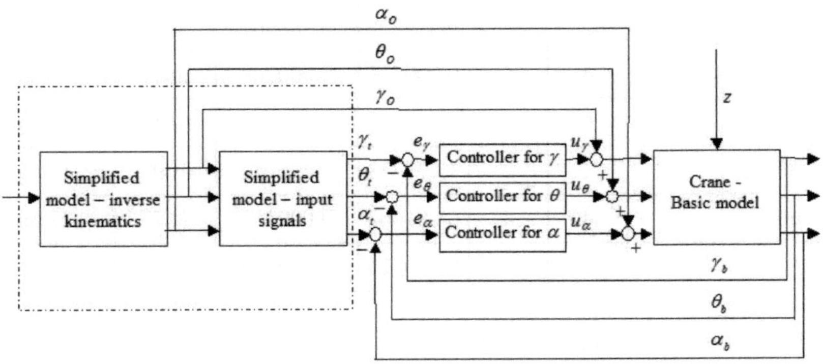

Fig. 10.30. Block diagram of control system: γ_O, θ_O, α_O – drive functions determined according the simplified model, γ_t, θ_t, α_t – input signals determined according the simplified model, γ_b, θ_b, α_b – current values of controlled signals, e_γ, e_θ, e_α – dynamic errors, u_γ, u_θ, u_α – output signals, z – disturbance

A block diagram of the considered control system is shown in Fig. 10.30. Three independent controllers, one for rotation of the drum and two for rotational motions of the stiff element of the auxiliary system, are used. The control system contains feed-back loops as well feed-forward loops. Time courses of the inputs signals ($\alpha_t = \alpha_t(t)$, $\gamma_t = \gamma_t(t)$, $\theta_t = \theta_t(t)$) have to be determined simultaneously with the drive functions ($\alpha_0 = \alpha_0(t)$, $\gamma_0 = \gamma_0(t)$, $\theta_0 = \theta_0(t)$) according to the simplified model. It is worth to be mentioned that in a standard stabilization task the following relationships are fulfilled:

$$\begin{aligned} \alpha_t(t) &= z_L^0 = const \\ \gamma_t(t) &= y_L^0 = const\,. \\ \theta_t(t) &= x_L^0 = const \end{aligned} \tag{10.72}$$

Time courses of load coordinates, x_L, y_L, z_L respectively, for waves increased about 10% in relation to the nominal defined in (10.57), are shown in Figs. 10.31, 10.32 and 10.33. Numerical calculations were carried out according to the basic model, drive functions were determined from chaper 10.3.3. Two cases were considered: with and without control system. The P controller was used in analyzed control system. The obtained results are very promising.

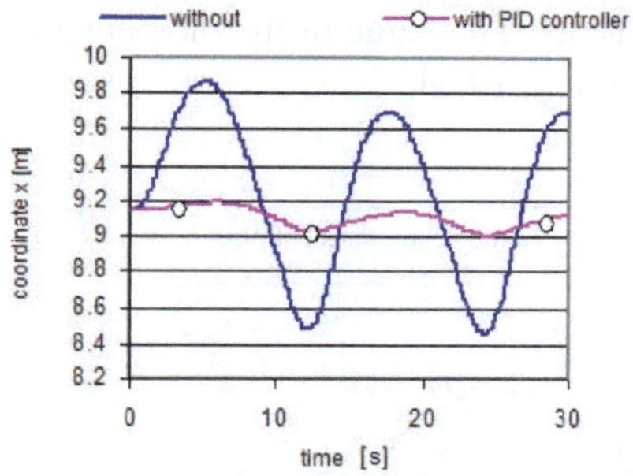

Fig. 10.31. Time course of xL coordinate

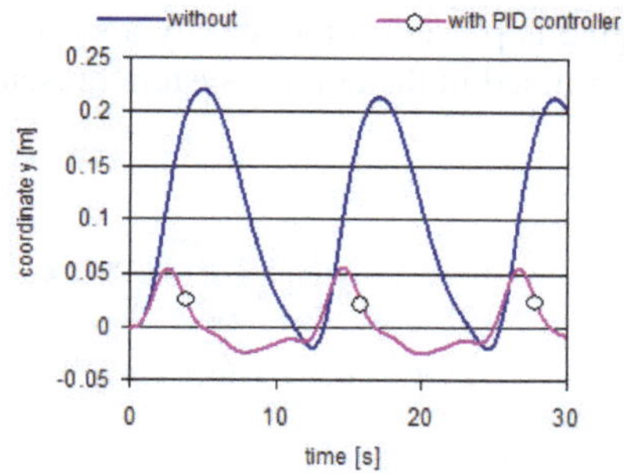

Fig. 10.32. Time course of y_L coordinate

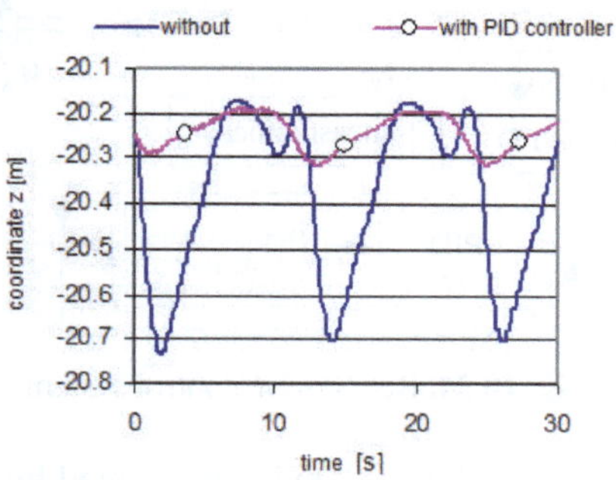

Fig. 10.33. Time course of z_L coordinate

10.4 Active Waves Compensation System for the Reel's Drive

Practice and the results of numerical simulations (chapter 9.3) show that during operation of a reel device for laying pipelines under sea waves its uneven work may occur. To reduce this undesirable effect, a modification of the reel's drive system is proposed. In place of a passive system (in which the braking force of the reel is a constant set by the operator) an active system may be installed with controllable value of the force. That value would be chosen so that the moment maintains the assumed velocity of the reel and constant tension in the pipeline despite the occurrence of additional dynamic loads caused by sea waves. That would be a new solution, as yet unseen in existing drives, however, it would require a change of the way the drive system is designed: the engines (hydraulic or electric) would need the ability of exerting moments in both directions (classical passive systems are only capable of braking the reel during normal operation). The concept is discussed in details by [Szczotka M., 2010], [Szczotka M., 2011b].

10.4.1 Model of the Control System

Frequency converters are commonly used to control velocity and drive moment of electric engines [Olsson G., Piani G., 1998]. The drive moment M_E created by the engine is passed to the reel by a system of gears. The value of the moment M_E is determined in the control system presented in Fig. 10.34.

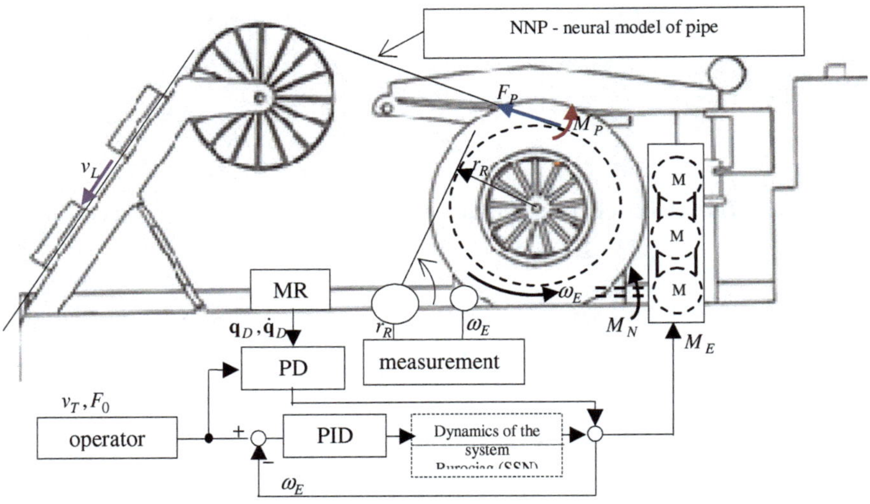

Fig. 10.34. Reel's drive control system

In the computer model, a drive system is represented by a differential equation describing the motion of the reel with pipeline wound onto it and a block performing the calculations of forces and moments caused by the action of the pipeline on the reel (denoted as F_P, M_P). In the concerned application, the operation of this block is based on a model using an artificial neural network. One of the advantages of this approach is short computation time due to the fact that the programme executes only simple operations of multiplication of matrices (containing weight coefficients of the trained network) whose dimensions depend on the numbers of layers and neurons [Osowski S., 1996], [Żurada J., et al., 1996]. To collect data necessary to train the network, a series of simulations was performed in which the equations (9.95) were integrated for different parameters describing waves. The obtained results were saved as a data set. It is worth to note that the training data was obtained for a model taking into account nonlinear material and geometric models. Details of the mathematical model, architecture of the neural network used, the process of generating training data and its application to determining forces and moments exerted by a pipeline being unwound from a reel are presented in [Szczotka M., 2010].

Practical realization of the proposed control system requires, in addition to replacing the drive with an electric one, measuring the winding diameter r_R of the pipe, the reel's velocity ω_E and the vessel's motion (components of the vector $\mathbf{q}_D$ and its derivative). Control is performed in a feedback loop with a PID controller. The system features also an additional PID controller which introduces an adjustment to the previous PID's response. The additional PID controller enables quick reaction of the system to changes input by the operator and excitation caused by waves. It is an example of a feedforward system which in many cases improves stability and accuracy [Olsson G., Piani G., 1998].

As mentioned, further calculations assume an equivalent model of a pipeline implemented with an artificial neural network. To verify the correctness of obtained results, a comparison is made in Fig. 10.35 between simulations using the full model (a pipeline discretized with the RFE method, discussed in chapter 9.3) ant the functional one. On the graphs, the following denotations are present: SSN – results yielded by an artificial neural network, SES – a pipeline discretized with the RFE method. Computation time for a motion lasting 50 s equalled 5 s approximately, whereas calculations according to the full RFE model with the number of RFEs being 100 take roughly 15 min.

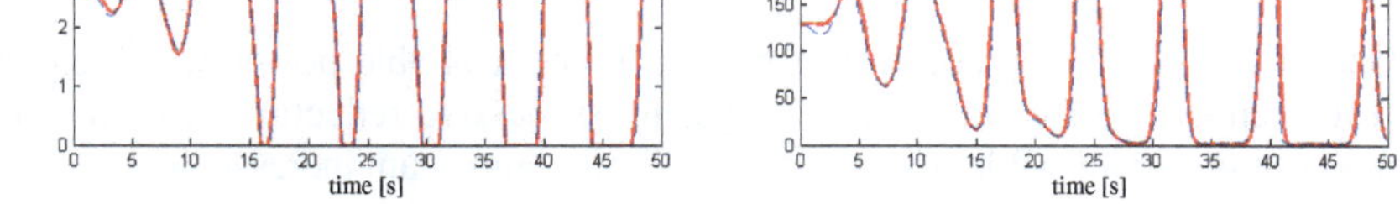

Fig. 10.35. Results of simulation in a passive system: a) reel's velocity, b) tension

10.4.2 Installation of a Pipeline Using the Reel Method with Active Drive System

Sample results of calculations for devices with passive and active reel's drive system are shown in Fig. 10.36. Harmonic excitation of the swaying angle with amplitude $A_\psi = 4^\circ$ and period $T = 8$ s was assumed. Due to structural limitations, the proposed system cannot fully compensate for the influence of dynamic forces caused by waves in cases when occurring forces exceed some limit values. This shows as variations visible on the graphs of force and angular velocity of the reel.

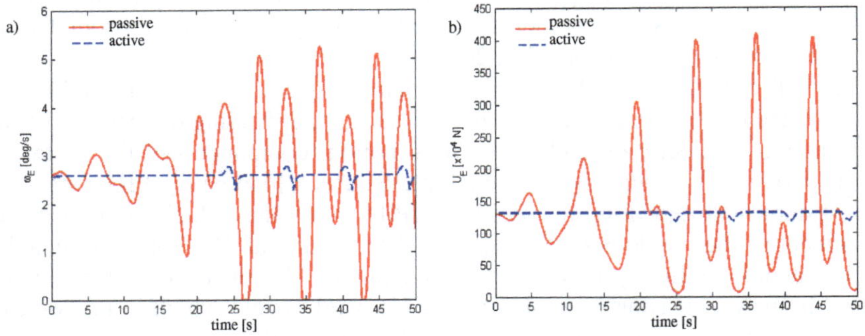

Fig. 10.36. Comparison for passive and active reel drive: a) angular velocity of the reel, b) axial force in the pipeline

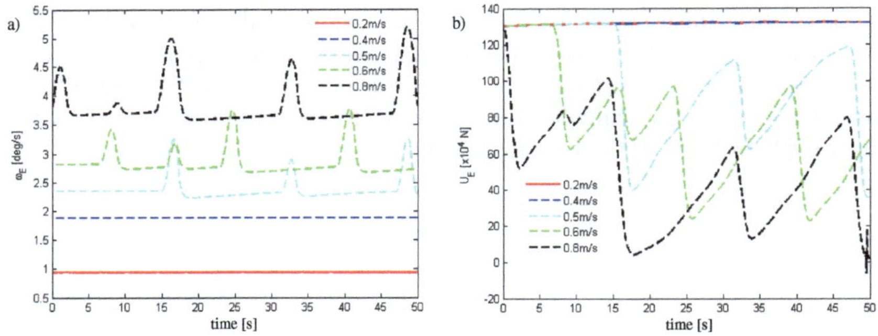

Fig. 10.37. Influence of the speed of laying the pipeline on the behaviour of the system with active compensation for waves: a) reel's velocity, b) tension

One of the limitations not to be forgotten is the available power which can be used to compensate for waves. In the system, successive reduction of drive force after exceeding admissible forces is used. This causes an increase in the reel's velocity (due to forces of inertia caused by the vessel's motion), and further decrease in the moment caused by limited power. The system is capable of returning to the nominal conditions (tension in the pipeline and velocity of the reel) if the operator decreases the tension or waves weakens. Sample results for different settings of the velocity of unwinding the pipeline are summarized by Fig. 10.37.

Note that the assumed power of the drive system (670 kW) is sufficient to operate with constant tension of about 1300 kN when the amplitude $A_\psi = 4^\circ$, that is under intense wave action. For the vessel's operator also calculating the range of safe, stable operation under variable waves conditions may be of interest.

The parameters of the device's operation (peak changes of the reel's velocity and the axial force in the pipeline) in the selected range of swaying amplitudes from 1° to 4° for periods of waves in the interval between 6 s and 12 s are shown in Fig. 10.38. The values of increment of the angular velocity $\Delta\omega_E$ are given as follows:

$$\Delta\omega_E = \frac{\omega_E^{(\max)} - \omega_E^{(\min)}}{\omega_E^{(0)}} 100\% , \qquad (10.73)$$

where $\omega_E^{(\max)}, \omega_E^{(\min)}$ – maximal and minimal velocity of the reel,

$\omega_E^{(0)}$ – nominal velocity of the reel.

Change in the axial force ΔU_E (Fig. 10.88 c and d) is expressed by:

$$\Delta U_E = \max\left\{\left|U_E^{(0)} - \max\left(U_E\right)\right|, \left|U_E^{(0)} - \min\left(U_E\right)\right|\right\}, \qquad (10.74)$$

where $U_E^{(0)}$ – axial force at the moment $t = 0$ (nominal value),

$U_E = U_E(t)$ – current value of the force.

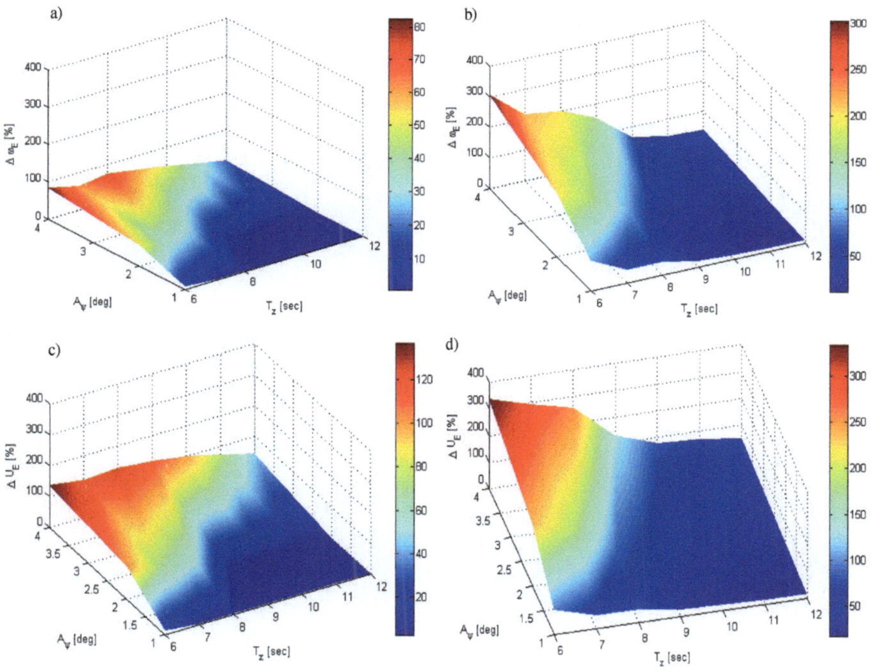

Fig. 10.38. Surfaces describing changes of dynamic parameters with passive and active reel drives a) percentage increments of the reel's velocity in an active system, b) percentage increments of the reel's velocity in a passive system, c) percentage increments of the tension in an active system, d) percentage increments of the tension in an passive system

As the presented graphs indicate, using a system for compensation of waves makes it possible to achieve reduction by a factor greater than three of the dynamic forces and limitation of oscillations of the reel's angular velocity a few times. The graphs were produced for one fixed speed of laying the pipeline equal to 2000 m/h and assumed limits (maximum force passed by the structure of the drive system being 2000 kN and available power 670 kW). The practical effect of employing such a system would be significantly increased efficiency (speed of laying the pipeline) which is a considerable improvement.

The analyses presented are only examples of possible applications of the developed models, algorithms and programmes. It seems particularly useful to employ the model of a drive with automatic adjustment of the force applying tension to the pipeline which allows one to entirely eliminate large overloads in the system with sufficient power. Measuring the vessel's motion is fairly simple. It requires a sensor (e.g. Konsberg Seatex's MRU) to be installed and the reel's angular velocity and winding diameter to be measured. Short computation time, especially when the programme is based on a neural network, is a significant advantage.

11 Summary

In the process of new designs of mechanical structures or systems and control strategy development, a great role for numerical modeling and simulation can evidently be identified. The most comprehensive verification provided by an experiment is naturally the best solution. It is however tedious and costly and in many cases difficult to perform. Offshore structures are often produced as single specimen for a specific order. Carrying out detailed empirical research would raise the final price of a device considerably. Therefore many design companies, including ones in the business of offshore engineering, are interested in access to appropriate calculation software. Such programmes have different purposes. Some of them are suited for strength analysis, others to simulate operation of a device or its control system. In addition to accurate calculations of precise values which are necessary when designing a given machine, companies also need quick and rough simulations, e.g. when preparing an offer (during initial negotiations with a counterparty). Calculations performed at the design stage are not significantly constrained by allowed duration of the simulation. On the other hand, control systems of devices must perform real-time calculations which requires using sufficiently numerically efficient models and methods. In many cases, to obtain satisfactory correspondence to reality flexibility of links must be taken into consideration by their discretisation. In some problems, nonlinear properties of the material or other specific conditions may be important. At present, different discretisation methods are used in calculations of machines' dynamics. The most widely known is the finite element method. The authors of this book have been involved for many years in the development of the rigid finite element method. Based on their experience, it is their position that this method allows developing models of structures adequately reflecting the actual features of the dynamics involved while keeping the number of generalized coordinates small. It is also fairly simple to implement on a computer. It furthermore enables quick and convenient changes of the number of rigid finite elements in the discretized links. This allows both the calculations in real time (for small numbers of RFEs) necessary for control and more time consuming ones which better reflect the flexibility of the system (assuming more RFEs) to be carried out.

In the book, the authors have discussed applications of the rigid finite element method in offshore technology. In addition to giving basic formulas and dependencies, a way has been presented to model added phenomena typical for operation of offshore structures. In particular, modelling of motion of the device's base caused by sea waves and interaction of the sea environment with the elements immersed in water (pipelines, cables) is of importance. Moreover,

E. Wittbrodt et al.: Rigid Finite Element Method, OEO 1, pp. 245–246.
springerlink.com

selected analyses concern nonlinear material characteristics, including deformations of elements in the plastic region. To describe the geometry of systems using the method of homogeneous transformations and joint coordinates is proposed. This method is characterized by conciseness of notation and simplicity of description of complex structures. Complex systems may be modelled with it, including ones in which rigid and flexible links are interleaved. The method also enables taking large deflections into consideration.

The models of offshore structures presented in this book vary in their complexity, therefore they reflect real objects in different ways. Some of them are suitable for quick calculations in real time, others enable more accurate analyses. Certain models have seen actual use in design practice (e.g. the model of an A-frame, the model of a device for unwinding pipes). Applications of the rigid finite element method are naturally not limited to offshore technology. The authors used this method before to model dynamics of spatial mechanisms and manipulators of robots with rigid and flexible links, passenger vehicles and lorries, power trains of vehicles and even satellite dishes. Contents of many of these works confirm the usefulness, effectiveness and correctness of the presented methods, documented by correspondence of the results of calculations and measurements on actual objects.

We hope that the readers will appreciate the two fundamental advantages of the rigid finite element method. First the simplicity of the physical interpretation of a system divided into rigid bodies connected by spring-damping elements, and secondly the ease both of the division into elements and combinations of the natural division into rfes and sdes with the virtual division, which is necessary for the discretisation of the flexible links of machines and mechanisms. We look forward to receiving feedback about the usefulness and applications of the methods presented.

References

Adamiec-Wójcik, I.: Dynamic analysis of manipulators with flexible links, Ph. D. thesis, Strathclyde University, Glasgow (1992)

Adamiec-Wójcik, I.: Modelling dynamics of multibody systems using homogenous transformations, Wydawnictwo Akademii Techniczno-Humanisycznej, Seria Rozprawy Naukowe, Nr 3, Rozprawa habilitacyjna, Bielsko-Biała (2003)

Adamiec-Wójcik, I., Maczyński, A., Wojciech, S.: Zastosowanie metody przekształceń jednorodnych w modelowaniu dynamiki urządzeń offshore (Application of homogeneous transformation method for modeling the dynamics of offshore structures), WKŁ, Warsaw (2008) (in Polish)

Ansys Documentation, Release 12, SAS IP

API-RP-2A-LRFD, Planning, designing and constructing fixed offshore platforms - load and resistance factor design, American Petroleum Institute (1993)

Arteaga, M.A.: On the properties of a dynamic model of flexible robot manipulators. Journal of Dynamic Systems, Measurement and Control 120, 8–14 (1998)

Bai, Y., Bai, Q.: Subsea pipelines and risers. Elsevier (2005)

Balachandran, B., Li, Y.Y., Fang, C.C.: A mechanical filter concept for control of nonlinear crane-load oscillations. Journal of Sound & Vibrations 228, 651–682 (1999)

Bathe, K.J.: Finite element procedures. Prentice-Hall (1996)

Berzeri, M., Campanelli, M., Shabana, A.A.: Definition of the elastic forces in the finite element absolute nodal coordinate formulation and the floating frame of reference formulation. Multibody System Dynamics 5, 21–54 (2001)

Bukowski, J.: Mechanika płynów (Fluid Mechanics). PWN, Warsaw (1968) (in Polish)

Cha, J.H., Roh, M.I., Lee, K.Y.: Dynamic response simulation of a heavy cargo suspended by a floating crane based on multibody system dynamics. Ocean Engineering (2010), doi:10.1016/oceaneng.2010.06.08

Chakrabarti, S.K.: Handbook of Offshore Engineering. Elsevier (2005)

Clauss, G.F., Weede, H., Riekert, T.: Offshore pipe laying operations - Interaction of vessel motions and pipeline dynamic stresses. Applied Ocean Research 14(3), 175–190 (1992)

Cosstick, H.: The sway today. Container Management, 42-43 (1996)

Craig, J.J.: Introduction to Robotics, 3rd edn. Prentice-Hall (2004)

Das, S.N., Das, S.K.: Mathematical model for coupled roll and yaw motions of a floating body in regular waves under resonant and non-resonant conditions. Applied Mathematical Modelling 29, 19–34 (2005)

Dean, R.G., Dalrymple, R.A.: WaterWave Mechanics for Engineers and Scientists, vol. 2. World Scientific, Singapur (1998)

Denavit, J., Hartenberg, R.S.: A kinematics notation for lower-pair mechanisms based on matrices. Journal of Applied Mechanics 2, 215–221 (1955)

DNV-RP-C205, Environmental conditions and environmental loads, Det Norske Veritas (2007)

DNV-RP-F105, Free spanning pipelines, Det Norske Veritas (2006)

Dokka, A., Midttun, O.: Facts - the norwegian petroleum sector, Oslo, Norwegian Petroleum Directorate, Ministry of Petroleum and Energy (2006)

Driscoll, F.R., Lueck, R.G., Nahon, M.: Development and validation of a lumped-mass dynamics model of a deep-sea ROV system. Applied Ocean Research 22, 169–182 (2000)

Dziubiński, I., Świątkowski, T.: Poradnik matematyczny (Handbook of mathematics). PWN, Warsaw (1982) (in Polish)

Eames, M.C.: Steady state theory of towing cables. Trans. of the Royal Inst. of Naval Architects 10 (1968)

El-Hawary, F. (ed.): The Ocean Engineering Handbook. CRC Press LLC, Danvers (2001)

Ellermann, K., Kreuzer, E., Markiewicz, M.: Nonlinear dynamics of floating cranes. Nonlinear Dynamics 27(2), 107–183 (2002)

Ellermann, K., Kreuzer, E., Markiewicz, M.: Nonlinear primary resonances of a floating crane. Meccanica 38, 5–18 (2003)

EN 13852-1, Cranes – Offshore cranes – Part 1; General – purpose offshore cranes (May 2004)

Faltinsen, O.M.: Sea Loads on Ships and Offshore Structures. Ocean Technology Series. Cambridge University Press, Cambridge (1990)

Fałat, P.: Analiza dynamiczna żurawia okrętowego typu A-rama (Dynamic analysis of an A-frame), PhD Thesis, Bielsko-Biała University, Bielsko-Biała (2004) (in Polish)

Fałat, P., Gancarczyk, T., Wojciech, S.: Program for static load analysis of an A-frame. In: Proc. Euroconference on Compatational Mechanics and Engineering Practice, Szczyrk, pp. 144–149 (2001)

Fałat, P., Adamiec-Wójcik, I., Wojciech, S.: Application of the rigid finite element method to modelling dynamics of an Aframe. In: Proc. 8th Conference on Dynamical Systems Theory and Applications, Łódź, pp. 277–284 (2005)

Fossen, T.I.: Guidance and control of ocean vehicles. John Wiley & Sons, Chichester (1994)

Frączek, J.: Modelowanie mechanizmów przestrzennych metodą układów wieloczłonowych (Modelling of spatial mechanisms using the multibody system metod). Warsaw University of Technology Press, Warsaw (2002) (in Polish)

Gerwick, B.D.: Construction of Marine and Offshore Structures, 2nd edn. CRC Press LLC, Wydanie (2000)

Gronowicz, A.: Podstawy analizy układów kinematycznych (Foundations of analysis of kinematic systems). Wroclaw Technical University Press, Wrocław (2003) (in Polish)

Grzegożek, W., Adamiec-Wójcik, I., Wojciech, S.: Komputerowe modelowanie dynamiki pojazdów samochodowych (Computer Modelling of Vehicle Dynamics). Cracow Technical University Press, Kraków (2003) (in Polish)

Harlecki, A.: Metoda analizy dynamicznej mechanicznych układów wieloczłonowych z tarciem suchym w parach kinematycznych (A method of dynamic analysis of multibody systems with dry friction). Bielsko-Biala University Press, Bielsko-Biała (2002) (in Polish)

Hasselmann, K., et al.: Measurements of wind-wave growth and swell decay during the Joint North Sea Wave Project (JONSWAP). Ergnzungsheft zur Deutschen Hydrographischen Zeitschrift Reihe 8(12) (1973)

Hoerner, S.F.: Fluid Dynamic Drag. S. F. Hoerner, New Jersey (1958)

Holthuijsen, L.H.: Waves in oceanic and coastal waters. Cambridge University Press (2007)

Huang, S.: Stability analysis of the heave motion of marine cable-body systems. Ocean Engineering 26, 531–546 (1999)

Huston, R.L., Wang, Y.: Flexibility effects in multibody systems. In: Proceedings of the NATO Advanced Study Institute on Computer-Aided Analysis of Rigid and Flexible Mechanical Systems, pp. 351–376. Kluwer Academic Publisher (1994)

Jezierski, E.: Dynamika robotów (Robots dynamics). WNT, Warsaw (2006) (in Polish)

Jordan, M.A., Bustamante, J.L.: Numerical stability analysis and control of umbilical-ROV systems in one-degree-of-freedom taut-slack condition. Nonlinear Dynamic 49, 163–191 (2007)

Jurewič, E.I. (ed.): Dinamika upravlenija robotami (Dynamics of robot controls), Science, Moskwa (1984) (in Russian)

Komen, G.J., Cavaler, L., Donelan, M., Hasselmann, K., Hasselmann, S., Jenssen, P.A.E.M.: Dynamics and Modelling of Ocean Waves. Cambridge University Press, Cambridge (1994)

Kręglewski, T., Rogowski, T., Ruszczyński, A., Szymanowski, J.: Metody optymalizacji w języku FORTRAN (Optimization methods in FORTRAN). PWN, Warsaw (1984) (in Polish)

Krukowski, J., Maczyński, A.: Wyznaczenie parametrów zastępczych uproszczonego modelu kolumny żurawia typu offshore (Assignment the alternative parameters of simplified model of the offshore crane pedes tal). Problemy Maszyn Roboczych (Engineering Machines Problems) z.33, 25–35 (2009) (in Polish)

Krukowski, J., Maczyński, A.: Efektywność obliczeniowa różnych sposobów modelowania ruchu ogólnego bazy urządzeń posadowionych na jednostkach pływających (Numerical efficiency of the modelling methods of a base's general motion in offshore equipments). Acta Mechanica et Automatica 4(2), 93–98 (2010) (in Polish)

Krukowski, J., Maczyński, A.: Dynamic analysis of an offshore crane with a shock absorber. Transport Przemysłowy i Maszyny Robocze 2(12), 28–31 (2011) (in Polish)

Kruszewski, J., Gawroński, W., Wittbrodt, E., Najbar, F., Grabowski, S.: Metoda sztywnych elementów skończonych (The rigid finite elements method), Arkady, Warszaw (1975) (in Polish)

Kruszewski, J., Sawiak, S., Wittbrodt, E.: Metoda sztywnych elementów skończonych w dynamice konstrukcji (The rigid finite elements method in dynamics of structures). WNT, Warsaw (1999) (in Polish)

Kruszewski, J., Wittbrodt, E.: Computer methods in vibrations of mechanical systems, T. I., WNT, Warsaw (1992) (in Polish)

Leyko, J.: Mechanika Ogólna (Mechanics), T. T. I i II, WNT (1996) (in Polish)

Li, Y.Y., Balachandran, B.: Analytical study of a system with a mechanical filter. Journal of Sound and Vibration 247, 633–653 (2001)

Lighthill, J.: Waves in Fluids. Cambridge University Press, Cambridge (1978)

Longuet-Higgins, M.S.: Bifurcation in gravity waves. Journal of Fluid Mechanics 151, 457–475 (1984)

Łomniewski, K.: Oceanografia fizyczna (Physical Oceanography). PWN, Warsaw (1969) (in Polish)

Maczyński, A.: Pozycjonowanie i stabilizacja położenia ładunku żurawi wysięgnikowych (Positioning and stabilization of the load's position in jib cranes). Bielsko-Biala University Press, Bielsko-Biała (2005) (in Polish)

Maczyński, A.: Stabilizacja położenia ładunku żurawia typu offshore z wykorzystaniem specjalnego układu pomocniczego (Application of the special auxiliary system to stabilization of load of offshore crane), Czasopismo Techniczne. Mechanika, Cracow Technical University, z. 1-M, 247–254 (2006) (in Polish)

Maczyński, A., Wojciech, S.: Dynamics of a mobile crane and optimisation of the slewing motion of its upper structure. Nonlinear Dynamics 32, 259–290 (2003)

Maczyński, A., Wojciech, S.: Stabilization of load position for offshore cranes. In: Twelfth World Congress in Mechanism and Machine Science, Besancon, France (2007), http://130.15.85.212/proceedings/WorldCongress07/articles/article_cd.htm (April 3, 2008)

Maczyński, A., Wojciech, S.: The influence of stabilization of load positioning in an offshore crane on taut-slack phenomenon in a rope. The Archive of Mechanical Engineering LVI(3), 279–290 (2009)
Masoud, Z.N.: A control system for the reduction of cargo pendulation of ship-mounted cranes, Virginia Polytechnic Institute and State University, Doctoral Thesis, Blacksburg, Virginia, USA (2000)
Masoud, Z.N., Nayfeh, A.H., Mook, D.T.: Cargo pendulation reduction of ship-mounted cranes. Nonlinear Dynamics 35(3), 299–311 (2004)
Massel, S.R.: Ocean Surface Waves; their Physics and Prediction. Advanced Series on Ocean Engineering. World Scientific (1996)
Mei, C.C.: The Applied Dynamics of Ocean Surface Waves. Advanced Series on Ocean Engineering. World Scientific (1989)
Mohitpour, M., Golshan, H., Murray, A.: Pipeline design and construction, 2nd edn., A practical approach. The American Society of Mechanical Engineers (2003)
Morecki, A., Knapczyk, J., Kędzior, K.: Teoria mechanizmów i manipulatorów. Podstawy i przykłady zastosowań w praktyce (Theory of mechanisms and manipulators. Fundamentals and examples of applications in practice). WNT, Warsaw (2002) (in Polish)
Morison, J.R., O'Brien, M.P., Johnson, J.W., Schaaf, S.A.: The force exerted by surface waves on piles. Petroleum Transactions 189, 149–154 (1950)
Mróz, Z.: On the description of anizotropic workhardening. Journal of the Mechanics and Physics of Solida 15, 163–175 (1967)
Nayfeh, A.H., Masoud, Z.N., Nayfeh, N.A.: A delayed-position feedback controller for cranes. In: IUTAM Symposium on Chaotic Dynamics and Control of Systems and Processes in Mechanics, pp. 385–395 (2005)
Nayfeh, A.H., Masoud, Z.N., Nayfeh, N.A., Abdel-Rahman, E.: Control of ship-mounted cranes. In: IUTAM Symposium on Vibration Control of Nonlinear Mechanisms and Structures, pp. 21–35 (2005)
Nayfeh, A., Masoud, Z.: Delayed position-feedback controller for the reduction of payload pendulations of rotary cranes ASME. In: Design Engineering Technical Conference and Computers and Engineering Conference, Pittsburgh (2001)
Newman, J.N.: Marine Hydrodynamics. MIT Press, Cambridge (1977)
Nowacki, W.: Teoria pełzania (Theory of creep), Arkady, Warsaw (1963)
Nowacki, W.: Teoria sprężystości (Theory of elasticity). PWN, Warsaw (1970) (in Polish)
Ochi, M.K., Hubble, E.N.: On six-parameters wave spectra. In: Proc. 15th Coastal Eng. Conf., vol. 1, pp. 301–328 (1976)
Olsson, G., Piani, G.: Computer systems for automation and control, 2nd edn. Prentice Hall Inc., London (1998)
Osiński, M., Maczyński, A., Wojciech, S.: The influence of ship's motion in regular wave on dynamics of an offshore crane. The Archive of Mechanical Engineering 51(2), 131–163 (2004)
Osiński, M., Wojciech, S.: Dynamic of hoisting appliances in marine conditions. Machine Vibration 3, 76–84 (1994)
Osiński, M., Wojciech, S.: Application of nonlinear opitmisation methods to input shaping of the hoist drive of an off-shore crane. Nonlinear Dynamics 17, 369–386 (1998)
Osowski, S.: Sieci neuronowe w ujęciu algorytmicznym (Neural networks in algorithmic terms). WNT, Warsaw (1996) (in Polish)
Ottosen, N.S., Ristinmaa, M.: The mechanics of constitutive modeling. Elsevier (2005)
Palmer, A.C., King, R.A.: Subsea Pipeline Engineering. Tetra Verlag GmbH (2008)
Parszewski, Z.: Teoria maszyn i mechanizmów (Theory of machines and mechanisms). WNT, Warsaw (1978) (in Polish)
Paul, R.P.: Robot Manipulators: Methematics, Programming and Control. MIT, Cambridge (1981)

Pedrazzi, C., Barbieri, G.: LARSC: Launch and recovery smart crane for naval ROV handling. In: 13th European ADAMS Users' Conference, Paris (1998)

Pierson, W.J., Moskowitz, L.: A proposed spectral form for fully developed wind seas based on the similarity theory of S. A. Kitaigorodskii. J. Geophys. Res. 69(24), 5181–5190 (1964)

Prager, W.: On isotropic materials with continous transition from elastic to plastic state. In: Proc. 5th Int. Congr. Appl. Mech., Cambridge, Mass, pp. 234–237 (1938)

Press, W.H., Teukolsky, S.A., Wetterling, W.T., Flannery, B.P.: Numerical Recipes in C++, The Art of Scientific Computing. Cambridge University Press, Cambridge (2002)

Ramberg, W., Osgood, W.R.: Description of stress-strain curves by three parameters. National Advisory Committee For Aeronautics, Washington DC, Technical Note No. 902 (1943)

Sarpkaya, T., Isaacson, M.: Mechanics of wave forces on offshore structures. Van Nostrand Reinhold Company, New York (1981)

Schaub, H.: Rate-based ship-mounted crane payload pendulation control syste. Control Engineering Practice 16, 132–145 (2008)

Schwartz, L.: Computer extension and analytical continuation of Stokes' expansion of gravity waves. Journal of Fluid Mechanics 62, 553–578 (1974)

Skrzypek, J.: Plastyczność i pełzanie. Teoria, zastosowania, zadania (Plasticity and creep. Theory, applications, tasks). PWN, Warsaw (1986) (in Polish)

Spathopoulos, M.P.: Pendulation control of an offshore crane. International Journal of Control 77(7), 654–670 (2004)

Spong, M.W., Hutchinson, S., Vidyasagar, M.: Robot Modeling and Control. John Wiley & Sons (2006)

Szczotka, M.: Modelowanie ruchu pojazdu przy uwzględnieniu różnych układów przeniesienia napędu (Modeling of movement of the vehicle, taking into account the different transmission systems), PhD Thesis, Bielsko-Biała University, Bielsko-Biała (2004) (in Polish)

Szczotka, M.: Pipe laying simulation with an active reel drive. Ocean Engineering 37(7), 539–548 (2010)

Szczotka, M.: Dynamic analysis of an offshore pipe laying operation using the reel method. Acta. Mech. 27(1), 44–55 (2011)

Szczotka, M.: Metoda sztywnych elementów skończonych w modelowaniu nieliniowych układów offshore (The rigid finite element method in modeling of nonlinear offshore system). Gdansk Technical University Press, Gdańsk (2011) (in Polish)

Szczotka, M., Wojciech, S., Maczyński, A.: Mathematical model of a pipelay spread. The Archive of Mechanical Engineering 54, 27–46 (2007)

Szuwalski, K., Życzkowski, M.: On the phenomenon of decohesion in perfect plasticity. Int. J. Sol. Struct. 9(1), 85–98 (1973)

Torsethaugen, K., Haver, S.: Simplified double peak spectral model for ocean waves. In: Proc. 14th International Offshore and Polar Engineering Conf., Toulon (2004)

Tucker, M.J., Pitt, E.G.: Waves in Ocean Engineering. Ocean Engineering Series. Elsevier (2001)

Urbaś, A.: Analiza dynamiczna i sterowanie maszynami roboczymi posadowionymi podatnie (Dynamic analysis and control of the engineering machines flexible supported), PhD Thesis, Bielsko-Biała University, Bielsko-Biała (2011) (in Polish)

Urbaś, A., Szczotka, M., Maczyński, A.: Analysis of movement of the BOP crane under sea weaving conditions. Journal of Theoretical and Applied Mechanics 48(3), 677–701 (2010)

Verley, R., Lund, K.M.: A soil resistance model for pipelines placed on clay soils. In: 14th International Conference on Offshore Mechanics & Arctic Engineering, Copenhagen (1995)

Webb, D.J.: Nonlinear transfer between sea waves. Deep-Sea Res. 25, 279–298 (1978)
White, F.M.: Fluid Mechanics, 6th edn. Series in Mechanical Engineering. McGraw-Hill (2006)
Wilson, J.F. (ed.): Dynamics of offshore structures. John Wiley & Sons (2003)
Wit, R.: Methods of nonlinear programming, Minimization of smooth functions. WNT (1986)
Wittbrodt, E.: Dynamika układów o zmiennej w czasie konfiguracji z zastosowaniem metody elementów skończonych (Dynamics of systems with changing in time configuration analysed by the finite element method), Monographs, vol. 343. Gdansk University of Technology Press, Gdańsk (1983) (in Polish)
Wittbrodt, E., Adamiec-Wójcik, I., Wojciech, S.: Dynamics of flexible multibody systems. The rigid finite element method. Springer, Berlin (2006)
Witz, J.A.: Parametric excitation of crane loads in moderate sea states. Ocean Engineering 22(4), 411–420 (1995)
Wojciech, S.: Dynamika płaskich mechanizmów dźwigniowych z uwzględnieniem podatności ogniw oraz tarcia i luzów w węzłach (Dynamics of planar linkage mechanisms with consideration of both flexible links and friction as well as clearance in joint), Monographs, vol. 66. Łodz Technical University Press, Łódź Polish) (1984) (in Polish)
Wojciech, S.: Dynamic analysis of manipulators with flexible links, Archiwum Budowy Maszyn (The Archive of Mechanical Engineering). T. XXXVII Z. 3, 169–188 (1990)
Wojciech, S., Adamiec-Wójcik, I.: Nonlinear vibrations of spatial viscoelatic beams. Acta Mechanica 98, 15–25 (1993)
Wojciech, S., Adamiec-Wójcik, I.: Experimental and computational analysis of large amplitude vibrations of spatial viscoelastic beams. Acta Mechanica 106, 127–136 (1994)
Wojnarowski, J., Adamiec-Wójcik, I.: Application of the rigid finite element method to modelling of free vibrations of a band saw frame. Mechanism and Machine Theory 40(2), 241–258 (2005)
Ylinen, A.: A method of determining the buckling stress and required cross-sectional area for centrally loaded straight columns in elastic and inelastic range, Mem. Ass. Int. Ponts Charp. 16, 529–550 (1956)
Zienkiewicz, O.C.: Metoda elementów skończonych (The finite element metod). Arkady, Warsaw (1972) (in Polish)
Zienkiewicz, O.C., Taylor, R.L.: The Finite Element Method, 5th edn. Butterworth-Heinemann, Oxford (2000)
Żurada, J., Barski, M., Jędruch, W.: Sztuczne sieci neuronowe. Podstawy teorii i zastosowania (Artificial neural networks. Basic theory and application). PWN, Warsaw (1996) (in Polish)

MIX
Papier aus verantwortungsvollen Quellen
Paper from responsible sources
FSC® C105338

If you have any concerns about our products,
you can contact us on
ProductSafety@springernature.com

In case Publisher is established outside the EU,
the EU authorized representative is:
Springer Nature Customer Service Center GmbH
Europaplatz 3, 69115 Heidelberg, Germany

Printed by Libri Plureos GmbH
in Hamburg, Germany